New Directions in Radical Cartography

New Directions in Radical Cartography

Why the Map is Never the Territory

Edited by
Phil Cohen and Mike Duggan

ROWMAN & LITTLEFIELD
Lanham • Boulder • New York • London

Published by Rowman & Littlefield
An imprint of The Rowman & Littlefield Publishing Group, Inc.
4501 Forbes Boulevard, Suite 200, Lanham, Maryland 20706
www.rowman.com

86-90 Paul Street, London EC2A 4NE

Cover Image: Inefficient mapping of the edges of privilege. 2020. Artist: Linda Knight. Linda Knight developed inefficient mapping as a methodological protocol, it offers a methodologic practice for attuning to phenomena in projects using speculative and 'post-' theories. More on Linda's work can be found at www.lindaknight.org

British Library Cataloguing in Publication Information Available

Library of Congress Cataloging-in-Publication Data

Names: Cohen, Philip, 1943- editor. | Duggan, Mike, editor.
Title: New directions in radical cartography : why the map is never the territory / edited by Phil Cohen and Mike Duggan.
Description: Lanham, Maryland : Rowman & Littlefield, [2021] | Includes bibliographical references and index. | Summary: "This collection explores the meaning of maps and of map-making in the modern world"—Provided by publisher.
Identifiers: LCCN 2021032095 (print) | LCCN 2021032096 (ebook) | ISBN 9781538147191 (Cloth) | ISBN 9781538147214 (eBook) Subjects: LCSH: Cartography.
Classification: LCC GA105.3 .N49 2021 (print) | LCC GA105.3 (ebook) | DDC 526—dc23
LC record available at https://lccn.loc.gov/2021032095
LC ebook record available at https://lccn.loc.gov/2021032096

Contents

Preface

Phil Cohen

All the contributors to this book are connected, directly or indirectly, to Livingmaps Network. They may have written for our online journal, taken part in one of our conferences or webinars, or been involved in one of our participatory mapping projects.

The network was set up in 2013 by myself, an ethnographer and a graphic artist, John Wallett, who had collaborated on a map of east London for a book about the impact of the 2012 Olympics on local communities.[1] I had been using a form of participatory mapping for many years in my work with young people in East London, exploring issues of identity, place, and belonging; John had been involved in a number of community arts projects in which map-making was an important element. From the outset, the aim was to move beyond purely academic critiques of cartography, to develop a range of activities in which issues of counter-mapping theory and practice could be addressed concretely. The focus has been on developing and supporting local initiatives designed to challenge the prevailing relations of knowledge-power and open up spaces of representation for groups marginalised in the urban planning process.

Initially we organised a programme of workshops, seminars, and conferences, bringing together academics, artists, and activists around specific themes. Quite a few of the early participants were post-graduate students who were writing theses on cartographic topics, but were isolated within their departments, where interest in counter-mapping was minimal. These events helped to provide these people with a context of mutual engagement and support lacking in their academic institutions.

With the advent of the pandemic, this programme has, of course, moved online, which has also widened its scope to include contributors from many parts of the world. The focus has increasingly been on exploring the role of mapping in relation to the environmental and public health crisis. The network does not have a specific political goal, unlike groups such as those involved in the Anti-Eviction Mapping Project based in San Francisco, which we support, but one of our key influences was William Bunge's expeditionary geography, with its emphasis on combining quantitative and qualitative cartography in addressing social inequalities based on class and race. We organise an annual lecture in his honour and have invited speakers such as Iain Sinclair, Tim Ingold, Hugh Brody, and James Cheshire, who have made significant contributions to the field.

Starting in 2017, the network launched its own bi-annual online journal, which is now edited by this book's co-editor, Mike Duggan. We made a deliberate decision that the journal would not be peer reviewed and would provide an accessible platform for a wide diversity of approaches, encouraging experimental types of writing and multi-media, as well as accounts of counter-mapping practices and theoretical essays.

At the same time, the network has launched a number of participatory mapping projects. 'Speaking out of Place' worked with young people living around the Queen Elisabeth Olympic Park in East London. Through this project we produced a 'Young Person's Alternative Guide to the Park', a video and photographic exhibition, examining how the 'Olympic Legacy' was actually being lived. This has been followed by 'Groundbreakers', a multi-media trail and guide to the Olympic Park. The on-going flagship project is a Young Citizen's Atlas, a toolkit of specially designed resources for working with children and young people around a range of urban issues of direct concern to them.

The core members of the network come from many disciplines and perspectives in the arts, humanities, and sciences, many of them young scholar-activists and visual artists. We have been conscious that the demographic of professional cartography is overwhelming white and male and that this is reflected in its cultural perspectives. The world of counter-mapping is much more diverse and has close links with various social movements including feminist, anti-racist, and environmentalist campaigns. Against this background, we are concerned to address the current divide between the two cultures of mapping, the world of professional cartography with its emphasis on digital mapping technology and the visualisation of big datasets, and the small scale participatory forms of mapping which create platforms for locally situated knowledge and draw on techniques of storytelling, drawing, and imaginative expression to represent people's cognitive maps and emotional geographies.

For more information on Livingmaps Network, please visit: www.livingmaps.org.uk

NOTE

1. See Phil Cohen, 2013. *On the Wrong Side of the Tracks: East London and the Post Olympics.* Lawrence and Wishart.

Acknowledgements

First and foremost, we would like to thank all the contributors for their time and efforts in working with us to produce this book. Edited collections are long projects, and this one has been no different. The fact that we have completed this during a global pandemic, with all the stresses and heartache that we now know this entails, is reason to celebrate. We were grateful for the many useful comments we received from readers of the initial proposal which helped us to avoid some of the pitfalls of an edited collection.

We would also like to thank all the contributors to the Livingmaps Review and to those that are part of the wider Livingmaps project. The book could not include all the material that we would like due to limits of space, but we are in no doubt that this wider network of knowledge and critical practice has contributed substantially to the formulation of this book project. We hope that in the future we can further develop some of the themes highlighted here through the pages of our online journal, our monthly seminars and bi-annual lecture series.

Introduction

Why the Map is Not the Territory

Phil Cohen and Mike Duggan

> It is significant that 'culture' is sometimes described as a *map*. It is the analogy which occurs to an outsider who has to find his way in a foreign landscape and who compensates for his lack of practical mastery, the prerogative of the native, by the use of a model of all possible routes. The gulf between this potential abstract space, devoid of landmarks or any privileged centre, and the practical space of journeys actually being made can be seen from the difficulty we have in recognising familiar routes on a map or town plan until we are able to bring together the system of axes linked unalterably to our bodies, and carried about with us wherever we go.
>
> *Pierre Bourdieu, Outline for a theory of practice, 1977*

Maps used to mean single boards stuck on walls by planners, pieces of old paper carried around in their backpacks by ramblers or stuffed into the side doors of cars. These kinds of maps are still in use, but today digital maps are routinely used by millions of people to find their way around and go about their business. The transition from analogue to digital and mobile mapping, if we can call it that, has transformed the public understanding and negotiation of space, but also heightened curiosity about what maps are for, the different forms they can take, and how they relate to wider patterns of change in contemporary culture and society.

The history of cartography is often still written as a one way trip driven by scientific enlightenment from the mythographies of the early maps with their inaccurate contours and erratic scales, their *terra incognita* populated by monstrous races, to the current technologies of geovisualisation which generate cartograms showing with enormous precision the spatial distribution of big datasets.[1]

We might admire and indeed collect old maps for their aesthetic qualities, or as important historical documents – indeed map-collecting is a booming industry – but we probably would not actually use them to get about. This is not only because of their topographic inaccuracy and the fact that so much has changed on the ground since they were made, but also because the locally situated knowledge which produced them and enabled them to be interpreted and used for navigational purposes is no longer available to us.

It is true that maps of the world produced by orbiting satellites have symbolically replaced the original mappa mundi produced in the Middle Ages, but there is little evidence to suggest that these changes have followed a direct path of translation over time. Mappa mundi, which showed Jerusalem or Mecca to be the axis of a concentric universe around which everything else revolved, were not considered navigational devices in the same way that the digital maps of today are.[2] And yet, the presentation, replication, and preservation of *Mappa Mundia* as historical artefacts that today sit behind glass screens remain a point of comparison for today's maps in the public's lexicon.[3] In reality, they share very little in terms of form or use.

At the same time, we have to reckon with the fact that despite the radical de-centering of our view of the world accomplished by its navigation through the application of scientific calculation – the so-called Copernican revolution in map-making – ethnocentrisms and nationalisms have continued to flourish and indeed have come to exercise a dominant influence on contemporary geopolitics in a way that neither a global pandemic, a global economic recession, or the global environmental crisis can do more than amplify. Equally as Laura Kurgan has shown, the smooth zoom from the macroscopic to microscopic view, from an image of planet earth from outer space to the tiniest particle of matter, conceals the operations of a whole technology of surveillance that sustains an apparently frictionless circulation of information but also reproduces largely invisible networks of knowledge/power on a global scale.[4]

These shifts in the scoping of cartographic reason cannot therefore be rendered into a linear or teleological narrative of historical progress, driven by scientific discoveries. Equally, any account which does not register the impact of these developments, or regards them as mere iterations of some fundamental knowledge/power matrix, does not recognise the significance they have had in the shaping of society.

Once we abandon the linear narrative of cartographic progress, the risk is that we get trapped in a form of circularity in which every new development becomes a cause of itself as an effect in yet another form of technological determinism. The conditions that made possible the invention of GIS or GPS are read off from the forms of its application without introducing any socio-economic mediation. Equally, tracing the changing theory and practice of mapping cannot be a simple exercise in the historical sociology of knowledge, given that the territory being mapped is itself partly made up of mappings, and these mappings increasingly put in question both the common sense principles of representation that are applied to them *and* the status of the knowledge claims which the maps or their makers themselves implicitly or explicitly make.

In putting together this collection of work by writers associated with the Livingmaps Network, we diverge from the seeming linearity of cartographic progress and meander through different forms of map-making and spatial representation.[5] The writing and maps shown in these chapters will likely irk readers looking for new developments in scientific cartography. This *is* intentional. It will mean that the maps are registering in different ways that counter the normative ideals of cartography that are all too often striven for in map-making. What we want to show here is that contemporary map-making is a diverse practice with a multiplicity of meanings and uses. By showing this, we hope to encourage others to explore alternative mappings and to experiment with cartographies of their own making. There is merit in formalised cartography and we do not want to dismiss it – formalised and professional maps have no doubt been useful for all of us at some stage or another – but it does not mean we have to stick to these maps; we can and should look beyond, because, as the contributors to this book make very clear, rarely do the maps we usually encounter reflect the realities of everyday life.

This collection brings together chapters on what we see as four themes of contemporary radical cartography. The first section, Are We that Map? examines debates and new approaches to mapping in contemporary cartography and sets them in their political and socio-cultural context. It develops the theoretical strands of this introduction and situates them in specific case studies. The second section, Reclaiming the Territory, moves on from the theory to highlight the innovatory-mapping projects by practitioners opening up new spaces of representation in relation to marginalised communities and the environment. In particular, it shows how maps are being used to (re) assert ownership and control of lost lands, urban culture, and community life. Continuing this theme in the third section, Watch this space, artists, activists, and academics discuss ground-breaking approaches to counter-mapping including walking maps, audible maps, and deep maps. In the final section, New Scopes, New Scales, we look to the cutting edge of radical cartographies to highlight the experimental edges of analogue and digital-mapping across a range of platforms and cultural contexts. First, however, we must lay out our claims for cartography and put forward our position on what we understand a truly radical cartography to be.

SCIENTIFIC FICTIONS

The reader will notice that the subtitle of our book is phrased as if it offers an answer to its own rhetorical proposition: the map is not the territory (and this is why). It deliberately refers back to the proposition made famous by a now otherwise largely forgotten Polish American philosopher of science called Alfred Korzybski who coined the phrase 'The Map is not the Territory'.[6] By this, he meant that the world which we think we know directly through our senses is in fact created by our beliefs and language games which come to be embodied in what he called 'the nervous system'. These common-sense constructs do not at all correspond to the maps of the material world discovered by the natural sciences.

Korzybski thought that this disjuncture between everyday experience and scientific reason was as dangerous for human well-being as the commonsensical belief that map and territory, words and things, the world and its representation, always and already coincide. In fact, this disjuncture is a matter of common, though often disregarded, knowledge. After all, the word 'sugar' is not itself sweet and indeed may leave a bitter taste in view of its historical association with the slave trade not to mention its impact on public and personal health. What counts as 'sweet' may thus vary greatly according to taste, those with a sweet tooth will take one view and those with an abhorrence of sugary sentimental prose will take another, but as Korzybski would have been keen to point out, and as the parents of young children who are indulged in this way know to their cost, irrespective of standpoint (or how it is mapped), a sugar rush is an undeniable bio-chemical phenomenon within the nervous system with definite behavioural consequences irrespective of moral or cognitive standpoint. In other words, the territory is relatively autonomous in its effects.

Korzybski was no postmodernist *avant la lettre*, he was a neo-Kantian and what he actually wrote was: 'the map is not the territory it represents, but if correct, it has a similar structure to the territory, which accounts for its usefulness'. So he is no budding Baudrillard, who believes, quoting Gertrude Stein, 'there is no there there', that there is no reality outside its representation, and who is likely to have difficulty in dealing with situations in which material reality strikes back, perhaps in the form of a lamppost you walk into when following a mobile map.

However, the utility of the map is not limited to its efficacy as a technical affordance, whether as a mobile centre of calculation or in its capacity to calibrate relevant and reliable scopes and scales. Maps are good to think with precisely because they pose important questions about the nature of scaling and scoping as human activities and their relation to both external and internal realities. And here we think we could do worse than take a leaf out of the book of another neo-Kantian, a contemporary of Korzybski, and today also largely forgotten, Hans Vaihinger.[7] He wrote:

> Many thought processes appear to be based on consciously false assumptions, which either contradict reality, or are even contradictory in themselves. But they are intentionally thus formed in order to overcome difficulties of thought, and reach the goal of thought by roundabout ways. These artificial constructs are called scientific fictions, and distinguished, as conscious creations, by their 'as if' character.

We think it might be interesting as a thought experiment, to explore the possibility of reading maps as scientific fictions of this kind rather than treating them, as many critical cartographers do, simply as mystifying ideological constructs or instrumentalities of governance and power.[8] Maps, whether the maps in our heads, or the maps we learn to read and use as navigational devices, or the maps we create to give flesh to other possible worlds, all serve to create a set of *imaginary* correspondences with the territory they depict, they enable us to act *as if* the structure of the map and the territory were homologous and so enable us to find our way around both the material and the symbolic worlds we co-habit. It is the *as if-ness* of what we are going to call the 'cartographic pact' and its taken-for-granted character that this book is seeking to put in question in a new way, a way that goes beyond simply de-constructing the map.[9]

For Korzybski, map and territory belong to mutually exclusive realms, that is, domains which stand in a purely external relation to one another. He offered a deficit model of their relation: only science can produce adequate knowledge of the material world while common sense entertains a purely phantasmagoric relation to the real as a projection of its own operations. For Cartesian cartography, however, map and territory are two sides of the same mimetic story – they stand in a relation of internal adequation to one another. Nothing is lacking in either the real or its representation. From this standpoint, the map, like language itself, is always indexical. It mirrors, and ever more accurately records, the terrain it covers.[10]

In one of his short stories J.L. Borges exposes the nature of the map as a scientific fiction by pushing the correspondence theory to a reductio ad absurdum when the map becomes spatially coextensive with the territory it covers:

> Let us imagine that a portion of the soil of England has been levelled off perfectly and that on it a cartographer traces a map of England. The job is perfect; there is no detail of the soil of England, no matter how minute, that is not registered on the map; everything has there its correspondence. This map, in such a

case, should contain a map of the map, which should contain a map of the map of the map, and so on to infinity.[11]

Borges is here taking a leaf out of Lewis Carroll's book, *Bruno and Sylvie*, where he cites the Anglo-Saxon passion for cartography in the following terms:

'We actually made a map of the country, on the scale of a *mile to the mile*!'
'Have you used it much?' I enquired.
'It has never been spread out, yet,' said Mein Herr: 'the farmers objected: they said it would cover the whole country, and shut out the sunlight! So we now use the country itself, as its own map, and I assure you it does nearly as well'.

In Borges, inspired by Carroll, carried the reductio ad absurdum a stage further in his short story 'On the Exactitude of Science':

In that Empire, the Art of Cartography attained such Perfection that the map of a single Province occupied the entirety of a City, and the map of the Empire, the entirety of a Province. In time, those Unconscionable Maps no longer satisfied, and the Cartographers Guilds struck a Map of the Empire whose size was that of the Empire, and which coincided point for point with it. The following Generations, who were not so fond of the Study of Cartography as their Forebears had been, saw that that vast map was Useless, and not without some Pitilessness was it, that they delivered it up to the Inclemencies of Sun and Winters. In the Deserts of the West, still today, there are Tattered Ruins of that Map, inhabited by Animals and Beggars; in all the Land there is no other Relic of the Disciplines of Geography.

Baudrillard famously read this story as a postmodernist parable taking Korzybski's dictum in a one-sided way, to argue that the map not only precedes the territory epistemologically but ontologically and indeed can in some instances become its own territory. As he puts it:

Today abstraction is no longer that of the map, the double, the mirror, or the concept. Simulation is no longer that of a territory, a referential being, or a substance. It is the generation by models of a real without origin or reality: a hyperreal. The territory no longer precedes the map, nor does it survive it. It is nevertheless the map that precedes the territory.[12]

Baudrillard tries to invert the relation between map and territory, reserving agency for the 'hyperreal' and its power of dis/simulation. He ignores the denouement of the Borges story, which is the physical decomposition of the map by the action of sun and weather. Even more significantly, the central issue addressed by both stories is completely overlooked by Baudrillard, namely that map/territory relations are fundamentally shaped by practices of translation, which involve both scoping and scaling and if the scale of the map is disproportionate to its scopic register.

In an attempt to disembed the map entirely from the territory, postmodern geographers abandoned the theory of representation in order to establish a new principle of correspondence between the unmappable and the de-territorialised, as equally lacking in any fixed anchorage.[13] This move provoked a riposte from critical geographers who wanted any model of the map to maintain the significance of its material referents (the territory), as a counterpoint or corrective to its function as a social imaginary, produced through an act of pure ideation, while at the same time rejecting the principle of mimetic correspondence.[14]

More recently, the issue has been addressed through the lens of digital and social navigation by actor-network theorists and others who argue that the map–territory relation should be rethought (again) as a single flat 'ontological plane' rather than a conceptual gap to be argued over.[15] This in turn has led to a return to the materiality of the map itself, in its mode of production as an object which necessarily withdraws its meaning from any territorial inscription and whose mode of being in the world we are now invited to consider in and for itself.[16]

THE ART OF TRANSLATION

For most people, maps are primarily navigational devices. Here prediction is subsumed more or less completely under performativity. The declarative role of road maps, military maps, town plans, or sea charts is to enable their readers to safely traverse the terrain in question – their 'realism' being an effect, but not a cause of this fact. Their paradox is that they necessarily privilege the mimetic, the fictive concordance of map and territory which we have argued is central to the cartographic pact, but their actual usage as interactive navigational platforms implies an altogether different model more alive to contingency. Disruptive roadworks, the fog of war, popular protest, and the work of weather see to that.

Lewis Carroll famously spoofed the scientific fiction of navigational maps in *The Hunting of the Snark* in which an unlikely crew who have nothing in common except for their ignorance of nautical matters and the fact that their occupations all begin with B (a Boot maker, a Baker, a Butcher, a Banker, etc.) set sail for an unknown destination. One of their number, the Bellman is sent off to find a chart:

He had bought a large map representing the sea,
Without the least vestige of land:
And the crew were much pleased when they found it to be
A map they could all understand.

'What's the good of Mercator's North Poles and Equators,
Tropics, Zones, and Meridian Lines?'
So the Bellman would cry: and the crew would reply
'They are merely conventional signs!'

'Other maps are such shapes, with their islands and capes!
But we've got our brave Captain to thank
(So the crew would protest) 'that he's bought *us* the best –
A perfect and absolute blank!'

Landlubbers of course, do regard the seascape as a blank space, a vast expanse of nothingness on which to write their fear of the unknown, whereas for mariners it is full of significance, densely populated with features, stories, and memories. Every sandbank, every channel has its name and with that name, a history. Landlubbers looking at a sea chart are quite unable to translate between map and territory. It literally means nothing to them, while sailors who do know how to translate nevertheless have to take into account the specific conditions of wind and tide in order to navigate by it.[17] Even the most operational map, like a maritime sat nav, remains a scientific fiction and cannot substitute itself for the 'real deal' – the ability to read and interpret the environment in real time as it happens.

Bruno Latour and others have drawn our attention to the paradoxical way in which navigational maps, by fetishizing their mimetic function, actively supress or mystify the actual navigational process which they enable. They point out that digital-mapping technologies have *rematerialized* the whole chain of production that makes them possible and that what is now involved in navigation is not so much a toing and fro-ing between map and territory, checking that the topographical features of mental or paper map correspond to what is happening geographically on the ground, but a more intricate and dialectic interaction between different kinds of information, both digital and analogue, which involves learning to identify and select appropriate, context-specific, cues from a mass of heterogeneous data.

The example they give is the interaction between a skipper at the helm of a yacht and a member of the crew below deck operating a GPS chart plotter, and possibly using a back-up paper chart. This is how they characterise the relation:

> [It] is based not on some resemblance between the map and the territory but on the detection of relevant cues allowing her team to go through a heterogeneous set of datapoints from one signpost to the next: some signposts are made visible from the cockpit in the hurly burly world (for instance, a roaring red buoy that the crew was desperately trying to tack), and some are visible in the no less hurly burly nauseating world of the cabin (for instance, a dark spot on the map with a red tip, which is just at the right angle expected by the navigator since the last beacon has been safely recognized and pinpointed with a blue pencil).[18]

One thing that their model, and indeed Latour's version of actor-network theory does not sufficiently allow for is that such translations are only as effective as their human relays, and this is far from a frictionless process.[19] To take their example, there may well be situations at sea in which the skipper over-rides the information they are getting from the chart plotter, in the light of their knowledge about how the boat behaves in different weather conditions. That information is not something that can be programmed into a GPS system or depicted on a paper chart. In other words, the boat itself is a relatively autonomous player in the ecological network that comprises the socio-material practices of navigation. It is only by excluding such entanglements from consideration they can end up declaring that the territory is a spurious referent, as they put it:

> We understand in retrospect that the very notion of territory is nothing but the 'virtual image'-to use an optical metaphor-of a paper map *interrupted* in its navigational usage to answer a mimetic interpretation, after all its real-life users and makers have been all but deleted.[20]

It is this an *a priori* de-materialisation, if not deletion, of the boat as an agent of its own navigation that enables Captain Latour to attribute materiality exclusively to the map and to treat territory as a retrospective illusion. Outside this scientific fiction, if the skipper ignores the boat's behaviour and consequently makes a wrong navigational judgement, then the result may be quite disastrous, and the boat may even capsize, in which case he will discover that there is nothing at all illusory about the consequences.

As Tim Ingold insists, following Bourdieu's original insight, navigation always involves a process of wayfaring. As he puts it 'an embodied performance in which we feel our way towards a goal, drawing on past experience and selectively or progressively integrating new information about the environment as we go."[21] The forces that transform that environment, whether local weather systems, Covid-19 or global recession, are material inscriptions which in many instances initially resist mapping or else provoke new mapping practices, new thought experiments, new scientific fictions. It is precisely because map and territory are never neatly synchronised that we have to allow each its own relative agency.

Mapping is always an act of translation, and much is inevitably lost in attempting to reduce a three-dimensional multisensory landscape to a set of highly selective two-dimensional features. But when this process of abstraction is embedded in a wider apparatus of cultural exploitation or political oppression, it becomes lethal. This is an issue brilliantly explored by Brien Friel in his aptly named play *Translations*, set in rural Ireland in the mid-nineteenth century. The play features a pragmatic cartographer and a young, idealistic orthographer who are working together on a map for the Ordnance Survey. Much of their work consists in translating local place names from Gaelic into English for purposes of the map: *Poll na gCaorach*, meaning 'hole of the sheep' in Irish, becomes Poolkerry in English. While the Irishman, Owen, has no qualms about anglicising the names of places that form part of his heritage, Yolland, the English orthographer who has fallen in love with the country is unhappy with what he perceives as a destruction of the indigenous culture and language. This is their dialogue:

> *Yolland*: I'm not sure. But I'm concerned about the past. It's an eviction of sorts.
> *Owen*: We're making a six inch map of the country. Is there something sinister in that?
> *Yolland:* I'm not sure
> *Owen*: And we're taking place names that are riddled with confusion . . . .
> *Yolland*: Who's confused? Are the people confused?
> *Owen*: And we're standardising these names as accurately and as sensitively as we can
> *Yolland*: Something is being eroded.[22]

ARE WE THAT MAP?

Clearly then there is no single axiomatic version of the map–territory relation. For example, in the context of urban planning, the map does indeed precede, over determine, and largely governs the transformation of the territory, unless it is somehow resisted. But in military operations, this is rarely the case. Sophisticated aerial-mapping technology is no more effective than old fashioned sand tables in dispelling the fog of war. Just think of the Charge of the Light Brigade or the 'collateral damage' inflicted by military drones that are often unable to show a resolution fine enough to show the damage done by 'pinpoint' aerial strikes.[23] These catastrophic meetings of map and territory can also be seen where the promise of self-driving cars, developed as they are with Lidar and 'real-time' mapping technologies, has encountered the reality of the territory. There are many reasons why self-driving cars have yet to hit the street in the numbers predicted by technologists, but one of them is the failure of these technologies to grasp the messy realities of the territory and the things that happen there.

So it seems that here at least, territory trumps map. Yet Stanley Spencer's depiction of map reading in World War I trenches as a kind of cosy campfire story suggests otherwise.[24] So too does the renaming by the troops of the front line fortifications after familiar London streets, magically transforming a foreign country of hell into some approximation of home. There is no better illustration of Keil Miller's point about counter-mapping: 'We speak to navigate ourselves/Away from dark corners, and we become/Each one of us cartographers'.[25]

An equally variegated pattern of map/territory articulations can be found in more everyday contexts. The maps used as navigational devices in dystopian computer games may conform to the postmodern model, but, as we suggested, computer-generated maps of the world posted from orbiting satellites most certainly do not. Yet these are both scientific fictions, in Vaihinger's sense, albeit carrying very differently loaded meanings.[26]

One reason for the arbitrary *and* overdetermined relation of map and territory is their constitutive lack of synchrony. The cartographic pact may create the scientific fiction that they coincide in time as well as space, but in reality, they rarely do in the way we hope they will. In effect, they operate different time signatures. Traditionally, the map oscillates between prospective and retrospective registers – it is a device for planning a journey or for remembering it, and if it be consulted en route, it still interrupts the temporal trajectory of travel. Even the sat nav, which appears to offer 'real-time' feedback, and is perhaps more frequently viewed than the paper alternative given its prominent position on the vehicle dashboard, introduces a split-second interrupter. Territory in contrast is entirely present tense – it marks the here and now of our embodied encounter with the material world. And whereas maps fix topographies arbitrarily in time – your Google Map is never fully up to date – territories render them fluid. 'Territories' are formed through trajectories of movement which maps freeze into bounded spaces. Maps fix space and our understanding of spatiality in ways that Doreen Massey warned us to be wary of.[27]

For example, the mouth of the River Alde (Suffolk, UK) is characterised by sandbanks, which are continually shifting and constitute a permanent hazard to ships. The local river authority issues new aerial photographs at the beginning of every sailing season with the caveat that sailors still need to proceed with extreme caution as a sudden storm may alter the configuration. Similarly, the coastal town of Bournemouth on England's south coast is no longer situated at the mouth of the River Bourne, the river having shifted its course in the several centuries since the place was given its baptismal naming. Thanks to the map, this toponym serves as a kind of skeuomorph, a fixed and now obsolescent sign which no longer corresponds to its changed material referent but which continues to conventionally stand for it.

A further example of this chronic disjuncture is the occupied territories of the West Bank. The dispossession of Palestinians from their ancestral lands, and the illegal construction of new settlements by Israeli Jews, have transformed the physical, social, and political geography of the area and have called a whole new cartography into being in order to authorize the land grab, or alternatively, to resist it.[28] The Israeli maps not only erase the Palestinian's property rights but also retrospectively redraw the boundaries of the state in line with changes on the ground.[29] So here the territory immediately precedes the map, but only in far as another map – the ideological map of Zionism which entitles, and indeed sanctifies – the occupation as a reclaiming of the Jew's promised land – mandates these actions. Yet that map in turn is partly a response to the territorial confinement of European Jews, first to the ghetto and then to the Nazi extermination camps, and that too is driven and legitimated by an antisemitic map of the world in which the Jew is present only as dispersed threat or a hidden hand. In each of these distinct moments, the asynchrony between territory and map is *neutralised* and *naturalised* through the asymmetry of power relations (in the first instance between host communities and diasporic Jews, then between Israeli Jews and Palestinians), and this in turn is naturalised and neutralised, in part, through the means of its cartographic representation.

One way to identify how map and territory continually overwrite and underwrite one another while retaining, subliminally, the independent temporal traces of that process is to define the landscape they describe as a *palimpsest.* This is how memoryscapes are constructed, and which Freud likened to the operation of a device known in his day as a 'mystic writing pad' and which he used as an analogy of the way the mind works to selectively store and retrieve information:

> Thus the Pad provides not only a receptive surface that can be used over and over again, like a slate, but also permanent traces of what has been written, like an ordinary paper pad.[30]

Substitute territory for slate and mental map for paper pad and you have an interesting model of the dialectical process we are talking about. The territory is a surface that is continuously traversed and shaped by the non-linear forces of nature-culture but nevertheless always appears as a tabula rasa awaiting fresh impressions, while the traces of these trajectories of meaning are preserved on the cognitive maps we create in the process of wayfaring.

The cartographic pact, the belief that a particular map will be a good enough rendering of the territory to enable us to traverse it without losing our way, is not a question of blind faith in, say, the Ordinance Survey or Google Maps, but a matter of empirical research. And what this may show is that we cannot, after all, take the 'fit' for granted. Typically, the map may be out of date, and given the rapidly changing environment of many cities and even the countryside, accelerated by global heating, it is hard for even Google Maps to remain topically topographic. Equally we may misread the map, or our sat nav may misread the terrain and send us down a cart track to a dead end rather than back onto the main road. At which point, we realise that indeed the map is not the territory.

The question then is *how* does the map differ from the territory? One answer is that the map is necessarily inscriptive and prescriptive (and never purely descriptive) while the territory is always performative (and never purely informative). If you change a line on a map or add in a feature that was not otherwise recorded that alters the configurations on the map and, consequently, the directions you may follow from it, but it does not change the territory itself. However, if you block off a public footpath by building a wall across it, then that not only re-configures the territory (i.e., in this case privatises what was public space) but actively falsifies the map unless and until either the wall is knocked down or the map is changed.

Historically, of course, maps have been used as instruments of power and appropriation, for example to delineate property relations and de-legitimate customary land uses. Their power of abstraction, rendering three-dimensional space onto a flattened surface synchronised perfectly with the power of the commodity to abstract and flatten the substantial exchanges of pre-capitalist economies, effectively de-territorialising them.

Territory is created by staking forceful claims over amenity and resource and is sustained by a variety of material practices, patrolling, surveillance, physical presence. In geopolitical terms, such claims are usually made substantive by tanks, bulldozers, or by the threat of their use; alternatively mass mobilisations of people may occupy a site as a means of preventing things they do not want to happen there. These actions may be accompanied by legal statements and/or maps designed to legitimate them, often through proclamations of local or national sovereignty or historical entitlement. But this involves the exercise of rhetorical power, issuing in a prescriptive text or discourse which only becomes performative if there is a sufficient physical force on the ground to substantiate the claim.

Of course, such rhetoric's may spur people on to make the necessary moves, that indeed is their purpose, but what actually happens, or doesn't, is down to extra-discursive factors. If I stand in the middle of my local High Street and declare it a liberated zone, inviting startled passers-by to join me in building the People's Democratic Republic of Wivenhoe I can expect to be arrested for obstruction or referred for psychiatric treatment.

For a long time, practices of territoriality were de-historicised and naturalised as the expression of some basic instinct, shared by animals and humans, who were equally compelled to mark out physically bounded areas where they exercised control over material resources. Evolutionary psychologists and socio-biologists got a lot of mileage out of this notion; at the same time, it was taken up by political scientists to argue that statehood, and state craft, consists in the exercise of national sovereignty through the policing of physical borders.[31] Social ecologists applied the notion of territoriality at a more micro-level, to groups competing for symbolic resources of status and identity in civil society. In particular, the Chicago school of urban ethnography, in their study of local gang cultures and race relations advanced a model of demographic invasion, succession, and dominance of ethnic groups vis-a-vis housing and neighbourhood amenities.[32]

It was not surprising then, given its intellectual provenance, that cultural geographers and critical cartographers in the 1970s and 1980s should ignore the concept of territory, or relegate it to the status of a dependant variable, and concentrate instead on the inscriptive power of mapping. This move also corresponded to the de-territorialisation of social relations brought about by the global information economy, the growth of distributed social networks both on and offline, and the space–time compression accomplished by digital media. Proxemics shifted focus from place-based, face-to-face communication to digital media, and the local came to be seen as an immovable object vis-a-vis the irresistible force of the global. The push back against globalisation, in which counter mappers initially played a leading role, re-asserted the importance of locally situated knowledge and feeling, of embodied navigation over alienated forms of human traffic flow, the primacy of territory over map. But, it could be argued, this has resulted in an equally one-sided version of the map–territory relation.

The iffiness of the as-if formulation thus lies not just in its focus on the mimetic but in its fetishistic reduction of the multiple and variable articulations of that relation to that single over determining instance, whether as an object of affirmation or critique. But if we learn to see the cartographic pact as just one scientific fiction amongst many, one which underwrites the common sense or taken-for-granted understandings of how maps work and what they are for, then perhaps we can also admit other kinds of thought experiment into the account. We then might be able to move beyond the present great divide between the two cultures of cartography, one privileging (digital) technologies of ever greater scientific realism and the other the creative power of the imagination released by do-it-yourself mapping. But how can these two cultures be brought into a more productive conversation?

This collection (and the Livingmaps Network at large) begins to address this question by foregrounding the possibilities laid down by the diversity of maps and map makers out there in the world. In many ways, this collection responds to the recent call to rethink maps, by highlighting how they may enact different moments, have various modes of being and can be much more than a universal methodology for spatial representation.[33] This is where we situate this book; as a showcase of contemporary mapping practices that we hope will challenge how we look and engage with maps.

COUNTER CARTOGRAPHY

In the late 1960s, the dominance of the Cartesian model was challenged by critical cartographers and geographers influenced by the poststructuralist critique of Enlightenment values and scientific rationality. They had little difficulty in deconstructing maps to demonstrate that the history of scientific cartography was inextricably linked to the growth of capitalism, colonialism, and state regulation of civil society.[34] Numerous studies have showed how the land survey, the plan view, or the cadastral map work as practices of inscriptive power to performatively enact the process of dispossession on the ground, from the renaming of place to the enclosure of space.[35] To paraphrase the complaint of indigenous peoples faced with the advent of the colonial settler map maker, which Kei Miller explores in his collection of poems, 'you had the maps and we had the territory, now we have your maps and you have our territory'.[36]

Matthew Edney has argued that cartography ≠ map-making.[37] Cartography is a colonial turned scientific practice. Map-making refers to a far wider set a practices for representing space and place, which we'll come on to later. Cartography was developed by European colonisers as way to exert territorial power over people and lands that were deemed 'other'. Today, it continues to categorise, confine, and subjugate in (post)colonial settings as well as within coloniser countries, but now under the guise of scientific objectivity that has become accepted (by many, but by no means everyone) the world over. And yet, despite these associations, *Counter Cartography* has emerged to become an important part of the so- called 'anti-colonial' and 'decolonisation' processes that seek social justice from the inequalities produced by colonialism.[38] We now have a well-documented counter-mapping scene where cartographic practices are being used to (re)claim land, culture, and situated knowledges of indigenous people.[39] This school of 'indigenous cartography' now incorporates state-of-the-art cartographic methods to help threatened communities withstand land grabs and validate customary claims over local environmental resource.[40]

Nevertheless, there are some difficult questions and contradictions raised by using a colonial practice to fight against (post)colonial rule. Further to this, there is serious thinking to be done over what and who should be mapped and over who should be a map maker or map thinker in this context? As others have rightly pointed out, there is a danger in reproducing colonial geographical thought and praxis if counter cartography is theorised and practiced using the cartographic logics developed under settler colonialism.[41] As white geographers from Britain, who are invested in counter mapping practices, this is something we have to be acutely aware of in the following analysis. It is difficult not to discuss indigenous map-making in a book about radical and counter cartography, which is why we leave this section in, but we do acknowledge that we come at the issue from a problematic perspective. In light of this, we do not assume ourselves to be the authority on the topic and aim to do our best to be respectful to the mapping work of indigenous people, activists, artists, and scholars.

In examining the history of (post)colonial cartography, Raymond Craib has argued strongly that maps were used effectively as a decolonising device by indigenous people. He states that, "like novels, art, and music, maps became a means through which to re-present and develop a postcolonial identity liberated from colonial determinations. Maps and atlases thus helped perform the hard cultural work of decolonising the land, the past, and, in the famous phrase of Kenyan intellectual Ngugi wa Thiong'o, 'the mind'".[42] He argues alongside others that mapping land anew through changes to toponyms is a significant way for indigenous groups to distance themselves from a colonial past and reclaim a sense of (post)colonial identity. 'Toponyms and languages matter not solely for the imprint they leave and the impression they make but also for the knowledges they hold, the identities they evoke, the history they convey'.[43]

However, by subverting the same cartographic practices (and *ideal*) used to subjugate people and lands, counter cartography in this context does have a tendency to fall into coloniser epistemological discourses around how space and place are represented. As it is frequently pointed out by indigenous groups and scholars, there are *other* ways to represent and understand space and place that are different from the detached 'bird's eye' viewpoints developed by Western colonial cartographers.[44] Take for example the stories, songs, prayers, symbols, artworks, etchings, and three-dimensional objects, which are used the world over as a representational means to convey the spatiality of the land and its relationship to people. For example, the mapworks of Moshekwa Langa have been said to reinstate the textures of place, ritual, and social encounter in the everyday lives of black South Africans erased from the public record during the apartheid regime.[45]

As counter cartography develops as a practice, and especially as a digital practice, there is a danger that indigenous ways of knowing and representing space and place will be obscured by the (colonial) cartographic ideal.[46] Indeed, Craib notes how (post)colonial cartographic practices, as valuable as they have been culturally, rarely extend to changes to the borders drawn up by colonial powers. It was not uncommon for newly independent nation states in the twentieth century to reuse surveys and expertise (and existing personnel) indebted to colonial rule when creating new maps of the territory. Moreover, Craib notes that granted independence does not necessarily mean changes to the way's maps supported the worldviews of the powerful. New so-called decolonial maps are still likely to favour the views of a select few and not represent a democratic process of cartography.[47]

Others have argued that indigenous map-making – sometimes referred to as 'indigenising the map' – may be a more apt form of resistance to hegemonic rule.[48] This is because it tends to focus on specific place-bound knowledge that do not, and should not, adhere to the common idea that maps are designed to be universally legible. Reuben Rose-Redwood and others have argued that indigenous-mapping decentres Western imperial knowledge and therefore contributes to wider practices of decolonial thought and practice.[49] They argue that mapping indigenous knowledge is both a practice of representing indigenous culture and also a way to encourage thinking about space and place that disrupts and disturbs long-held colonial beliefs about scientific objectivity and sovereignty. Drawing on the decidedly non-scientific notions of spirituality and the sacred, Irène Hirt argues that the indigenous knowledges of Mapuche people in Chile cannot possibly be summed up by the normative cartographic map, because they do not correspond accurately (which is ironic considering cartography's preoccupation with accuracy) to the reality of their knowledge systems, which offer alternative understandings about human–non-human relationships to the earth.[50]

Nevertheless, then the question becomes one of representational power and whether indigenous maps that do not fit the cartographic ideal of the coloniser have the power for meaningful change in (post)colonial spaces, where the cartographic ideal remains the ontological and epistemological norm in planning decisions and policymaking. As Leah Meisterlin writes of counter-mapping in urban contexts, 'As thoroughly researched, contextually responsive, and vividly descriptive as they may be, uniquely framed and tightly individualized countermaps do not scale toward influencing policy and decision making with the same persuasive efficiency of authoritative, data-driven, quantitative, GIS based results'.[51]

Perhaps the important thing to remember here is that maps, no matter how inclusive of different knowledge systems, are always and already situated in wider knowledge-power assemblages that can dictate how, if, or when they are used and seen to be legitimate objects of representation. We see this most clearly in the legal cases put forward to reclaim indigenous land from settler colonisers, which are fraught with tensions and discrepancies over whose evidence of territorial ownership is legitimate and whose is not. Where settler coloniser legal systems favour the science of cartography, it is difficult for so-called 'non-scientific' indigenous spatial knowledge to compete.[52] This should not be regarded as a call to conform to the cartographic ideal in such acts of resistance, but to follow Meisterlin, a call to think more deeply about how the counter map can subvert established cartographies and be situated as evidence with weight, alongside other forms of indigenous knowledge claims, in the epistemologies of (post)colonial policymaking.[53]

Finally, there is also the question of whether counter cartography can ever be a decolonial practice when true decolonisation asks for more than acts of representational resistance. To decolonise the map, as is often called for, does not necessarily decolonise the land. As Eve Tuck and K. Wayne Yang have argued, *decolonisation is not a metaphor* and should not be used as the label for fighting for social justice in settler colonial contexts.[54] True decolonisation is nothing less than the repatriation of indigenous land, resources, and culture, and not a process of achieving social equality for indigenous people whose land, resources, and culture continue to be occupied and controlled by the coloniser. In 2021, where colonialism and coloniality remain bound up in the lives and lands of indigenous people around the world, it's difficult to see how counter cartography can act to achieve this alone.

RADICAL CARTOGRAPHY

> *...conformality to a [normative cartographic] schema is what divides the map from the not-map, and on the other hand such schemas are culturally bound and thus mutable. Here, then, is the space where cartographic radicality can be located. A truly radical cartography would be one where the accepted schema of mapicity, or significant parts of it, is broken down and replaced. Cartographic radicality, as a condition, would be connoted by a major paradigmatic shift: a change that introduces a new vocabulary, grammar, and syntax.*
>
> Mark Denil[55]

Linked to, but not always directly intertwined with settler colonialisation,[56] there is another form of counter cartography, which has been used as a way to put other marginalised people on the map, in many cases for the first time. This has been an especially valuable practice of resistance for low income, non-white, and LGBTQ+ communities in the United States fighting the inequalities imposed upon them by society and the state.[57] Equally, but perhaps less well known has been the counter cartographies of children (see chapters 8 and 9, this volume), the elderly, the disabled, and the homeless (see chapter 3, this volume), which have been used to shed light on experiences and perceptions of spaces that are notably different from those represented on most standard cartographic maps.[58] In some cases, the maps of these latter groups are not considered counter or radical at all, particularly when they are legitimised by local authorities which are increasingly interested in them as consultation devices amidst urban planning developments.[59]

If we accept Denil's articulation of radical cartography, then much of the counter cartography discussed above is not radical. It is 'a cartography of radicals and not a radical cartography'.[60] For sure it is subversive, but not radical, because it follows the schema of the cartographic ideal that many of us are accustomed to; the grid lines, scale, key, and symbols that make up the normative maps we have grown up with. To follow Denil, a true radical cartography calls for a rethinking of how spaces, places, territories, people, and processes can be represented, which in turn would require those who think we know what a map is, to (re)learn how spatialities can be understood and represented, as well as (re)considering the ethics of spatial representation and what should/should not be mapped.

Though Denil asserts that radical cartography must break free from normative schemas to be truly radical, he is, for the most part, happy to leave the map as an object of one kind or another intended to convey spatial information. Save for a few brief remarks, he is not particularly interested in the context of a map's use, where that might be considered radical and for who. We would like to follow this line of enquiry to suggest that the context of use is fundamental to the radicality of maps. Indeed, even the maps most detached from the cartographic norm can only be considered radical if they are put to use in a radical context. Without that context, the map has no meaning whatsoever. The context makes or breaks how the map works.

Readers will see the cracks in the normative cartographic schema in the maps presented in the chapters of this book. In some cases, the schema we are used to is completely upended and most definitely radical in the sense that Denil meant. Take, for example, Emma McNally's *Field Drawings*, which on first glance appear cartographic, but then on second inspection appear to be nothing of the sort. In other cases, the maps presented begin to show what the possibilities of an alternative normative schema could be; they offer potential for seeing beyond the grid lines and the usual suspects of scale.

For us, to engage with radical cartography means two things: first, to grasp the roots of why mapping in all its many forms; analogue and digital, material and immaterial, fictional and forensic is such a fundamental human activity and arouses such passionate curiosity and debate. And second to trace the many different routes and relays through which mappings enter the world and continue to transform it, even and especially if they merely claim to be reflecting what is already there.

TOWARDS A THEORY OF CARTOGRAPHIC GENRES

All too much writing about cartography assumes some normative definition of what a map is or does and then proceeds to develop its critique and the quest for alternative approaches on that basis. But what if we reversed the process or rather turned it on its head and onto its feet. In conclusion, let us look at all the different activities that are, or might be, construed as mapping and see in what ways they are similar or different in how they go about their business and what kind of relation to territory they denote, imagine, or imply.

The examples we have briefly discussed underline the fact that the same map may have many different readings, some of them greatly at variance from the map maker's original purpose. Indeed, in recent years, there have been many calls to focus on mapping practices – or 'mapping moments' as Clancy Wilmott describes them – rather than the map itself, as a way to draw our attention to the fact that any map is closely tied to the context in which it is used.[61] Michel Hoellbecq in his novel about the contemporary artworld which draws its inspiration and title from the Korzybski dictum portrays the following cartographic epiphany:

> Jed bought a 'Michelin Departments' road map of the Creuse and Haute-Vienne. It was then, unfolding the map, he had his second great aesthetic revelation. This map was sublime. Never had he contemplated an object as magnificent, as rich in emotion and meaning as this 1:150.000-scale Michelin map. The essence of modernity, of scientific and technological apprehension of the world, was here combined with the essence of animal life. The drawing was complex and beautiful, absolutely clear, using only a small palette of colours. But in each of the hamlets and villages, represented according to their importance, you felt the thrill, the appeal, of human lives, of dozens and hundreds of souls – some destined for damnation, some for eternal life.[62]

The sheer variety of maps, their different forms, styles, and functions resist being reduced to any simple analytic typology. The terms which professional cartographers use, like the choropleth or thematic map, the cartogram, and so on, tend to conflate technical description with social function, as if there could only be one way in which a particular kind of map could be used. Nevertheless, there are some distinctions between cartographic genres which can be usefully made, provided that we recognise these are ideal types and that empirically, many maps are hybrids.

So, for example, we can identify *conceptual* maps, or meta-maps, which may – or may not – take a diagrammatic form and whose aim is to provide a framework for defining a semantic or discursive field. Increasingly conceptual maps are operationalised as part of the toolkit of business planning and used to encourage 'blue sky thinking'. The space is primarily theoretical although in some kinds of expository map this may be given a geo-location, for example in explaining the distribution of power, poverty, or population. William Bunge's expeditionary geography in Detroit produced maps illustrating the transfer of capital to the suburbs or the traffic accident pattern as white commuters drive through black neighbourhoods on their way home. Here the map is a device, at once graphic and conceptual, for correlating economic and racial patterns in the city in terms of two separate but equally lethal flows.[63]

Metaphysical maps are also propositions about the world, albeit ones which usually depict a cosmology that can be configured in topographic terms as in the Christian *paysage moralisee* produced by Victorian evangelists, depicting a pilgrim's progress that follows the narrow, though not always straight, path to salvation. Heaven and Hell are real places for religious believer and also for their secular counterparts in the contemporary self-improvement industry. Such maps should be distinguished from those which depict explicitly imaginary geographies. Utopias and dystopias are popular subjects for map-making, if only to give a symbolic location to material dreams and nightmares that otherwise might lack credibility or anchorage in the social imagination.

Fictional maps come in many different shapes and sizes; some are purely illustrative like Arthur Ransome's maps for his *Swallows and Amazons* series which allow young readers to follow the adventure, while in the case of RL Stevenson's *Treasure island*, the map itself becomes a key protagonist in the unfolding plot. Maps can spell out the geo-politics of 'countries of the mind' and no fantasy novel is complete without its map. Cartography is central to Shuiten and Peeter's graphic novel *The Invisible Frontier*, while Geoff Dyer has written a short story which is meant to be read like a map.[64] The literary critic Franco Moretti has analysed the deep maps to be found in the plots of classic Victorian novels as indicative of the social geography of class and race which subliminally shapes the characters actions. The recent emergence of literary cartography as a distinctive field of study in its own write suggests that the intersection between arts, humanities, and social science is continuing to be productive and that the hybrid nature of maps as image/texts creates a platform for experimentation that further undermines the classic divide between the 'two cultures' of cartography.[65]

Iconographic maps which give scope to visual artists to create their own territories of meaning may still contain referents to the material world, but are not constrained by them (see Section 2, this volume). This shift in the cartographic imaginary from the topographic to the topological is brilliantly accomplished in Julie Mehretu's work. She writes:

> The characters in my maps are plotted, journeyed, evolved, and built civilizations. I chart, analyze, and map their experience and development: their cities, their suburbs, their conflicts, and their wars. The paintings occurred in an intangible no-place: a blank terrain, an abstracted map space. By combining many types of architectural plans and drawings I tried to create a metaphoric, tectonic view of historical structures.[66]

In a somewhat similar vein, Layla Curtis has deconstructed the 'United Kingdom' into a new and quite surreal geography by collaging together fragments cut from commercially produced road atlases. Scottish, Northern Irish, and Welsh roads, cities, and towns are redistributed within the English border and English, Northern Irish, and Welsh roads, cities, and towns now appear within the Scottish landmass. In the process, formerly coastal towns have become landlocked, while cities far from the sea have become coastal destinations.[67]

Traditional maps like gazetteers can also furnish props for giving artists or their work a site-specific connotation, as in the case of Gilbert and George's tongue-in-cheek portrayal of themselves as iconic figures in East London's cultural landscape,[68] while in Geoff Ryman and Andrew Wilson's graphic novel, *Lars Arrhenius - A-Z*, the London A-Z serves to both locate and shape the action.[69]

Genealogical maps, such as family trees or maps which trace the development of a particular phenomenon often draw implicitly on teleological models of history or on shared myths of origin to weave a narrative that connects people and places according to what Wittgenstein called 'family resemblances' between discrete events or environments. This can create a palimpsest effect and such 'over-writing' has become a popular device amongst contemporary artists – and television producers – who are using maps or map-like formats to visualise the hidden connections between past and present or near and far.

Rhetorical maps, maps which are overtly or covertly propagandist and concerned to persuade the reader about a particular state of affairs, often use heraldic devices or iconic landmarks as condensed narratives, like the classic maps of the British Empire. Such maps invite deconstruction and provide a ready reference point for 'detournement' or satire as in Yaslov Tsvetkos' *Atlas of Prejudice*.[70] Israel's illegal occupation of Palestinian territory already referred to has produced some strenuous counter-mapping by artists, such as Joshua Neustein, exploring the impact of the land grab on the ground.[71]

In contrast, *narrative* maps proper, in which autobiographical events are indexed to specific locations, and sometimes written directly onto the map use this as a realist device to literally 'authorise' the text and the memoryscape which it unfolds. We can see these maps in the pages of diaries as well as in the inscriptions etched onto personal pocket maps like the London A-Z (see chapter 2, this volume).

The realism of *operational* maps works in quite a different way. They can be differentiated into two types: *forensic* and *navigational.* The role of forensic maps is to predict, monitor, or regulate the phenomena they depict, for example accident or crime 'hot spot' maps, weather charts, epidemiological maps. Today every branch of government routinely issues forensic maps as part of their mission to reassure the public that there is no aspect of civil society in which the state's writ does not run and to which its powers of surveillance and precaution do not extend. The realism of such maps stems as much from their symbolic efficacy as their operational efficiency, in giving the data (viz. crime statistics) public visibility, and hence credibility. Their (disavowed) rhetorical role is to achieve consensual validation of the facticities they present.

The cartographic genres sketched out here are weaved throughout the following chapters. In some this is very obvious, but in others readers will have to make up their own mind about how the mapworks fit, or indeed break free from or further complicate, these genres. If there's one thing we have come to understand about maps by bringing this collection together, it's that they are slippery, surprising, and always challenging to categorise in ways that everyone can agree on. The following chapters were brought together to emphasise this point and draw attention to the diversity of maps, map makers, and map users, all of which, we think, constitutes a radical cartography and demonstrates that the map is never the territory.

NOTES

1. This is certainly true for books and articles written for a general audience.
2. J.B. Harley and David Woodard, *The History of Cartography Project: Volume 1, Cartography in Prehistoric, Ancient, and Medieval Europe and the Mediterranean,* (The University of Wisconsin Press, 1987).
3. Map exhibitions the world over are guilty of projecting this linear imaginary. Visitors are often guided around exhibitions using chronological time as a narrative for understanding how one type of map links to the next.
4. Laura Kurgan, *Close Up at a Distance: Mapping, Technology, Politics* (Zone Books, 2013).
5. Livingmaps network was established in 2013 as a network of researchers, community activists, artists, and others with a common interest in the use of mapping for social change, public engagement, critical debate, and creative forms of community campaigning. For further information, see the preface to this book and https//www.livingmaps.org.uk.
6. Alfred Korzybski, *Science and Sanity: An Introduction to Non Aristotelian Systems and General Semantics* (Institute of Semantics, 1994).

7. Hans Vaihinger, *The Philosophy of as If* (Kegan Paul, 1935). For an application of Vaihinger's philosophy in a fictional setting, see Phil Cohen, *Graphologies* (Mica Press, 2014). For an in depth analysis of Vainhinger's theory see Kwame Appiah, *As If Idealisation and Ideas* (Harvard UP, 2017).

8. This is an idea we have seen used elsewhere as way to open up new possibilities of thought on seemingly closed topics. See for example, William Davies, *Economic Science Fictions* (MA: MIT Press, 2018).

9. The concept of a cartographic pact is analogous Philip Lejeune's notion of an 'autobiographic pact', the assumption that the author of a memoir or life story is identical to its subject/narrator in the text. See Philip Lejeune, *On Autobiography* (University of Minnesota Press, 1989). For a discussion of this, see Phil Cohen, To seek in the inferno, that which is not: Some reflections on writing a memoir, *History Workshop Journal* 50 (Summer 2013).

10. John T Pickles, *A History of Spaces: Cartographic Reason, Mapping and the Geo-Coded World* (Routledge, 2012); also Jeremy Crampton, *Mapping: A Critical Introduction to Cartography and GIS* (Wiley Blackwell, 2010).

11. J-L Borges, The exactitude of science, in *The Universal History of Infamy* (Penguin, 1976).

12. Jean Baudrillard, *Simulations and Simulacra* (University of Michigan Press, 1994).

13. John Krygier, *Everything Sings: Maps for a Narrative Atlas* (Siglio Press, 2013); Gunnar Olsson, *Intimates: Lines of power, limits of language* (University of Minnesota Press, 1991); For an overview from an American perspective, see Brian Jarvis, *Postmodern Cartographies: The Geographical Imagination in Contemporary American Culture* (New York: St Martin's, 1998).

14. The most succinct programmatic statement of critical cartography is Dennis Wood, *Rethinking the Power of Maps* (The Guildford Press, 2010).

15. Sam Hind and Alex Gekker, On autopilot: Towards a flat ontology of vehicular navigation, in Chris Lukenbeal, Laura Sharp, Elisabeth Sommerlad and Anton Escher (eds), *Media's Mapping Impulse* (PA: Casemate Academic, 2019), pp. 141–160.

16. See Tania Rossetto, *Object-orientated Cartography: Maps as Things* (Abingdon: Routledge, 2019).

17. See Valérie November, Eduardo Camacho-Hübner and Bruno Latour, Entering Risky territory: space in the age of digital navigation, *Environment and Planning D* 28 (2010): 581–599.

18. November et al. (2010), pp. 585–586.

19. For a further example of failed translation and 'friction-full' mapping, see Matthew Edney, *Mapping an Empire: The Geographical Construction of British India, 1765-1843* (Chicago: The University of Chicago Press, 1997).

20. November et al. (2010), p. 590.

21. Tim Ingold, *Perception of the Environment: Essays on Livelihood, Dwelling and Skill* (London: Routledge, 2000). See also Pierre Bourdieu *Outline for a Theory of Practice* (Cambridge University Press, 1997).

22. Brian Friel, *Translations* (Longman, 1996). See also Fintan O'Toole, *The Lie of the Land: Irish Identities* (Verso, 1997).

23. Eyal Weizman, *Forensic Architecture: Violence at the Threshold* (Zone Books, 2017).

24. See http://www.nationaltrustcollections.org.uk/object/790190.

25. Another linguistic example of popular counter-mapping from this period is 'Blighty', a term coined by homesick British soldiers to refer nostalgically to England and still used by some expatriate communities. According to the Oxford English Dictionary, the word derives from 'bilayati', a regional variant of the Urdu word 'vilayati', meaning 'British', 'English' or 'European'. So here a 'foreign' word used by colonial natives to refer to the English as foreigners is reworked, its significance reversed, so that it becomes a synonym for English attachment to a homeland in which Urdu and its speech community would undoubtedly be regarded as wholly 'other'. Not all counter-mappings are counter-hegemonic, especially in the area of race.

26. L. Kurgan (2013).

27. Doreen Massey, *For Space* (London: Sage, 2005).

28. At the same time, these cartographic actions have made an attempt to erase Palestinian cartographies, which are only now being reclaimed with the use of digital-mapping technology (see the Palestine Open Maps project. URL: https://palopenmaps.org).

29. For a detailed analysis of this processes and how digital-mapping technologies have further shaped them, see Jess Bier, *Mapping Israel, Mapping Palestine How Occupied Landscapes Shape Scientific Knowledge* (MA: MIT Press, 2017).

30. Sigmund Freud, A note on the 'Mystic Writing Pad', Vol 19 Complete Works (Hogarth Press, 1961). For a discussion of Freud's use of this device as metaphor or model of the Unconscious vis-a-vis perception and memory see Jacques Derrida and Jeffrey Mehlman, Freud and the scene of writing, *Yale French Studies* 48 (1972).

31. See the work of Saskia Sassen, Territory and territoriality in the global economy, *International Sociology* (June 2000) and When territory deborders territoriality, *Territory, Politics, Governance* 1, no. 1 (2003).

32. See the critique in Phil Cohen, Discursive rules and rituals of territoriality, in *Rethinking the Youth Question* (Palgrave, 1992).

33. Martin Dodge, Rob Kitchin and Chris Perkins, *Rethinking Maps: New Frontiers in Cartographic Theory* (Abingdon: Routledge, 2009).

34. The classic statement is: Brian Harley, Deconstructing the map, in T. Barnes and J. Duncan (eds), *Writing Worlds: Discourse, Text and Metaphor in the Representation of Landscape* (Routledge, 1992) also Brian Harley, *The New Nature of Maps: Essays in the History of Cartography* (John Hopkins University Press, 2001). A good introductory overview is Geoff King, *Mapping Reality: An Exploration of Cultural Cartographies* (London: Macmillan, 2003) and D. Denis Wood, *Rethinking the Power of Maps* (The Guilford Press, 2010). For a sample of the current debate see Martin Dodge, Rob Kitchin and Chris Perkins (eds), *The Map Reader: Theories of Mapping Practice and Cartographic Representation* (Wiley-Blackwell, 2011).

35. *Gerrymandering* and *Redlining* are two common examples of this. See the cartographic exposition of these processes in *The Detroit People's Atlas* published by Wayne University Press 2019.

36. Kei Miller, *The Cartographer Tries to Map a Way to Zion* (Carcanet, 2015). This collection won the Forward Prize.

37. Matthew Edney, *Cartography: The History and Its Ideal* (USA: Chicago University Press, 2019).

38. See for example Paul Carter, *The Road to Botany Bay* (University of Minnesota Press, 2010); *Map* INIVA 2000. The Institute of International Visual Art (INIVA) has pioneered the development of an arts based critical cartography in relation to the politics of race and post-colonial studies, with a series of exhibitions and educational projects. For further information see www.iniva.org.

39. See *This is Not an Atlas* book (2018) and online project (https://notanatlas.org/book/); Cartograhica's special issue on *Indigenous Cartographies and Counter-Mapping* (2012) Vol. 47, No. 2 URL: https://utpjournals.press/toc/cart/47/2; Alexis Bhagat and Lize Mogel, *An Atlas of Radical Cartography* (USA: Journal of Aesthetics and Protest Press, 2007); Martin Dodge, Rob Kitchin and Cris Perkins, *The Map Reader: Theories of Mapping Practice and Cartographic Representation* (Oxford: Wiley-Blackwell, 2011); The Decolonial Atlas (https://decolonialatlas.wordpress.com).

40. The pioneering study in indigenous cartography is Hugh Brody, *Maps and Dreams: Indians and the British Columbia Frontier* (Douglas and McIntyre, 1981); See also Barbara Mundy, *The Mapping of New Spain: Indigenous Cartography and Maps of the Relaciones Geograficas* (University of Chicago Press, 1996); John Rennie Short, *Cartographic Encounters: Indigenous Peoples and the Exploration of the New World* (Reaktion Books, 2009); The Native Land Map project (https://native-land.ca).

41. Reuben Rose-Redwood, Natchee Blu Barnd, Annita Hetoeve, hotohke'e Lucchesi, Sharon Dias, and Wil Patrick, Decolonizing the map: recentering indigenous mappings. *Cartographica* 55, no. 3 (2020): 151–162; see also the doctoral work in progress of David Garcia (@mapmakerdavid on Twitter).

42. Raymond Craib, Cartography and decolonisation, in: James R. Akerman (ed), *Decolonising the Map: From Colony to Nation* (Chicago University Press, 2017), pp. 18.

43. Craib (2017), pp. 19.

44. A process that cannot be separated from colonialism itself, whereby cartographic practices and techniques were developed with and by colonial subjects as well as by colonisers themselves, albeit not equally.

45. See http://www.andrewkreps.com/artists/moshekwa-langa.

46. M. Edney (2019).

47. Craib (2017).

48. Margaret W. Pearce and Stephen J. Hornsby, Making the *Coming Home* map. *Cartographica* 55, no. 3 (2020): 170–176.

49. Reuben Rose-Redwood et al. (2020).

50. Irène Hirt, Mapping dreams/dreaming maps: bridging indigenous and Western geographical knowledge. *Cartographica* 47, no. 2 (2012): 105–120. See also Gay'wu Group of Women, *Song Spirals* (Sydney: Allen & Unwin, 2019); and on similar lines, the discussion between Western understanding of navigation and wayfinding, and

its other conceptualisations, e.g. Fetaui Iosefo, Stacy Holman Jones, Anne Harris, *Wayfinding and Critical Autoethnography* (London: Routledge, 2020).

51. Leah Meisterlin, Cartographies of distance, in Laure Kurgan and Dare Brawley (eds), *Ways of Knowing Cities* (New York: Colombia Books on Architecture and Cities, 2020), pp. 253.

52. Though this is not necessarily a one way street. See, for example, V. C. Cruz and D.A. Oliveira (eds.), *Geografia e Giro Desco- lonial: Experiências, Ideias e Horizontes de Renovação do Pensa- mento Crítico* [Geography and the Decolonial Turn: Experiences, Ideas and Horizons of Renewal of Critical Thought] (Rio de Janeiro: Letra Capital Press, 2017); Reuben Rose-Redwood et al. (2020).

53. For an overview of cases that demonstrate the strategies and tactics of indigenous counter-cartography, see Bjørn Sletto, Joe Bryan, Alfredo Wagner, and Charles Hale, *Radical Cartographies Participatory Mapmaking from Latin America* (TX: University of Texas Press, 2020).

54. Eve Tuck and K. Wayne Yang, Decolonization is not a metaphor, *Decolonization: Indigeneity, Education & Society* 1 (2012): 1–40.

55. Mark Denil, The search for radical cartography, *Cartographic Perspectives* 68 (2011): 7–28.

56. For cases where there is a direct correlation, see the Anti-Eviction Mapping Project, *Counterpoints: A San Francisco Bay Area Atlas of Displacement and Resistance* (Oakland, CA: PM Press, 2020).

57. See Rebecca Solnit's collection of counter maps in U.S. cities (http://rebecca-solnit.net/atlases/); The antieviction mapping project (https://antievictionmap.com); The counter cartographies collective (https://www.countercartographies.org); An Everyday Queer New York (http://jgieseking.org/AQNY/).

58. Nancy Duxbury, W.F. Garrett-Petts and Alys Longley, *Artistic Approaches to Cultural Mapping: Activating Imaginaries and Means of Knowing* (UK: Routledge, 2018); Jin-Kyu Jung and Ted Hiebert, Imagining the details: happy places and creative geovisualisation, *Livingmaps Review* 7 (2019).

59. See also a typology of different critical/radicalmapping approaches in Sam Hind, Disruptive Cartographies: Manoeuvres, Risk and Navigation (PhD Thesis, 2016), p.75. Available at http://wrap.warwick.ac.uk/89673/1/WRAP_Theses_Hind_2016.pdf.

60. M. Denil (2011), pp. 19.

61. Clancy Wilmott, *Mobile Mapping: Space, Cartography and the Digital* (Amsterdam University Press, 2020); Cris Perkins, Cultures of map use, *The Cartographic Journal* 45, no. 2 (2008): 150–158; Rob Kitchin, Justin Gleeson and Martin Dodge, Unfolding mapping practices: a new epistemology for cartography, *Transactions of the Institute of British Geography* 38 (2013): 480–496.

62. Michel Houellbecq, *The Map Is Not the Territory* (Cape, 2012).

63. William Bunge, *Fitzgerald: The Geography of a Revolution* (Schentman, 1971).

64. Francois Schuiten and Benoit Peeters, *The Invisible Frontier Vol 1 and 2* (NEM, 2013); Geoff Dyer, The boy out of Cheltenham, in Geoff Dyer (ed), *Where You Are* (Picador, 2014).

65. For a good example of this trend see High Lewis-Jones (ed), *The Writers Map* (Thames and Hudson, 2017).

66. July Mehretu Interview in *Art Quarterly* (June 2008).

67. See http://www.laylacurtis.com.

68. Gilbert and George, Fournier street, 2008.

69. Geoff Ryman and Andrew Wilson, *Lars Arrhenius - A-Z* (London: Peer, 2002).

70. Yaslov Tsvetkov, *Atlas of Prejudice Vol 1* (Alpha Designer, 2013).

71. Eyal Weizman, *Hollow Land: Israel's Architecture of Occupation* (Verso, 2007); See also http://joshuaneustein.com.

BIBLIOGRAPHY

Anti-Eviction Mapping Project. 2020. *Counterpoints: A San Francisco Bay Area Atlas of Displacement and Resistance*. PM Press. Oakland, CA.

Appiah, K. 2017. *As If Idealisation and Ideas*. Harvard UP. Cambridge.

Baudrillard, J. 1994. *Simulations and Simulacra*. University of Michigan Press. Chicago.

Bhagat, A. and Mogel, L. 2007. *An Atlas of Radical Cartography*. Journal of Aesthetics and Protest Press, USA.

Bier, J. 2017. *Mapping Israel, Mapping Palestine: How Occupied Landscapes Shape Scientific Knowledge*. MIT Press. MA

Borges, J. 1976. *'The Exactitude of Science' in The Universal History of Infamy*. Penguin. London.

Bourdieu, P. *Outline for a Theory of Practice*. Cambridge University Press. Cambridge.

Brody, H. 1981. *Maps and Dreams: Indians and the British Columbia Frontier*. Douglas and McIntyre. BC.

Bunge, W. 1971. *Fitzgerald: The Geography of a Revolution*. The University of Georgia Press. Georgia.

Carter, P. 2010. *The Road to Botany Bay*. University of Minnesota Press. MN.

Cohen, P. 1992. *Rethinking the Youth Question*. Palgrave. London.

Cohen, P. 2013. To seek in the inferno, that which is not: Some reflections on writing a memoir. *History Workshop Journal*, 50, Summer.

Cohen, P. 2014. *Graphologies*. Mica Press. Wivenhoe.

Craib, R. 2017. Cartography and decolonisation. In: James R. Akerman (Ed) *Decolonising the Map: From Colony to Nation*. Chicago University Press. Chicago, pp. 18.

Crampton, J. 2010. *Mapping: A Critical Introduction to Cartography and GIS*. Wiley-Blackwell. Oxford.

Cruz, V. C and D.A. Oliveira, eds. 2017. *Geografia e Giro Desco- lonial: Experiências, Ideias e Horizontes de Renovação do Pensa- mento Crítico [Geography and the Decolonial Turn: Experiences, Ideas and Horizons of Renewal of Critical Thought]*. Rio de Janeiro: Letra Capital Press.

Davies, W. 2018. *Economic Science Fictions*. MIT Press. MA.

Denil, M. 2011. The Search for Radical Cartography. *Cartographic Perspectives*, 68: 7–28.

Derrida, J. and Mehlman, J. 1972. Freud and the scene of writing. *Yale French Studies*, 48: 74–117.

Dodge, M., Kitchin, R., and Perkins, C. 2009. *Rethinking Maps: New Frontiers in Cartographic Theory*. Routledge. Abingdon.

Dodge, M., Kitchin, R., and Perkins, C. 2011. *The Map Reader: Theories of Mapping Practice and Cartographic Representation*. Wiley-Blackwell. Oxford.

Duxbury, N. Garrett-Petts, W.F. and Longley, A. 2018. *Artistic Approaches to Cultural Mapping: Activating Imaginaries and Means of Knowing*. Routledge. London.

Dyer, G. 2014. The boy out of Cheltenham. In: Geoff Dyer (Ed) *Where You Are*. Picador. London.

Edney, M. 1997. *Mapping an Empire: The Geographical Construction of British India, 1765-1843*. The University of Chicago Press. Chicago.

Edney, M. 2019. *Cartography: The History and Its Ideal*. Chicago University Press. Chicago.

Freud, S. 1925. *'A note on the 'Mystic Writing Pad' in Vol 19 Complete Works*. Hogarth Press. London.

Friel, B. 1997. *Translations Longman 1996 See also Fintan O'Toole The Lie of the Land: Irish Identities*. Verso. London.

Gay'wu Group of Women. 2019. *Song Spirals*. Allen & Unwin. Sydney.

Harley, B. 1992. Deconstructing the map. In: T. Barnes and J. Duncan (Eds) *Writing Worlds: Discourse, Text and Metaphor in the Representation of Landscape*. Routledge. London.

Harley, B. 2001. *The New Nature of Maps: Essays in the History of Cartography*. John Hopkins University Press. Baltimore.

Harley, J.B. and Woodard, D. 1987. The History of Cartography Project: Volume 1, Cartography in Prehistoric, Ancient, and Medieval Europe and the Mediterranean. The University of Chicago Press. Chicago.

Hind, S. and Gekker, A. 2019. On autopilot: towards a flat ontology of vehicular navigation. In: C. Lukenbeal, L. Sharp, E. Sommerlad and A. Escher (Eds) *Media's Mapping Impulse*. Casemate Academic. PA, pp. 141–160.

Hind, S. 2016. Disruptive Cartographies: Manoeuvres, Risk and Navigation. PhD Thesis (p. 75). Available at http://wrap.warwick.ac.uk/89673/1/WRAP_Theses_Hind_2016.pdf.

Hirt, I. 2012. Mapping dreams/dreaming maps: bridging indigenous and Western geographical knowledge. *Cartographica*, 47(2): 105–120.

Houellbecq, M. 2012. *The Map Is Not the Territory*. Cape. London.

Ingold, T. 2000. *Perception of the Environment: Essays on Livelihood, Dwelling and Skill*. Routledge. London.

Iosefo, F., Jones, S. H. and Harris, A. (2020) *Wayfinding and Critical Autoethnography*. Routledge. London.

Jarvis, B. 1998. *Postmodern Cartographies: The Geographical Imagination in Contemporary American Culture.* St Martin's. New York.

Jung, J-K. and Hiebert, T. 2019. Imagining the details: happy places and creative geovisualisation. *Livingmaps Review*, 7.

King, G. 2003. *Mapping Reality: An Exploration of Cultural Cartographies.* Macmillan. London.

Kitchin, R., Gleeson, J. and Dodge, M. 2013. Unfolding mapping practices: a new epistemology for cartography. *Transactions of the Institute of British Geography*, 38: 480–496.

Kollektiv Orangotango. 2018. *This is Not an Atlas.* Transcript Verlag. Bielefeld.

Korzybski, A. 1994. *Science and Sanity: An Introduction to Non Aristotelian Systems and General Semantics.* Institute of Semantics. New York.

Krygier, J. 2013. *Everything Sings: Maps for a Narrative Atlas.* Siglio Press. New York.

Kurgan, L. 2013. *Close up at a Distance: Mapping, Technology, Politics.* Zone Books. New York.

Lejeune, P. 1989. *On Autobiography.* University of Minnesota Press. MN.

Lewis-Jones, H. 2017. *The Writers Map.* Thames and Hudson. London.

Massey, D. 2005. *For Space.* Sage. London.

Meisterlin, L. 2020. Cartographies of distance. In: Laure Kurgan and Dare Brawley (Eds) *Ways of Knowing Cities.* Colombia Books on Architecture and Cities. New York, pp. 253.

Miller, K. 2015. *The Cartographer Tries to Map a Way to Zion.* Carcanet. Manchester.

Mundy, B. 1996. *The Mapping of New Spain: Indigenous Cartography and Maps of the Relaciones Geograficas.* University of Chicago Press. Chicago.

November, V., Camacho-Hübner, E. and Latour, B. 2010. Entering Risky territory: space in the age of digital navigation. *Environment and Planning D*, 28: 581–599.

Olsson, G. 1991. *Intimates: Lines of Power, Limits of Language.* University of Minnesota Press. MN.

Pearce, M. W. and Hornsby, S. J. 2020. Making the coming home map. *Cartographica*, 55(3): 170–176.

Perkins, C. 2008. Cultures of map use. *The Cartographic Journal*, 45(2): 150–158.

Pickles, J. 2012. *A History of Spaces: Cartographic Reason, Mapping and the Geo-Coded World.* Routledge. Abingdon.

Rose-Redwood, R., Barnd, N. B. Hetoeve, A., Lucchesi, H., Dias, S. and Patrick, W. 2020. Decolonizing the map: Recentering indigenous mappings. *Cartographica*, 55(3): 151–162.

Rossetto, T. 2019. *Object-orientated Cartography: Maps as Things.* Routledge. Abingdon.

Ryman, G. and Wilson, A. 2002. *Lars Arrhenius - A-Z.* Peer. London.

Sassen, S. 2000. Territory and territoriality in the global economy. *International Sociology*, June 2000.

Sassen, S. 2003. When territory deborders territoriality. *Territory, Politics, Governance*, 1: 21–45.

Schuiten, F. and Peeters, B. 2013. *The Invisible Frontier Vol 1 and 2*. NBM. New York.

Short, J. R. 2009. *Cartographic Encounters: Indigenous Peoples and the Exploration of the New World*. Reaktion Books. London.

Sletto, B., Bryan, J., Wagner, A., and Hale, C. 2020. *Radical Cartographies Participatory Mapmaking from Latin America*. University of Texas Press. TX.

Hale, Janice, et al. 2020. *A People's Atlas of Detroit.*. Wayne University Press. Detroit. The University of Wisconsin Press. Madison.

Tsvetkov, Y. 2013. *Atlas of Prejudice Vol 1*. Alpha Designer. London.

Tuck, E. and Yang, K. W. 2012. Decolonization is not a metaphor. *Decolonization: Indigeneity, Education & Society*, 1: 1–40.

Vaihinger, H. 1935. *The Philosophy of As If*. Kegan Paul. London.

Weizman, E. 2007. *Hollow Land: Israel's Architecture of Occupation*. Verso. London.

Weizman, E. 2017. *Forensic Architecture: Violence at the Threshold*. Zone Books. New York.

Wilmott, C. 2020. *Mobile Mapping: Space, Cartography and the Digital*. Amsterdam University Press. Amsterdam.

Wood, D. 2010. *Rethinking the Power of Maps*. The Guildford Press. New York.

Part 1

ARE WE THAT MAP?

Chapter 1

The Cultural Life of Maps

Everyday Place-Making Mapping Practices

Mike Duggan

TORI

Like many of us, Tori now uses a range of digital maps as she navigates her way through contemporary life in the city. It was not uncommon to find her using a Sat-Nav device while driving, examining maps within popular smartphone applications or exploring the planet on Google Earth via her laptop at home. And yet, in her journey-making practices through London, the popular A-Z pocket atlas held a permanent place within her day bag; thrown in alongside her smartphone, purse, notepad, pens, jewellery, fresh and fading receipts.

Tori had grown quite attached to the atlas. It was an object she had come to like, not only for its practical purposes in navigation but also as a material artefact with a tactile and aesthetic quality that she found appealing. She saw no reason to replace it, for it had been her trusty guide in getting to know London for the last ten years, long before digital maps had become the popular choice of city map. She had learnt how to use it by cross referencing what she saw out of the top deck windows of buses with what she saw on the page. Over time she had used this method to build up a navigational system and spatial imaginary of London, which was very specific to her own experiences of living in the city. She did not consider her method any better than using digital alternatives. On the contrary, she often said that her method was time consuming and would certainly be made easier if she bothered to learn the digital way. Rather, she had simply become accustomed to this way of use; it had become a habit, which she saw no reason to change at this point.

Tori's affection for the A-Z map was made clear by its state of appearance. It was a map made through many years of use. Taking it from her bag, 'where it lives', she showed me its dog-eared corners, tatty pages, its out-of-date tube map printed on the back page, and where she had scribbled annotations across many of its pages. Personal meanings were inscribed all over the map, many of which only made sense to her during the time they were made. Whilst Tori could easily recall what some of the arrows, numbers and circles penned on the map referred to – places of work, bus stops, tube stations, estimated journey times – in most cases she had forgotten their significance. These pages were littered with a permanent record of forgotten journeys. Although digital maps may be similarly annotated, this particular type of personalisation is only made possible by the form of the pocket atlas. There is a freedom to this type of personalisation which is not available to her when using the digital map.

(Notes taken from field diary, June 2015)

Maps are constituted of a collection of lines, symbols, words, and numbers. These attributes are layered and meshed together to form the common appearance of a map. Gridlines intermingle with contour lines, place names, street names, road numbers, lines representing rivers, roads, railways, footpaths, boarders, buildings, and symbols representing stations, churches, pubs, and parks. As Denis Wood and John Fels suggest, these graphical forms wield great power, which feed into our understanding and experience of space, and how we imagine the territories, terrains people, and places that are represented on maps.[1] It is often said that when we use maps in everyday life, we are confined by the cartographic logics of these forms, that there is no way to escape the gleaming grid lines of cartographic reason.[2] There is a widespread belief in the map as neutral and transparent, which is made all the more believable by the fact that what is represented on a map has a striking resemblance to what we see in the world and a remarkable ability to work in the world.[3] These cartographic logics are understood, produced, and reproduced as they circulate through the material and cultural practices of everyday life.[4] Indeed, it is a belief that is reinforced and verified when maps works; when we reach our destination on time and without getting lost and when that road on the map corresponds with that road ahead.[5] Despite the misgivings of some maps, we are not put off by the idea that cartographic logics are the best way of ordering and navigating the world in which we live. This goes for the layperson as well as those schooled in subverting the power of maps.[6] We are all susceptible to the lure of the map.

Nevertheless, as the opening vignette suggests, maps may also be constituted through practices of everyday place-making. Maps don't simply go out into the world as finished graphical forms wielding great power; rather they also become part of cultural practices that are often far removed from the uses envisioned by their makers.[7] Maps have a cultural life. Or rather, maps and mappings become deeply embedded into cultural life as artefacts *and* practices.[8] In the case of Tori, the A-Z atlas has become bound up in her daily practices of navigating and getting to know London. Her subjective practices of annotating the map have become an important part of this process. Indeed, it could be said that she made the map her own through acts of inscription, for such scribbles and marks may be useless to anyone else not familiar with her particular way of working. These annotations offer her a personalised way of knowing and documenting her place-making in the city.

This chapter offers a commentary on the cultural life of maps by describing how they become intertwined with everyday place-making practices. It also provides an insight into how the cultural life of maps is changing in world increasingly perforated by digital maps. I will demonstrate how the map is both an artefact to be worked with to produce knowledge of place, and a 'quasi-object' (following Michael Serres) [9] used to inform social relations in place. In describing what and why everyday users inscribe onto and into specific types of map, I build on the notion that knowledge of place is produced and accumulated *through* and *with* the map in a social context,[10] as a performance,[11] and that uses of the map must be understood by examining the cultural contexts in which they take place.[12] Moreover, I highlight how repeat practices of annotating the map may constitute a sense of place for the user, which further confirms the notion that we come to know a place by returning over and again to specific locations and undertaking specific practices.[13]

I develop these arguments through an ethnographic study of contemporary mapping practices in London and the South East of England between October 2013 and May 2016. Focusing on participants from this research, I give descriptive snapshots of place-making mapping practices in order to show how contexts of use matter in practices of annotation and to highlight how maps extend beyond their envisioned uses and become folded into the cultures of daily life. In the first of these cases, I build on the ethnographic account of Tori, outlined above, by describing how and why the popular London A–Z mapping atlas is used and annotated in the everyday navigational practices of another London resident, Sally (aged twenty-nine).[14] The second case, that of a road cyclist named Barbara (approximately forty-five), illustrates how place-making practices unfold with digital maps. Focusing on popular cycling mapping software I show how digital maps are used to produce routes and social experiences for her and her cycling club mates.

Collectively, these empirical examples seek to demonstrate the diversity of place-making mapping practices and offer some insight into the different ways in which paper and digital maps come to be used and annotated in everyday life. Moreover, I use these examples to illustrate how the textual and technical capacities of each form have an impact on how users may annotate the map. These differences, I argue, are important to note for they show that place-making with paper and digital maps are fundamentally not the same. Users of digital maps are far more restricted in terms of the annotations they can make, which I suggest can have an impact on how place-making practices unfold.

The chapter is split into two primary sections. The first section explores how maps are used beyond the original use envisioned by their author in the context of place-making. I focus on paper maps, describing further how marking up an A–Z map constitutes a repeat place-making performance. In the second section, I focus on how digital maps are used in practices of place-making. Following Sybille Lammes,[15] I make a claim about the textual and technical composition of maps as artefacts and argue that there are fundamental differences to the ways that paper and digital maps may be annotated and put to use in everyday practice. In laying out this argument, I make a clear distinction between annotations made to paper maps and those made to digital maps, suggesting that the former offers a more permanent juxtaposition to the common cartographic form than the latter.

LINES, MAPS, MOVEMENT: PLACE-MAKING PAPER-MAPPING PRACTICES

Tim Ingold has noted that the lines of sketch maps are the etchings of movements from the past; previous experiences of journeys within the world retold in a graphical form.[16] For Ingold, every line on a sketch map is the trace of a gesture that is a retelling of an experience of movement through the world, which he calls *mapping*. Nevertheless, not all mappings are necessarily inscribed as artefacts, with most remaining only gestural or fleeting as they are retold through storytelling (for example, in giving directions). What Ingold does so brilliantly here is to put maps into contention with movement. Maps, he suggests, cannot be separate from the movements (mappings) we make throughout the world because they are representative of these movements and because the making of sketch maps is a performative act of movement itself as gestures are traced or etched. We can, however, take this claim further and suggest that maps, sketch maps or otherwise, are never finished, their lines are never finally drawn. Instead, I contend, alongside others, that maps, like mappings, are always coming into being as they a put to use in everyday life.[17] Thus, practices of mapping are produced anew each and every time a map is called upon. These practices unfold as assemblages defined by specific contexts of use, which I suggest are deeply embedded in specific cultural and technological practices. In the following, I show how specific these contexts can be, beginning with unpacking Sally's relationship to her London A–Z.

KNOWING LONDON THROUGH THE MAP: SALLY

Sally's pocket-sized A–Z atlas (2005 edition) is another well-worn artefact that has been used in place-making practices around London. Like Tori's atlas, it tells a number of social and material stories of how Sally came to know and make sense of London's sprawling landscape. Unlike Tori's continued use, however, Sally's atlas sits upon her shelf in a way that suggests that it doesn't get much use any longer. And it doesn't; like many people, Sally migrated long ago to using digital formats – mostly Google Maps – on her smartphone or laptop. Looking at the atlas, having reached down to take it off the shelf, she tells me it holds great sentimental value; not for now, but as a material reminder of how she came to know this place (London). Many pages are littered with scribbles, crosses, small paper markers, and loose scraps noting addresses and inserted on specific pages (see figure 1.1). They

Figure 1.1a and b Sally Pocket London A–Z Map Atlas, Pictured with Permanent Annotations in Pen and Individually Stuck Down Pieces of Paper, Some of which Are Stuck Down with Glue While Others Are Attached by Post-it Notes. Author's Image.

point to work places, shops, art galleries, friends' houses, social spots and nightclubs, all of which speak to the daily practices of her early adulthood in the city. As she flicks through it, she tells me a little about what each annotation represents, which shops and galleries she was going to and who was playing at that gig, *that* night. At first she is reluctant for me to take a look, and frequently reminds me to be careful not to pull out pages as I make my way through its narrative. It is clearly precious to her now, but I do question if that would have been the case at the time of its daily use as it made its way in and out of bags into hands and onto tables and laps in preparation for a journey or in a hurry during a journey. She tells me that it was, but that it is more so now, like a series of diary entries from the past, all of which evoke a sense of nostalgia about a place not especially present in my discussion with Tori.

By writing and sticking points of personal interest on to the map, Sally and Tori have effectively defaced the map. Pen and paper obscure places all over these maps, making them irrelevant and unimportant, at least for them at these times. Their practices literally reorganise the map. Through these material practices of place-making, they have inadvertently renegotiated the authority of the map's gleaming grid lines in ways that would not be permissible or even possible on other maps. This is partly down to the maps and contexts in question. The pocket-sized A–Z atlas encourages personalised use in ways that other maps do not. They are sold on the premise that they are for individual use and their paper form allows for annotations to be made with ease. Scribbling addresses over public street maps or sticking pieces of paper onto London's tube maps would not be so easy nor would it be a socially acceptable alternative.[18] Such maps are displayed in public spaces and made of a different material, where different rules apply. One may trace routes and pinpoint locations with their fingers without any trouble, but to apply ink or paper would be socially inappropriate. This goes to show that the cultural life of maps is heavily dependent on the spaces in which they are used.

Like Tori, Sally made the A–Z her own over time in the daily cumulative acts of getting to know London. The slow accumulation of this knowledge, presented on the same pages over time had become bound up in their practices of place-making. By adding notes and points bit by bit, they both layered their own spatial understanding of London in a way that could now be done instantly using a digital map. However, it was exactly the speed in which these annotations could now be added which was a problem for Tori. To add everything at once is to add nothing. Tori assumed the clarity of her spatial understanding was only made possible by the slow accumulation of knowledge, which was added to the map over a long time, after multiple experiences navigating the city. These insights reflect Tim Ingold's notion that we build up a knowledge and understanding of places in practices of repetition over time.[1]

Sally's annotations have become part of a map of her own making. They build upon a familiar pre-existing cartographic form but do not fit so neatly within the same cartographic style. In effect, they offer a juxtaposition, a disruption to the cartographic norm. Moreover, having made these additions in pen and with paper stuck down by glue, they have become permanent features of this map and cannot be removed without any further defacing. Such a record no longer exists in the same way now that Sally uses digital maps on her phone and laptop. Nor is such personalised permanence technologically permissible in the same way. Any annotations or notes Sally makes on Google Maps, which is possible but not often done by Sally owing to the convenience of making repeated searches with this technology, must be recalled computationally each and every time she uses them. The algorithmic, codified, and electronic processes that come to produce the representation of digital-map-with-annotation must be re-collected from where they are stored (a data centre) and reproduced each and every time they are requested to load.[19]

All cartographic information may be stored and recalled for use. Sally and Tori both made notes on their maps and then recalled this information each and every time they used them, albeit in a specific context of use that would have shaped how this information was asked to be used. As Rob Kitchin et al. note, maps have a great many uses, any number of which may emerge depending on the contexts of their use.[20] A–Z maps and Google Maps are no different in this respect, for it is the context of use which determines how and why each map is used in any given practice. However, what *is* different about the way in which Sally brings the cartographic form of her annotated A–Z into being and the way she brings Google Maps into being can be aligned with the textual and technical properties of each mapping artefact. The paper A–Z as annotated by Tori and Sally is immutable in ways that Google's digital maps are not.[21] Pages may tear, get a little damp and lose colour in the rain, and her stuck down inserts may begin to peel in the heat, but within a relative time frame, her particular A–Z can be expected to remain more or less the same, allowing Sally to revisit and recall the same information time and again. As Sybille Lammes has suggested, digital maps work differently.[22] They are mutable in ways that offer something different each time they are called to use. The interfaces on which they are represented shape different performances, have a capacity to produce different events, and fundamentally mediate place-making practices in different ways.

When Sally opens Google Maps on her phone prior to or during a journey she cannot expect to view the same information as when she last opened the application. While the map aesthetic is likely to be the same,[23] giving her an impression that the map is indeed the same as when she left it, all manner of changes to the software may have been made in the code used to produce the map in this time. The capacity of the paper surface to retain information is therefore not equivalent to the capacity of digital maps to do the same. Paper surfaces needn't be refreshed or reloaded upon every new encounter. For instance, yesterday's map and today's Google Map could be very different as changes were made by software developers looking to add new points of interest, cartographic styles or test new routing algorithms. Were Sally to look back upon Google Maps in five year's time in an effort to remember her place-making mapping practices of today, the personal details afforded to her by the annotated A–Z would not be accessible and therefore not likely to produce similar feelings of nostalgia for past journeys made. Besides her memories of these events, all that would be left would be digital traces of searches made, routes taken, and places visited, available only to Google, used to feed future product development with the ultimate goal of producing further profit for the company.

The mutability of digital maps is often bound up with the profit-driven models of capitalist logics.[24] While maps have long been produced and sold to make a profit, the complex computational systems behind today's popular for-profit digital maps ensure that profits from maps are generated in different ways. Let's take Google Maps as an example. Revenue is not generated by user purchases (though it is through selling its Application Programming Interfaces (API) as a business service). Instead, Google Maps are freely given away to everyday practitioners in the form of a download or web service. Using this model, revenue is generated through a computational system of advertising in which the user becomes targeted by increasingly personalised adverts in their use of the map rather than by direct payment for the map. Most simply, what users see are digital maps littered with icons indicating the location of businesses paying the map provider to be included on the map.

DIGITAL TRANSITIONS

Sally continues to use maps on a daily basis as she discovers new places and new routes through the city, but crucially, the mapping technologies she now uses offer her different place-making possibilities. As digital maps and mobile technologies made possible previously latent capacities of place-making, Sally's experience of using maps in London has changed. Discussing her use of Google Maps and City Mapper she tells me that digital maps allow her to make judgements on the move-about journey times, modes of travel, and places of interest in ways that her paper A–Z could not. Previously, Sally added notes and paper inserts onto her map before making a journey, making searches online using a laptop and adding details onto the pages of the map, before using these permanent features as a static map when making journeys across town. Her mapping practices were constituted by a hybrid of analogue and digital technologies in this respect. Now, while she continues to make online searches before taking a trip, she has access to mapping technology that allows her to make adjustments to these plans on the move, easily search for new points of interest and provide her with time and travel information as and when she needs it. Her current mapping practices have an added layer of dynamism and temporality which have affected her previously hybridised system. This is to say that digital-mapping practices have long replaced the majority of her analogue mapping practices. The analogue which remains, and will perhaps always remain, is embedded in the contexts of her journey-making practices, which are assembled by a multiplicity of digital and non-digital maps that constitute contemporary experiences of being in the city. Like many she remains a user of tube maps on station platforms and those slotted into carriage interiors, as she does a user of analogue you-are-here street maps and gallery maps.

Sally is, what some people call, an *early adopter* of technology, and therefore it is perhaps little surprise that she has made this shift away from paper to digital. The digital maps she is now familiar with allow her to make searches, create routes, and mark places in far more convenient manner than she used to. According to Sally, the digital maps, available to her, do a better job than those of old. One example she gives relates to finding bus stops in areas of town that have multiple stops. Using digital maps she is no longer required to correlate her A–Z with that of a localised bus stop map, which indicates what buses stop at which stops. Google Maps does much of the work for her, for which she is grateful, having considered it a great success when she had managed to navigate London's buses previously. Nevertheless, it is clear from our conversations about the A–Z that Sally considers something lost after making this switch, albeit something that is draped in a sentimentality for the past which is not entirely bound up in her use of maps, but instead in her life's history. What is clear is that maps once played a different role in her place-making practices than they currently do. What is not clear is how she will recall her place-making mapping practices in the future, for she may well have fond memories of this current juncture, which are brought into being in different ways, off the map.

LINES, MAPS, MOVEMENT: PLACE-MAKING DIGITAL-MAPPING PRACTICES

Planning Cycle Routes with Barbara

Tori and Sally make for specific case studies for they don't much bother with making annotations to digital maps. In other cases, drawing on the digital map is a daily practice undertaken by many. Journey trackers, fitness apps, and location-based mobile games are all common examples of digital products which trace movements onto maps that may be viewed and analysed by users during or after the fact. Much of this software allows users to also produce their own routes to be downloaded onto mobile devices to be followed by themselves or others. These actions have been said to be both serious and leisurely efforts to train bodies and reach goals using quantitative biotechnical practices.[25] In others contexts, these actions have been described as playful mapping practices, for instance, in artistic practices and community mapping projects.[26]

Barbara, a keen road cyclist, is one such user of digital maps. Bound up in her training plans are precise journey-planning practices in which she produces cycling routes in accordance with specific training goals. When we meet, she is training for an epic non-stop 600km ride across Norway. The digital technology she uses (a computer and mouse) permits her to draw on a base map where she wants to go. This then gives her precise data about how long a route is, how much elevation gain/loss there is in the topography of the land, how long this route is expected to take based on her previous average speeds uploaded onto the system and how many calories she is expected to burn. The base map itself provides the cartographic form and styles one expects from a map, giving her place names, road types, and points of interest, all of which she works into her calculations (see figure 1.2).

The process by which Barbara creates and updates routes using various software and websites available is something that can be extremely fiddly, frustrating, and time consuming. We frequently discuss the pros and cons of .GPX (digital map) files, storage capacities, the limits of Garmin's Connect software and the advantages of Strava and Google's route-planning services.[27] Each platform offers some advantage over the other and each has its own specific environment that takes time to master. And then there is the task of getting it on the device; a case of plugging in, updating firmware, and freeing up the disk space needed for storage. All in all, route-planning is 'a fine art' says Barbara, 'not as easy as you might think'. And like an artist, it has taken her a while to hone her craft. After a number of years practicing, she is now quite capable and has a large number of routes (or courses as they're known on the Garmin software) stored on her home computer, which she can then upload, put into rotation, edit, and share with others on her cycling club's online forum or via social media. Barbara tells me this sharing has become an important part of her cycling practices in recent years, as it has fostered new social relationships with other members of the club. Through the exchanging of routes, Barbara and her maps have become part of a sub-culture of her cycling club that regularly discuss, question, and build home-made digital routes with the aim of sharing and riding them together.

Figure 1.2 A Screenshot Showing One of Barbara's Training Routes Laid Over a Digital Base Map on the Garmin Connect Website. Author's Image.

The purpose of making these routes at home is tied up in Barbara's training plan, but also in her practices of place-making from a distance. Like Sally, Barbara chooses to plan routes ahead of making journeys, albeit for different reasons. She doesn't want to repeat the same routes week in week out, and enjoys the fact that she can discover new places using this technology. She purposefully produces routes that will direct her through unfamiliar places on unfamiliar roads in the hope that she will discover something new and interesting while riding. It is the routes that she has found to be particularly interesting that make it in to the sharing that goes on within the online and offline spaces of her cycling club. She is especially proud to contribute routes that take in places of historic interest as well as good training on quiet roads.

Barbara's annotations to the map are different to that of Tori and Sally. As I suggested above, inscriptive practices may appear ethnographically similar across different mapping platforms, but digital and paper maps are textually and technologically distinctive due to the different ways in which they have been produced and the different ways that they are called into use. Barbara's route-making practices are bound up in a computational system whereby all her annotations are subject to the parameters set by the software she is using. If one was to look at a finished route, it may appear as though she has drawn freehand onto the map using a computer mouse, but what is represented is not the same, for her actions have been processed and represented through software and an interface. Tori and Sally's practices escape these systems of software control and therefore can be said to offer a different kind of annotation, one they are both free to choose and not one predetermined by software. Following Clancy Wilmott,[28] I suggest that Tori and Sally's mapping practices sit within *and* beyond cartographic logics of software, whereas Barbara's are contained within software.

This becomes clear when she demonstrates how she plots a route on the map. Where Sally and Tori are able to annotate a route in a free-flowing manner, Barbara must click and drag the screen to move around the base map and then intermittently click on points on the map and wait for the software to generate a route between them. During these brief load-times, the software is processing all manner of things unbeknown to Barbara. It is processing the routes she is making based on a pre-formulated system of parameters and not simply on her own judgements about where to draw on a route. In doing so, the map emerges into being in a textually and technically different way to that of Sally and Tori's marked-up A–Z atlases. The ethnographic result is that her route-planning practices stutter along as she is made to check each segment of the route to see if its been marked properly and redo certain sections if she notices something that needs addressing, such as an indication that she is to cycle on a busy A road. The advantage of this laborious process, she says, is that when finished, she can quickly upload it onto a device to be used as a route map when cycling rather than having to stop too often to consult a paper map to check if she is on course. Moreover, she tells me it will give her a route with added contextual data about her ride which she can incorporate into her training plan. Nevertheless, it cannot be said that Barbara's practices of inscription are altogether equal to those of Sally and Tori, precisely because of the different maps that they are using. Unlike Sally and Tori, Barbara is not free to choose how to annotate the map, for her inscriptions are confined to the rules set by the route-planning software.

CONCLUSION

Place-making practices are constituted within complex socio-technical, spatial, and temporal assemblages. In this chapter, I have sought to demonstrate how just one common everyday mapping practice, that of annotating the map, contributes to this complexity. I have argued that maps don't simply go out into the world to be used as their makers intended them to be. Instead, I demonstrated how maps have cultural life whereby they become bound up in everyday navigational place-making practices. Using ethnographic examples taken from a study of mapping practices, I showed how map users accumulate knowledge of place, in part, by working *with* the malleability of graphical maps, whether this be working with paper materials or within digital systems. In this regard, place-making mapping practices are said to be co-produced by material, social, and technical practices. In the case of two London A-Z users, it was shown that maps were marked up with pens, pencils, sticky notes, and folded pages in order to inform and keep a note of journey-making practices that could be revisited and revised for future trips. Similarly, in the case of a popular route-planning software for cyclists, routes were shown to be digitally etched onto (or into) base maps as a way of planning and sharing cycling trips which suited specific cycling needs. In these cases, maps were inscribed as means to discover new places, revisit past experiences, and create a sense of social cohesion between a group of riders. Together, these cases illustrate how place-making in everyday life may be continually made and remade by repetitive and culturally specific practices of inscribing the map. This reasserts the notion that maps are 'quasi-objects' (after Michael Serres), for they enact social relations and inform everyday experiences of practice. To quote Serres:

> Our relationships, social bonds, would be airy as clouds were there only contracts between subjects. In fact, the object, specific to the Hominidae, stabalises our relationships, it slows down the time of our revolutions. For the unstable

> bands of baboons, social changes are flaring up every minute . . . . The object, for us, makes our history slow.[29]

The cases of Tori, Sally, and Barbara all illustrate how the map itself has become enfolded into their social and place-making practices. In describing these cases, I have also drawn attention to the performative, textual, and technological differences between annotating paper maps and annotating digital maps. I argued that while maps are indeed brought into being anew each and every time they are used, the ways these emergences must be understood differs fundamentally between paper and digital maps. This, I suggested, has an effect on how the cultural life of maps is folded into the context of everyday life. As we saw with the two A–Z users, their practices were evidence of minor disruptions that went beyond the original intentions of the map maker. Although these disruptive practices were not always (or ever) intended to be against the maps, they did constitute a permanent change to the map that would not be possible if using digital maps. As was shown with the cyclist using route-planning inscriptions, digital annotations are always subject to the control of the dynamic systems of software used to produce them. Although annotations can be made to both digital and paper maps, the freedom as to what can be added is determined by the mapping form.

This is to say that the way in which digital maps and paper maps are brought into being in the contexts of our everyday place-making practices are not the same if we compare the technical and textual make-up of these maps. Nevertheless, as my accounts of inscriptive practices show, they may be ethnographically equivalent, for users of maps are often not aware of nor do they particularly care about the composition of maps. What they are concerned with is whether the map will work effectively for them in their everyday contexts of use, and as such they will pick whichever map works best for them for a particular task. This goes to show that the cultural life of maps extend beyond the textual and technical composition of maps. Ultimately, the differences lie in how the relations between the map and the user emerge.[30] It is therefore the contexts of a maps' use which are key to highlight, for it is context that will tell us why a map is brought into being and the reasons for why it is being annotated.

NOTES

1. Denis Wood and John Fels, *The Power of Maps* (New York: The Guilford Press, 2002).
2. Jeremy Crampton, *Mapping: A Critical Introduction to Cartography and GIS* (Oxford: Wiley-Blackwell, 2010).

3. ibid; John Pickles, *A History of Spaces: Cartographic Reason, Mapping and the Geo-Coded World* (London: Routledge, 2004).

4. D. Cosgrove Apollo's Eye: A Cartographic Genealogy of the Earth in the Western Imagination (Baltimore: The Johns Hopkins University Press, 2003).

Gunnar Olsson, *Abysmal: A Critique of Cartographic Reason* (Chicago: University of Chicago Press, 2007); John Pickles, *A History of Spaces: Cartographic Reason, Mapping and the Geo-Coded World* (London: Routledge, 2004).

5. Nevertheless, such a belief is sometimes challenged when the map fails in our expectations of it, such as when the Sat-Nav leads us down a dead end or when the new road layout has not yet been updated on our map. Although such cases ask us to question the map and its worth, our scepticism is short lived if we find another map to work for us.

6. J. Crampton, *Mapping: A Critical Introduction to Cartography and GIS* (Oxford: Wiley-Blackwell, 2010)

7. Chris Perkins, Cultures of Map Use, *The Cartographic Journal*, 45, no. 2 (2008): 150–158.

8. See also Sébastian Caquard and William Cartwright, Narrative Cartography: From Mapping Stories to the Narrative of Maps and Mapping, *The Cartographic Journal*, 51, no. 2 (2014): 101–106.

9. Michael Serres, *Genesis* (Ann Arbor: Michigan University Press, 1995), p. 87.

10. Barry Brown and Eric Laurier, Rotating Maps and Readers: Praxiological Aspects of Alignment and Orientation, *Transactions for the Institute of British Geographers*, 33 (2008): 201–216.

11. Veronica Della Dora, Performative Atlases: Memory, Materiality and (Co-) Authorship, *Cartographica*, 44 (2009): 241–256.

12. Chris Perkins, Cultures of Map Use, *The Cartographic Journal*, 45, no. 2 (2008): 150–158.

13. Tim Ingold, *Lines* (London: Routledge, 2007).

14. The London A-Z has been the subject of much public and academic interest in recent years. See, for example: Richard Hornsey, The Cultural Uses of the A-Z London Street Atlas: Navigational Performance and the Imagining of Urban Form, *Cultural Geographies*, 23 (2016): 265–280; Do You Still Love You're A-Z? Londonist (11/2/11). Available at: http://londonist.com/2011/02/do-you-still-love-your-london-a-z; Tern Productions for BBC Scotland, 'Mrs P's A-Z (1936)', Map Man, series 2, episode 7. First aired on BBC2, 17 October 2005.

15. Sybille Lammes, Transmitting Location: Digital Cartographical Interfaces as Transformative Material Practices, *Aether: Journal of Media Geography*, 11 (2011).

16. Tim Ingold, *The Perception of the Environment: Essays on Livelihood, Dwelling and Skill* (London: Routledge, 2000); Tim Ingold, *Lines* (London: Routledge, 2007).

17. Martin Dodge, Rob Kitchin and Chris Perkins, *Rethinking Maps: New Frontiers in Cartographic Theory* (New York: Routledge, 2009).

Rob Kitchin, Justin Gleeson and Martin Dodge, Unfolding Mapping Practices: A New Epistemology for Cartography, *Transactions of the Institute of British Geographers*, 38 (2013): 480–496.

18. Nevertheless, acts of defacing public maps do occur, and have a political and historical context. See, for example: Joe Austin, *Taking the Train: How Graffiti Art Became an Urban Crisis in New York City, 1970–1990* (New York: Columbia University Press, 2001); Andrea Ballatore, Defacing the Map: Cartographic Vandalism in the Digital Commons, *The Cartographic Journal*, 51 (2014): 214–224.

19. Sybille Lammes, Transmitting Location: Digital Cartographical Interfaces as Transformative Material Practices, *Aether: Journal of Media Geography*, 11 (2011).

20. Rob Kitchin, Justin Gleeson and Martin Dodge, Unfolding Mapping Practices: A New Epistemology for Cartography, *Transactions of the Institute of British Geographers*, 38 (2013): 480–496.

21. Sybille Lammes, Transmitting Location: Digital Cartographical Interfaces as Transformative Material Practices, *Aether: Journal of Media Geography*, 11 (2011).

22. Ibid.

23. That is unless any significant changes to the product have been made. For instance, if one was to compare the aesthetic qualities of Google Maps when launched in 2005 and Google Maps in 2017 they would notice some significant changes to the cartographic form. During this time, gradual changes are frequently made. Of significance is the changes made in 2013, which saw the form of Google Maps to change distinctly overnight.

24. Sybille Lammes, Transmitting Location: Digital Cartographical Interfaces as Transformative Material Practices, *Aether: Journal of Media Geography*, 11 (2011).

25. Deborah Lupton, *Digital Sociology* (London: Routledge, 2014).

Deborah Lupton, Foreword: Lively Devices, Lively Data and Lively Leisure Studies, *Leisure Studies*, 35 (2016): 709–711.

26. The Playful Mapping Collective, *Playful Mapping in the Digital Age* (Amsterdam: Institute of Network Cultures, 2017).

27. These are all popular mapping technologies that may be used to plot routes on, which may then be transferred onto mobile devices such as smartphone and portable GPS units.

28. Clancy Wilmott, Living the Map: Mobile Practices in Postcolonial Cities. PhD Thesis (University of Manchester, 2017).

29. Michael Serres, *Genesis* (Ann Arbor: Michigan University Press, 1995).

30. Chris Perkins, Plotting Practices and Politics: (Im)mutable Narratives in OpenStreetMap, *Transactions of the Institute of British Geographers*, 39, no. 2 (2014): 304–317.

BIBLIOGRAPHY

Austin, J. (2001). *Taking the Train: How Graffiti Art Became an Urban Crisis in New York City, 1970–1990*. New York: Columbia University Press.

Ballatore, A. (2014). Defacing the Map: Cartographic Vandalism in the Digital Commons. *The Cartographic Journal*, 51, 214–224.

Brown, B and Laurier, E. (2008). Rotating Maps and Readers: Praxiological Aspects of Alignment and Orientation. *Transactions for the Institute of British Geographers*, 33, 201–216.

Caquard, S. and Cartwright, W. (2014). Narrative Cartography: From Mapping Stories to the Narrative of Maps and Mapping. *The Cartographic Journal*, 51, 2.

Cosgrove, D. (2003). *Apollo's Eye: A Cartographic Genealogy of the Earth in the Western Imagination*. Baltimore: The Johns Hopkins University Press.

Crampton, J. (2010). *Mapping: A Critical Introduction to Cartography and GIS*. Oxford: Wiley-Blackwell.

Della Dora, V. (2009). Performative Atlases: Memory, Materiality and (Co-) Authorship. *Cartographica*, 44, 241–256.

Dodge, M., Kitchin, R. and Perkins, C. (2009). *Rethinking Maps: New Frontiers in Cartographic Theory*. New York: Routledge.

Hornsey, R. (2016). The Cultural Uses of the A-Z London Street Atlas: Navigational Performance and the Imagining of Urban Form. *Cultural Geographies*, 23, 265–280.

Ingold, T. (2000). *The Perception of the Environment: Essays on Livelihood, Dwelling and Skill*. London: Routledge.

Ingold, T. (2007). *Lines*. London: Routledge.

Kitchin, R., Gleeson, J. and Dodge, M. (2013). Unfolding Mapping Practices: A New Epistemology for Cartography. *Transactions of the Institute of British Geographers*, 38.

Lammes, S. (2011). Transmitting Location: Digital Cartographical Interfaces as Transformative Material Practices. *Aether: Journal of Media Geography*, 11.

Lupton, D. (2014). *Digital Sociology*. London: Routledge.

Lupton, D. (2016). Foreword: Lively Devices, Lively Data and Lively Leisure Studies. *Leisure Studies*, 35, 709–711.

Olsson, G. (2007). *Abysmal: A Critique of Cartographic Reason*. Chicago: University of Chicago Press.

Perkins, C. (2008). Cultures of Map Use. *The Cartographic Journal*, 45(2), 150–158.

Perkins, C. (2014). Plotting Practices and Politics: (Im)mutable Narratives in OpenStreetMap. *Transactions of the Institute of British Geographers*, 39(2), 304–317.

Pickles, J. (2004). *A History of Spaces: Cartographic Reason, Mapping and the Geo-Coded World*. London: Routledge.

Serres, M. (1995). *Genesis*. Ann Arbor: Michigan University Press.

The Playful Mapping Collective. (2017). *Playful Mapping in the Digital Age*. Amsterdam: Institute of Network Cultures.

Wood, D. and Fels, J. (2002). *The Power of Maps*. New York: The Guilford Press.

Wilmott, C. *Living the Map: Mobile Practices in Postcolonial Cities*. PhD Thesis (University of Manchester, 2017).

Chapter 2

This Noise Matters

Participatory Soundmapping and the Auditory Experience of Homelessness[1]

Paul Tourle

INTRODUCTION

In late May 2018, I had the opportunity to co-host a workshop with the Museum of Homelessness (MoH) that brought together twelve participants to think about sound and auditory culture. Co-founded by Matt and Jess Turtle in 2015, and guided by a small and organically evolving 'core group' of members, MoH is, in its own words, a 'community-driven social justice museum, created and run by people with direct experience of homelessness'.[2] It 'tackles homelessness and housing inequality by amplifying the voices of its community through research, events, workshops, campaigns and exhibitions'.[3]

The May workshop, *This Noise Matters* (*TNM*) squeezed three linked sessions into half a day. Following a round of introductions, I led the group in a broad discussion around the themes of sound and listening and introduced an analytical framework based on the work of Georges Perec (see below) for describing and thinking about particular sounds.

This session led into a silent listening walk,[4] tracing a circular route centred on the Old Diorama Theatre in Central London, an organisation MoH has worked with previously and which was generous in providing us free-of-charge a large rehearsal space in which to work for the day. As well as offering a means of re-energising the group part way through the day, the walk also served the purpose of situating our developing conservation relative to a collective experience of an exemplar urban soundscape.

In the final and longest portion of the workshop, the group split into smaller clusters of three, with each attendee given the opportunity to donate a sound of their choosing to the museum, guided by the above-mentioned Perecian framework. These donations took the form of verbal descriptions and reflections upon particular sounds, their social and biographical significance, and on the auditory content and experience of both home and homelessness, with participants taking it in turns to record each other's contributions.

At the conclusion of the workshop, I collected those recordings, reviewed their content, and then, over a period of several weeks, used the opensource editing package Audacity to organise excerpts of participants' contributions into a short audio piece. As well as attempting to shape the collected fragments of conversation produced through the workshop into a loosely coherent narrative (grouping contributions to the greatest extent possible by theme), I also mixed into the piece a selection of environmental sound recordings. Several of the recordings were sourced directly from workshop participants after the event; others were taken from open access archives.

Having completed production of the *TNM* recording by mid-summer, the next stage in the project process was to circulate the piece to those who had been present at the workshop. Attendees were given three months in all to raise comments or concerns about the piece or to make suggestions for edits, and they were also asked to consent to the recording being placed online as a part of the MoH archive.

Today, the recording still forms a part of the museum's digital collection and can be accessed at https://museumofhomelessness.org/collection-and-archive/noise-matters/. I intend it to be listened to as a companion to this chapter. Below, as well as describing further how the workshop unfolded, I also set out the circumstances in which the project arose and consider its value as a kind of socially engaged mapping practice. My primary objective, considering in the round the project's strengths and weaknesses, and its history (a confusion of different origins, ends, and means), is to share a case study that can both inform debate on the purpose, efficacy, and future of countermapping practices and to highlight the potentials and limits of listening as a tool in mapping.

STARTING POINTS

Soundmapping

The idea of staging and producing *TNM* developed as I tried to untangle and think through a knot of different themes, case studies, concerns, and questions that had emerged through my doctoral research. Situated primarily within the field of critical heritage studies, that research had taken as its starting point the publication of a series of online soundmaps within heritage institutions—most notably two maps: the British Library's *UK Soundmap* (2010–2011) and its *Sounds of our Shores* map (2015), created in partnership with the National Trust and the National Trust for Scotland.[5] Both projects used crowdsourcing as a mechanism for amassing large collections of soundscape recordings produced by members of the British public. The recordings sourced through the projects were presented as composite, digital maps-cum-archives, intended to be preserved in perpetuity as lasting statements on the sounds of the nation.

Although, undeniably, it is a marginal pursuit, soundmapping nevertheless can claim both a rich and modestly radical history and a committed and critically engaged community of practitioners. Rooted notably, though certainly not solely, in the works of the World Soundscape Project (founded in the late 1960s by Canadian composer and environmental activist R. Murray Schafer), the practice has often been deployed as a means of calling attention to and advocating against industrial noise pollution. Equally, and in its more expansive forms, it has been celebrated for its capacity to disrupt and subvert the confidence, impersonality, and lifelessness of traditional visual cartographies. As artist Salomé Voegelin argues, where visual cartography tends to deal in certitude, sonic geographies of the kind exemplified in online soundmapping are contrastingly 'anxious and affective: full of doubt, uncertainties, and the pathology of who we are'.[6] For Angus Carlyle, similarly, soundmaps promote:

> [. . .] affective attentiveness to what goes on behind the windows and walls, to the domestic beyond the architectural façade; intensified sensitivity to magnetic fluxes, to the internal vibrancy of matter, to shifts in heat, in wetness and wind, to the racket of the cicada and the buffalo's breath, to the dangerous [. . .] and to the precarious (in whatever language it speaks).[7]

The practice might be understood, Carlyle adds, as a form of 'collective counter-mapping that makes audible that which struggles to be seen'.[8]

Having written elsewhere on the two British Library projects referenced above, I do not want to dwell especially on them here.[9] That said, I would note that they appeared to me to exhibit several limitations, both in terms of the listening they brought to bear on the world, and in terms of the way they convened and represented the public (and, by extension, the nation). Where historically, soundmapping has often been driven by a will to critique and/or supplement visual mapping, at the British Library, in both projects, it seemed more an expression of a simple, indiscriminate compulsion to collect and conserve traces of an ever-changing world.[10] Both maps claimed to capture sonic diversity; yet confusingly, they used the visual index of the map as their primary means for measuring that diversity (aiming for visual completeness, with pinned recordings drawn from all corners of the UK, and as many points as possible in between). The strong emphasis the projects placed on the production of high-quality, naturalistic recordings as the end goal of listening implied a corresponding lack of interest both in the intimately relational experience of listening itself (e.g. the ways in which sounds bind themselves to memory) and in the historical and socio-political contexts out of which different sounds and ways of listening emerge. Finally, and corresponding to the concerns of writers including Jacqueline Waldock[11] and Milena Droumeva,[12] the projects (i.e. the media that framed and promoted them) tended to dwell on the importance of outdoor (especially natural) sounds, to the expense of the domestic. For Droumeva, commenting on a wide range of soundmapping projects that have appeared online since the late 1990s:

> One entire typology of sounds that are often missing are everyday, mundane, domestic sounds, sounds of strife, inequality, alterity and resistance. Those voices do not make it into these semi-insitutionalized portals of sound, and that is an ideological absence we must highlight and shift.[13]

In part, allied to an intensifying and uncomfortable familiarity I had developed with the silence often characteristic of my own encounters with people forced to sleep on the streets (registering homelessness as an issue with, perhaps, a unique auditory character of its own), it was this recognition of an 'ideological absence' within sonic cartography that led me, in 2015, to begin to develop projects exploring listening and homelessness in tandem, aiming both better to understand the one via the other (and vice versa), and, if possible, to harness the former in order to contest the conditions that shape and give rise to the latter. Here, crucially, my research also began to connect with recent work in homelessness studies aimed at unsettling and countering overly statistical, unfeeling, and monolithic accounts of the problem and those affected by it.

SENSORY APPROACHES TO THE STUDY OF HOMELESSNESS

Building upon the foundation of an extensive body of literature that tries to define and respond to homelessness by interrogating the (sensuous, gendered, rhythmic, etc.) nature and experience of 'home' itself,[14] the past twenty years have witnessed a slow proliferation of forms of research attentive to sensorial and emotional aspects of homelessness, and to how the senses may function as key loci through which homelessness is experienced. Concerned with documenting homelessness 'felt and lived',[15] with understanding homelessness (and becoming homeless) as an often slow, incremental, and corporeally and emotionally cumulative *process* rather than as an isolated event[16] and asking not 'what' but rather *how* homelessness is,[17] the research generated out of this qualitative turn is marked by a common refusal of 'normative distinctions between "the homeless" and "everyone else"',[18] with its overriding logic being thus to show the extraordinary effects that the process and experience of homelessness exert on ordinary (that is to say, most pointedly, non-deviant) bodies.

This work has developed in a wide range of directions, looking, for example, to account for the physical and psychological effects of the long periods of boredom often associated with homelessness,[19] to trace emotional cartographies of fear and comfort in temporary shelters,[20] to register the cumulative effects of assault and abuse that build up in individuals continually exposed to street violence,[21] to document how homelessness is experienced as a symbolic (but nonetheless embodied, affective) burden—a kind of 'feeling homeless', made manifest as a sense of shame and low self-worth,[22] and to show how homeless bodies are made subject to exclusionary urban design.[23]

The research often has highly practical ramifications. For example, work exploring individuals' sensorial experiences of the processes they must navigate in applying for housing relief has informed calls for welfare reform. One issue subject to considerable scrutiny in the United Kingdom at present is the role that local authority homelessness officers play – and the personal discretion they are able to exercise – in assessing the respective degrees of individual homeless applicants' 'vulnerability' (e.g. in terms of mental or physical health), their honesty (registered in judgements about individuals' voices, bodily comportment, and their overall appearance of well- or ill-being) and hence their eligibility for housing support under the category of 'priority need'.[24] Writing for Shelter Wales in a briefing against priority need assessments, Jennie Bibbings describes how, required to 'prove' vulnerability in order to access support, homeless applicants are placed in the 'humiliating' position of having to present a 'sob story' to councils in order to get to the front of the queue.[25] Significant in promoting an understanding of how contact with welfare services may reinforce trauma and undermine self-worth, Bibbings' analysis connects, too, with Daniel Edmiston's study of the formation of 'welfare' subjectivities, as well as with Andrew Woolford and Amanda Nelund's assessment of the emotional burden placed on individuals required to perform 'neoliberal citizenship', in order to be deemed deserving of state support.[26]

Finally, amid this growing corpus of sensorily attuned works, Catherine Robinson's *Beside One's Self* and Robert Desjarlais' *Shelter Blues* also demonstrate unusual attention to the role of sound in moulding experiences of homelessness.[27] In Desjarlais' book, an ethnographic account of the lives of residents at the Station Street Shelter in Boston, Massachusetts, sound is a pervasive material presence that seems continually to interrupt, defer, or displace residents' sense of self. His informants lament a lack of 'peace around the ears', find themselves 'distracted' and 'disturbed', or made 'nervous' by environmental noise; they are 'engulfed by the elements', 'their ability to sense or make sense of the word [. . .] often overwhelmed by the harshness of physical and social environments'.[28]

Robinson, meanwhile, in striving to express the intense vulnerability to pain and trauma suffered through by people experiencing homelessness, is sensitive to often neglected or unwritten sonic and auditory dimensions of research processes themselves. An encounter with a young girl fleeing domestic abuse is recounted partly in terms of Robinson's own physical incongruence in the refuge setting – her raised voice, 'the loudness of my whole body – its health, its power, its intentness'. In a shattering narrative, the girl's contrastingly 'quiet voice', her 'little shaky pauses', and 'held down sobs' are not gratuitous detail; rather they are a statement of a fundamental emotional and bodily reality so often masked by research that 'knows' homelessness only in terms of the layered abstractions of housing needs' assessments and statistical models of causation.[29]

As Robinson argues, such sensuous scholarship is urgently required, for:

> Felt evidence extends a potentially radical call for public recognition of and responsibility for homelessness. Knowledge of the felt begins to open 'those who are not themselves in pain' to 'those who are'.[30] If the felt dimension of homelessness remains unnamed, it remains foreign, effaced, and without a response.

For researchers, then, who are seeking to engage with homelessness effectively, a fundamental challenge is to refuse, resist, and complicate those ways of knowing people experiencing homelessness that deny or abstract from their essential humanity, presenting them merely as serial examples of an unindividuated, unfeeling, and malfunctioning type apart from the rest of society. Conducting research that acknowledges sensation – the pain and discomfort, and equally the pleasure and joy – experienced by individuals living through homelessness may be a key strategy in confronting and reworking public and policy attitudes to the issue. Though Robinson concedes that empathy and compassion alone will not solve homelessness (and certainly they can do little of lasting value divorced from a critical understanding of the diverse causes of the problem), it may yet be the case that creating complex, sensitive, and even sensual accounts of individuals experiencing homelessness can serve to suspend public prejudice and invite and promote recognition of, and curiosity about, individuals' circumstances. This in itself can be a valuable route to encouraging a critical public discussion of the issue. And it was with this aim in mind that my research into sound and homelessness began.

NOISE

> How much noise must be made to silence noise? And what terrible fury puts fury in order?
>
> Michel Serres[31]

Having begun to establish the context that my research into sound and homelessness grew out of, it is also important to touch upon the project that preceded *TNM* during my studies. For it, too, contributed greatly to shaping my collaboration with MoH, and moreover it brought into focus a series of problems associated with participatory mapping practices – and my own assumptions about their value – that I have come to think of collectively under the umbrella of 'noise'. That is, a phenomenon which, depending on one's point of view, speaks either to the very possibility of communication – to the differences that make communicative exchange possible and meaningful,[32] and to the resistance of ossifying truths that holds dialogue open;[33] or alternatively to its near impossibility within our contemporary media and intellectual environment – to the way in which our attention falters under the strain of informational overload,[34] and to the difficulty of discerning and defending truth in the context of a postmodern relativisation of all value.[35]

That first project, *Our Sounds*, comprised a series of workshops about sound, undertaken with guests at a London homeless shelter to which I was already connected as a volunteer. Comprising a series of workshops in which I and a group of shelter guests attempted to map sounds collectively at different scales and with different focal points (the shelter, the neighbourhood, London, voices, and rhythms) the project proper straddled a period of around four months before disintegrating into a longer, informal process of participant observation, or participant listening.

As a means of generating an understanding of the ways in which sound and listening could be said to feature as part of differing experiences of homelessness, the project was informative. Through conversations with workshop participants and (in the longer term) other shelter guests, it became possible to sketch a very partial typology of certain sounds and forms of listening, which, though I would stress they may pervade contemporary everyday life generally, nevertheless seemed to manifest more starkly in contexts of homelessness. These were sounds of displacement and social atomisation: stiff silences during mealtimes shared with an endlessly transforming assembly of strangers, or the drone of a television left untouched for hours by inattentive watchers, that provided, nevertheless, a kind of sonic shelter permitting mass solitude. They included the sounds made by makeshift, make-do architecture: crowded rooms and creaking bedsprings in a hastily converted industrial warehouse. There were sounds that evidenced individuals' attempts at spiritual and technological self-maintenance: ostentatious prayer routines and the tinny rattle of a dozen sets of headphones leaking at any given moment. And, as above, there were sounds that spoke to a certain form of calculated, bureaucratic disregard: several guests described growing tired of explaining their circumstances to multiple government agents without ever feeling listened or responded to, and without their cases being dealt with. Required to stay calm, to be polite to authorities, to modulate their voices, they experienced the discipline of the state without (in many cases) receiving its support.

As a means of deepening and broadening my own personal familiarity with the subject matter at hand, then, *Our Sounds* proved invaluable to me. Beyond that, however, from the perspective of advocacy or wider public impact, the project achieved little if anything. Throughout its lifespan, plans to co-produce materials to publish or even campaign around never came to fruition. Instead, the work ended with the publication of a sole ethnographic chapter within my dissertation, destined to fill a niche in heritage discourse that few looking to comprehend homelessness will ever have cause to visit. There are four key factors that I see as central to the project's collapse.

The first such factor was the disconnect between the theme of sound and listening and participants' typically highly unsettled, precarious, and stressful everyday lives. In short, in the context of individuals' struggles to find work or stable housing, or to rebuild health, a project about sound (as I had configured it) felt extraneous at best and frivolous at worst. It was 'noisy' in the sense that it provided little meaning within everyday life.

Second and third, there were complications relating to the nature (or absence) of community at the shelter. Most tangibly, gradually, over the months the project ran, each of the participants who had first signed up to be part of it left (happily, in all but one case, moving onto more permanent accommodation), and in this context, our work together simply lost momentum.

More importantly, however, the more time I shared with participants, the clearer it became how little we had in common with one another, and therefore how difficult it would be to formulate the shared narrative or sense of purpose that might underpin the kind of collective, community work that I had envisaged undertaking together. Reflecting on his experiences as an ethnographer investigating municipal men's shelters in New York, Anthony Marcus emphasises the permeability of the shelter environment, as well as the high degree of variance both in the ways residents spent their time and perceived themselves, and in the extent to which they connected socially with one another. The shelter, he writes, is 'more like a beach on a crowded day than a unified institution, community, or small town [. . .] the human networks among shelter residents mirrored the dispersed and atomized surrounding city'.[36]

In planning *Our Sounds*, failing at first to grasp the noisy reality of a social dynamic generated through multiple, distinct individuals' serial displacement, and proceeding instead on the strength of externally developed assumptions about project group members' likely shared grievances and suffering, I had invented a community – a common identity – where none existed. This misrecognition meant, further, that I invested unfair hope in community action arising from a context of severe social atomisation. Had my efforts instead (e.g. in the manner of formal participatory action research), centred on first negotiating questions of community, working to surface and articulate shared concerns, the project might have ended differently. But of course, that was not the approach that I took.

This brings me to a fourth and final issue, and perhaps the project's founding flaw. If one attempts, as I did, to understand and frame *Our Sounds* as a form of counter-mapping, then it is worth questioning what precisely it was supposed to counter. Several possible answers present themselves, some of them highlighted above: unfeeling accounts of homelessness, for example, or particular stereotypes of people living through homelessness. Yet, through the project, I failed notably ever to define specific narratives that needed to be resisted, specific institutions or organs whose work required refuting or contesting, or specific audiences who might be supposed to benefit from exposure to its published product. One set of projects *Our Sounds* did clearly look to critique and complicate (regrettably instrumentalising the time, energy, and emotional labour of its participants for that purpose) was the aforementioned family of soundmapping initiatives launched by the British Library, albeit in cryptic fashion. On the whole, however, despite good intentions, the project failed to counter anything.

And this, above all, is what I am getting at when I refer to the problem of noise. During a period in which relativistic attitudes to cultural value and experience allow researchers, cartographers, artists, and academics to conjure projects, objects, commentary, and commodities from even the merest, most ephemeral scraps of our social fabric (e.g., building noise), content proliferates. If due care is not taken (if contexts, audiences, networks of communication, cause and effect are not scrutinised), it proliferates without purpose. The upshot of that process is a mass of atomised media products that speak almost to no one. As in Jodi Dean's analysis of 'communicative capitalism', communication is transformed, losing the qualities of (potentially antagonistic, and therefore destructive/creative) dialogue.[37] Messages are dispatched into a saturated mediascape where they elicit no response. Ideas circulate, but never land. We feel the quantitative mass and weight of the media; the labour required to produce it is real; its inputs are captured and capitalised within a series of industries (notably, academic publishing); the attention it drains from us significant. Yet the qualitative differences inscribed in any given message are rendered increasingly irrelevant: we produce noise and fail ever to reach or influence those with the power to effect widespread material change.

THIS NOISE MATTERS

To summarise the above, through my initial attempt to map aspects of the auditory experience of homelessness, I reached two conclusions. First, research into sound and listening has some potential to inform and enrich a shared understanding of homelessness. But, second, that potential is easily squandered without careful planning of the research process, most particularly in terms of the groups, communities, and audiences it builds from (or aims to build) and speaks to, and in terms of the way in which it connects to the everyday realities, priorities, and needs of participants and audiences. In working to organise and produce *TNM*, I looked, accordingly, to refine my research process. Through the remainder of this chapter, building on the summary methodology provided in the introduction, I will discuss the changes I made, the extent to which they worked or did not, and how – if only modestly – they might inform future approaches to counter-mapping. Overall, in that spirit, there are four aspects of the project that I will highlight before offering some final analysis and concluding thoughts.

REDUCED BURDEN

First, and most simply, aiming to counter the gradual disintegration of the research group I had assembled at the shelter previously, the workshop I held in conjunction with MoH was compressed into a single half-day. After our meeting in person, as described above, I remained in contact with participants, via the museum's core team, for a period of several months, allowing each the opportunity to comment on, propose changes to, and ultimately sign off on my initial cut of the audio produced on the day, in a process approximating ethnographer Steve Feld's 'dialogic editing'.[38] In general, however, I aimed to reduce the level of commitment required of attendees and to take better account of the varying stressors in their everyday lives that might make serial participation difficult or impossible.

A COMMUNITY OF PRACTICE

Second, and perhaps chief among the changes I made, was the very decision to work with MoH, and thus to connect with and yoke my own efforts to an existing community of practice whose goals I shared, which had already forged a strong identity around those goals and which was seeking independently to develop and mobilise forms of cultural activism in response to the issue of homelessness. Often working in partnerships, and always adhering to an internally defined three-part 'Theory of Change' directed to (i) empowering individuals; (ii) influencing institutions, and (iii) increasing public awareness and understanding of homelessness,[39] MoH's first five years have seen it embark on a range of projects. Its two 'State of the Nation' shows, hosted respectively by the Tate Modern and Tate Liverpool galleries shared artworks relating to homelessness and gave speaking platforms to activists including squatters' rights groups. Two further co-curated displays in London, both entitled 'This Stuff Matters', presented works the museum's members had derived from archival collections. 'Objectified', an interactive exhibition designed to explore the cognitive science behind processes of dehumanisation, was developed in collaboration with the Wellcome Trust and hosted at Manchester Art Gallery. Since April 2019, meanwhile, firmly underscoring the activist nature of its work, MoH has taken on stewardship of the 'Dying Homeless' project: an initiative begun by the non-profit Bureau of Investigative Journalism to collect data, share stories about, and commemorate the more than 800 people who have died while homeless in the United Kingdom since October 2017.[40] And in early 2018—having begun to engage in more direct forms of protest (e.g. organising a vigil at Downing Street to highlight the issue of deaths on the streets), the Museum launched its 'Catalyst' programme; a month-long training initiative designed to support the individuals involved to develop their skills as creative campaigners.

Albeit working with MoH could provide no guarantee of my own specific auditory methods having impact, our collaboration did, then, as a minimum, allow me to learn from, participate in, and contribute to a community already well advanced in developing a range of communicative strategies. As I looked to mitigate to whatever extent possible the noise of communicative capitalism, the museum provided a formal structure in which participants' energies and thoughts might more effectively be captured.

A PROBLEM SHARED

The third key change I implemented in the design of *TNM* also responded to concerns about the nature of community and processes of community building. It went further, however, in drawing on above-mentioned analyses of the governance of homelessness within neoliberalism, and an identified tendency for policymakers to favour individualising 'cures' and processes of self-responsibilisation over structural, social remedies in combatting the problem.[41]

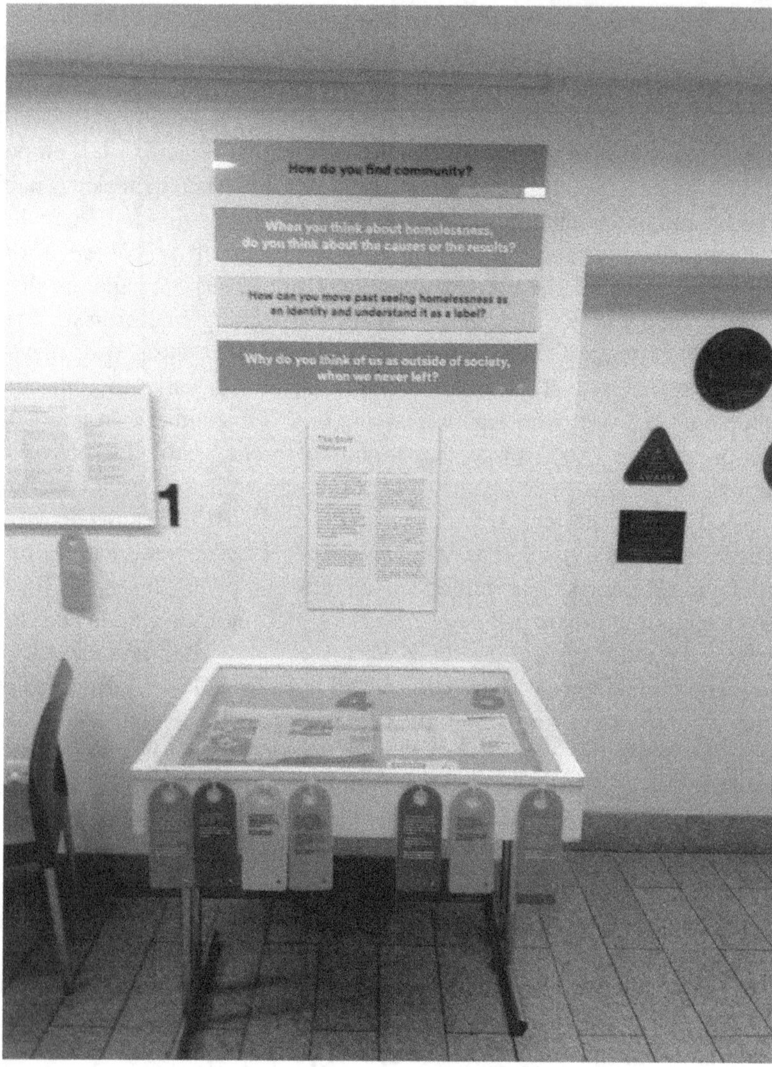

Figure 2.1 Displays at MoH Second 'This Stuff Matters' Exhibition at St Martin in the Fields Church. Author's Image.

During *Our Sounds*, as we have seen, I struggled to foster a coherent research community. Moreover, I began to question the legitimacy of a process that asked people experiencing homelessness, somehow, with minimal prompting, to adopt a shared (typically unwanted) identity and organise around it.

This was a critical question: to what ends is community constructed, and how and with what force could members of atomised and marginalised 'groups' be expected to act to improve their own situations? In response, I concluded that to ask homeless participants to stand alone in a shared act of representation solely by virtue of mutual hardship would be to replicate and promulgate a part of the founding logic of entrepreneurial neoliberalism: that the poor and the marginalised must serve themselves, be their own voice, and fight their own battles. Over-reliance on homeless people themselves to generate the materials that speak to and resist the conditions of their marginalisation becomes a way of relieving non-, less- (or at least, differently) marginalised others from shouldering a part of the burden of representation and action.

As such, together with the MoH team, and in attempting to articulate homelessness as a shared problem (an aspect and product of our common everyday life), it was agreed that *TNM* should be a public workshop. It would invite responses to the issue from all those who wanted or felt able to speak to homelessness, regardless of the extent and nature of their experience of it. And it would reflect the aspiration to develop a representation and interpretation of homelessness that synthesised traces of the vulnerability, complicity, care, ambivalence, helplessness (etc.) experienced and exhibited by those who are not directly affected by homelessness (but who are entangled inevitably in the society that produces it) with testimonies of those who do know homelessness personally and have lived through it.

In concrete terms, of course, it is one thing to aspire to create a context for a polyvocal, multi-perspectival discussion of homelessness engaging the whole of the shared substance of everyday life, and quite another to achieve the feat in practice. Though I might have speculated as to the benefits of bringing together, for example, welfare officers, politicians, and police officers, all to discuss the sounds of homelessness (their experiences of, responses to, and agency in reproducing homeless soundscapes), the reality of the workshop was shaped by the tangible limits of my own and MoH's communications networks and by the limited appeal of the event.

Free to book and attend, the workshop was publicised initially via an email sent out to the Museum's existing mass of friends and supporters, and to those who have registered for updates on its website. After that initial call had prompted only a handful of sign ups, however, and with time before the day of the event running short, I was forced into a second effort to drum up interest, resorting in this regard, rather ironically, to that noisiest of all social media platforms: Twitter. Tweets requesting retweets, dispatched to homelessness charities, to sound archives, to individual radio producers, and museum and heritage junkies; in short to those people and organisations, whom I suspected would already be receptive to the project, and crucially with whom I already shared some personal or professional interest or connection (however tenuous). In the realm of Twitter the radical possibility of communicating directly with any of its 330 million users spread worldwide is undermined both by the difficulty of knowing to whom one might most productively address a targeted appeal, and, contrastingly, by the futility of firing off general broadcast messages, fully aware of the odds against them ever meeting with interested eyes amid a torrent of other information.

On the one hand, then, that *TNM* assembled a group of highly energised strangers with varying degrees of direct experience of homelessness (from none to all too much) to exchange perspectives on sound and social marginalisation represented an important evolution of my practice as I understood it. On the other hand, however, that group was comprised almost entirely of politically active, left-leaning sound recordists, radio and heritage professionals, and members of the MoH core group, reflected a truism articulated by archaeologist John Carman. Namely

> that in conducting [in Carman's context] public archaeology (whether we call it 'outreach' or 'community participation' or 'democratic archaeology' [. . .]) we are always and inevitably – and despite any desire to the contrary – dealing with people like ourselves. This of course is neither what we imagine we are doing nor what we would prefer to do: what we intend is to 'reach out' to those who otherwise do not have access to us and our work. But in the end all we can do it talk to those who already speak in our language and share our values. Moreover, and equally inevitably, the process by which we create the community with which we engage is grounded not in processes of inclusion (which we would like) but in processes of exclusion.[42]

A COMMON FRAME: THE POSSIBLE

Turning to the final major development *TNM* marked in my research process relative to *Our Sounds*, I come to the question of framing. During *Our Sounds*, my approach to exploring sound and listening had been relatively unstructured, and many participants, quite understandably, had struggled to find meaning in the subject matter. Sessions tended to take the form of open-ended brainstorming and free association. And while this led to some lively discussions, it also generated a sort of noisy inertia, in which, to paraphrase Bauman, the fragments of our conversations became hegemonic, and the task of drawing out common threads and narratives, certainly at the moment, became almost impossible.[43] By inviting in potentially a broader range of participants to *TNM*, meanwhile, it was foreseeable that that problem would only grow and become more intractable.

My response to that challenge was to try to devise a common framework (referred to above in the introduction to this chapter) and set of conceptual tools with which to structure the new workshop, and, I hoped, give it more purpose and relevance. To this end, drawing on two key influences – Henri Lefebvre's *Critique of Everyday Life* and his overarching mission to create 'a different everydayness'[44] and Phil Cohen's formulation of counter-mapping as 'the postulation of other possible words'[45] – I settled on the future, and the question of how to shape it in order to make it more equitable, pleasurable, and so on, as key, critical frames. Instead of taking the form of co-ethnography, then, or a group oral history session, I intended that the session would feel and function more like a design workshop, interrogating the present (i.e. present soundscapes, and present ways and experiences of listening) in order to understand future possibilities, to surface and negotiate differences in the perceived desirability of those futures, and then discuss how they might best be realised. Finally, looking to turn that broad approach into a concrete plan, and bringing me now back again to a closer discussion of method, I took inspiration from a third key influence, Lefebvre's contemporary and sometime collaborator, Georges Perec.

In his classic short essay *Approaches to What?*, Perec sets out an agenda for studying everyday life, anchored ultimately in a concern to understand and contest social inequality and injustice. Here, though I am sure the text will be familiar to many readers, I will quote at length:

> In our haste to measure the historic, significant and revelatory, let's not leave aside the essential, the truly intolerable, the truly inadmissible. What is scandalous isn't the pit explosion, it's working in coalmines. 'Social problems' aren't 'a matter of concern' when there's a strike, they are intolerable twenty-four hours out of twenty-four, three hundred and sixty-five days a year. [. . .]
>
> What's really going on, what we're experiencing, the rest, all the rest, where is it? How should we take account of, question, describe what happens every day and recurs every day: the banal, the quotidian, the obvious, the common, the ordinary, the infraordinary, the background noise, the habitual?
>
> To question the habitual. But that's just it, we're habituated to it. We don't question it, it doesn't question us, it doesn't seem to pose a problem, we live it without thinking, as if it carried within it neither questions nor answers, as if it weren't the bearer of any information. This is no longer even conditioning, it's anaesthesia. We sleep through our lives in a dreamless sleep. But where is our life? Where is our body? Where is our space?
>
> How are we to speak of these 'common things', how to track them down rather, flush them out, wrest them from the dross in which they remain mired, how to give them a meaning, a tongue, to let them, finally, speak of what is, of what we are.[46]

Coiled tightly into a few dense paragraphs, Perec's impassioned appeal resonates strongly with the works of Robinson, Desjarlais, and others researching everyday experiences of homelessness. Furthermore, it offers a neat summary of many of the concerns that animate Lefebvre's *Critique*: a frustration and fascination with the varied forms of alienation that grip everyday consciousness, with daily rhythms, and the manifold contortions and humiliations they demand of and impose upon ordinary bodies. Perec fixes on the mundane detritus of quotidian existence and implores us to regard it all slowly, searchingly: 'Question your teaspoons'; 'What is there under your wallpaper?' If there is a truth to the everyday ('our truth'), he suggests, it will not be found in the exceptional, in history books alone, or in tabloid sensations, but in 'bricks, concrete, glass, our table manners, our utensils, our tools, the way we spend our time'.

As a basis for investigating sounds, and for framing the potential value of doing so, the above excerpts, shared and discussed with the workshop group, provided the theoretical foundation for the introductory third of *TNM*. Additionally, I also extracted from Perec's essay a final key passage in which he describes a modest method for grasping the everyday:

> Make an inventory of your pockets, of your bag. Ask yourself about the provenance, the use, what will become of each of the objects you take out.[47]

Within this simple instruction, I found, could be traced the essence of the counter-cartographic project as described by Cohen: presencing and elaborating other possible worlds. To shape it to the interrogation of auditory experience, I proceeded by breaking down the three elements listed above in Perec's instruction – provenance, use, and becoming – and recasting these as a set of eight short questions for the group to ask of the sounds we would discuss throughout the day (see figure 2.2).

When it came to the afternoon session of the workshop and the task of 'donating' sounds to the MoH archive, this again was the schema that I

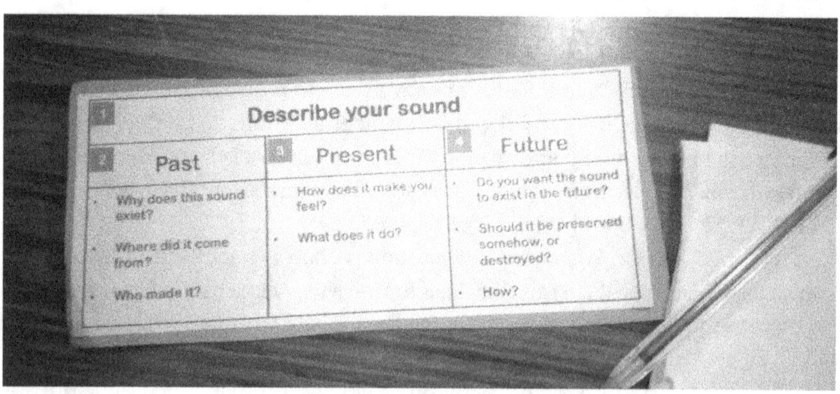

Figure 2.2 Primer for Exploring Sounds. Author's Image.

proposed attendees adopt in describing and explaining their various contributions. Prior to that concluding part of the day, meanwhile, aiming both to put flesh on the bones of that rather abstract model, and to seed a range of different ideas about the potentially vast scope of an enquiry into the auditory, I used the framework as a means of analysing a series of exemplar sounds in two phases.

First, individually, I presented a sound drawn from my own experience (a smoker's cough) to demonstrate the range of different connections and designs for a future society that might be drawn out from that sound and from my own desire not to have to listen to or produce it anymore. To begin with, what was the history of such a cough? How did it come into being? The growth of tobacco? The greed of corporations that foist their noxious product on consumers regardless of the known ill-effects? Personal weakness? Cursed Sir Walter Raleigh? Second, what did the cough do? What feelings did it evoke? Shame, guilt, fear for one's lungs and blood, an equal fear of being discovered by family members hitherto unaware of one's vice. Finally, how could the cough be destroyed? And would the task of destruction entail individual or collective action?

Though it bordered on the frivolous, the purpose of this exercise was to foreground the question of how the future could be re-made and whose responsibility it was to see to it that change occurred. Most pressingly here, though it began from an essentially banal phenomenon, the example also suggested, on the one hand, that behind every event/object/phenomenon one might care to devote one's attention to lies a complex, multi-scalar history; and, on the other, that behind every individual step taken to remedy a given present found to be unfavourable lies the question of how far larger social structures might need to be navigated or reworked in order to achieve personal and collective goals.

In the second phase of this preliminary discussion, workshop attendees worked in small groups analysing a series of video clips I had selected in order to focus attention on the range of ways in which sounds and varied processes of listening might be bound up with experiences of home and belonging, exchanges and flows of power, the production of social space, and so on. Here, we spent the majority of a plenary discussion analysing three clips: (i) a recording of the two-minute silence held regularly at the Cenotaph on Armistice Day; (ii) a video of the then speaker of the UK House of Commons, John Bercow, bellowing at his colleagues for silence during prime minister's questions, struggling to overcome the braying, hooting, and howling of members intent on drowning out one another's speech; and (iii) an excerpt from Director Ken Loach's 2016 film *I, Daniel Blake*, a film grounded in Daniel Edmiston's and others' above-mentioned research into welfare subjectivities, and loosely illustrative of a range of personal stories recounted to me during *Our Sounds*.

Although regrettably I opted not to document fully this early stage of the workshop (limiting audio recording to the afternoon, and treating the morning as a prelude to group work, rather than a part of my own formal research process), the image below, of notes jotted down by a participant during the video session, gives a strong sense of the flavour and content of our discussion.

Overall, as becomes clear in listening to the *TNM* recording, and as I discuss further below, my attempt to route the group's joint exploration of sound through the Perecian framework just described met with only a limited degree of success. To conclude on this point, however, it seems important to reiterate the rationale that led me to try to code a critical concern with the past, present, and future – the origins, efficacies, and destinies of everyday objects and experiences – into the archive and the recording itself.

For by asking participants to structure their donations according to that same schema, my objective was clear: to seed and promulgate a series of questions, and a mode of perception, as much as to share a selection of sounds. Bearing in mind the notion of the possible, and the task of designing into the future, the decision to structure the workshop so as to foreground sounds' prospective futures (or the futures of the conditions of their production and reception), reflected an aim to train listeners in a specific mode of

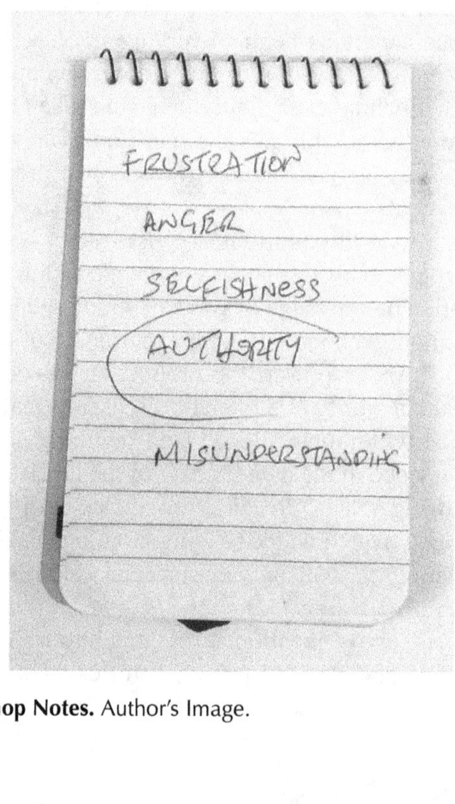

Figure 2.3 Workshop Notes. Author's Image.

attention to the acoustic environment; a listening, that is, that would recognise the possibility of that environment's reshaping and prompt reflection on the best means of achieving desired change.

REFLECTIONS: NOISE, NARRATIVE, AND THE CARTOGRAPHIC COMMONS

Having set out the process through which the recording was produced, and rather than providing a written summary of its content here, I want to invite readers now to take a moment to listen to *TNM* at the URL provided above. Though, on the one hand, this strategy might make for a strange, bodyless chapter, on the other, it feels important to foreground workshop participants' own direct descriptions of the sounds and experiences they donated through the project. Moreover, in discussing the project's material outputs with group members (including the possibility of masking their voices in order to preserve anonymity), it was unanimously agreed that precisely the weight of experience and feeling engrained in those voices was too important and resonant a form of evidence to omit. Together, then, albeit anonymously, we wished to speak for and as ourselves, and so here I preserve that possibility. I will, however, draw upon parts of the recording (referenced by timestamp), as below I offer some final analysis of the project. This I will undertake with a dual perspective, looking to reflect on the project first, more pragmatically, as an event, and second, somewhat more critically, as a form counter-mapping.

EVENT/PROCESS

First, then, to the simple mechanics of the research. While, to be sure there were elements of the workshop and post-production process that could have been improved upon, the limited feedback provided by attendees following the workshop supported my own impression that *TNM* held together as a coherent and enjoyable event. Eleven of the twelve participants felt comfortable to donate a sound to the MoH archive, and, equally, the one remaining participant felt comfortable not to. All attendees were happy to consent to the day's product being made public.

Attendees noted that they appreciated the opportunity the workshop had given them to share perspectives. It was, one said, 'a really special day full of respect, space to talk, space to listen and space to think', while a further participant valued having had 'some time to reflect on the topic and be mindful of [their] place in these issues'.

In terms of the aims I brought personally to the event, a number of the comments left by attendees also suggested that the workshop succeeded, at least partially, in delivering its key message (that sounds are products of socio-historical processes, and as such, to contemplate changing them is also to contemplate changing society). The workshop was 'interesting', according to another participant, in that it 'makes you think about sound(s) and dialogue and their place in history and how (if any) it affects other event(s)', while a fourth comment reflected on the ways in which the 'thoughts and sounds of a city can inspire change'.

One of the more experimental elements of the workshop format I adopted, and that I also want to focus briefly upon here, was my reliance on attendees to themselves document the contributions they each made; handling the recording process in groups of three following a short crash course on how to use the devices provided. Listening back to the recording as a whole, albeit my own editing has smoothed over some of the inconsistencies, interruptions, and stylistic idiosyncrasies this process of delegation spawned and preserved, it is nevertheless possible to identify both particular advantages and disadvantages that approach offered.

Beginning with the latter, a clear challenge I encountered by abdicating responsibility for the recording process was a relative loss of control over the shape of contributions and of the ability, on the one hand, to offer clarification and guidance to participants, and, on the other, to ask follow-up questions in line with my original Perecian framework. To me, most frustrating in this respect was that I could not be present to prompt further, deeper reflection on the question of the future; of the possible; of social change, and how to achieve it.

While in certain of the contributions, it is evident that participants are responding loosely to the framework I provided (thus, from 02m59s to 04m40s, traffic noise is considered to some extent in terms of its past [corporate greed], present [environmental and health impacts], and future [a reference to the potency of 'Critical Mass' style collective actions in which cyclists converge to disrupt traffic]), elsewhere participants either struggled to adapt to the imposed format or abandoned it altogether. At 11m50s for example, we hear an older male participant describe the contrasting feelings of depression and relaxation he associates with hearing the sound of a coffee machine. Though, to my ear, this brief testimony still carries a particular force; nevertheless, I struggle to hear it without also remembering the several minutes of confusion, stripped away in post-production, during which the participant in question struggled to come to terms with the framing questions being foisted upon him: 'What do you mean – the future of coffee? It's coffee'.

In terms of addressing that particular problem (notwithstanding the challenges attendant longer term group project formats), one solution would have been to stage the project over a series of days or weeks, thus allowing more time for participants to formulate responses to specific questions and equally accommodating the possibility of undertaking relevant supported research. Arguing against that adaptation, however, and coming to the benefits of the recording process *as was* are the relative informality and spontaneity that I feel was achieved within and promoted by both the single-day workshop format and by my deployment of a delegated recording process.

One of the participants whom a listener encounters several times through the course of *TNM* is John.[48] Where two of the four groups that split off to document their donations to the archive opted for a more individual approach – taking it in turns to record each other in isolation, John's group adopted a different approach, setting the recording rolling and capturing the full duration of a round table conversation. And it was this wholly unplanned, organic deviation from the process I had charted that yielded what to me stands among its most important outputs: John's reflection (at 15m05s, cutting away from talk of fireside circles) on the power dynamics involved in focus groups and circle-seated listening exercises of the very kind I had devised.

Whichever way one chooses to interpret or take on that contribution (to me it speaks powerfully to the risk of instrumentalisation touch upon above), what is certain from a practical perspective is that it would have been far less likely to have emerged either through a more rigid process or via the mechanism of a one-to-one interview. Perversely then, in the context of a workshop format that I had worked so hard to control, it was precisely at that moment that externally imposed structures gave way to relatively more natural conversation that some of the most resonant materials and opportunities for learning surfaced and could be documented.

COUNTER-MAPPING

As a process and tangible expression of counter-mapping, finally, there are numerous ways of apprehending and evaluating *TNM*. Here I want to adopt two principal perspectives, considering the project both as an investigation into and attempt to represent homelessness and as a contribution to the tradition and practice of critical and community cartography.

Beginning with the recording as a representation, then, there are three points I would make regarding the portrayal of homelessness arrived at through *TNM*. The first, and perhaps the most urgent concerns the failure of the project to foreground an explicit critical understanding or questioning of the causes of homelessness, and hence, too, its failure (in large part) to broach the question of the possible based on a concrete understanding of the past and present. In designing the workshop framework, my intent had been in part to encourage, and create space for, individuals to speak in depth about the origins of (their experiences of) social inequality and marginalisation. Plainly, however, that did not work.

Part of the explanation for this absence in the recording lies, I think, in the power and almost mechanical familiarity and flow of personal narrative: it seems far easier to tell and retell our own well digested and rehearsed stories than to try account for social conditions more broadly – at least in the context of a public workshop, where time and opportunities for slow reflection are scarce. In part, too, that absence stems from the near impossibility of accounting for homelessness as a single phenomenon. Finally, and equally unavoidably in my research context, this limit of the recording seems to be born of the unusual emphasis on sound and auditory experience upon which it is founded. Though, I would propose, listening functioned through the workshop as an effective means of defamiliarising the problems and experiences discussed on the day, it is also true that my exclusive privileging of sound worked at times to undermine the research. In the end, for example, although certain sounds might well be identified as stemming from economic inequality, it is important to concede that simply listening to them offers little as a means of grasping critically the root causes of that inequality.

The second point I would make about the recording as a representation of homelessness concerns doubt, and here I find far more cause for encouragement. In *TNM*, I would argue, two of the strengths of the finished piece are the doubt it generates as to the identity of its contributors and the degree of their experience of homelessness. Throughout the workshop, as part of creating a supported environment in which to work, it was agreed that participants need not disclose anything of their personal history that they did not wish to. As such, even having led the day's events, I remain uncertain as to who among the group had or had not experienced homelessness, either in the sense of rooflessness or in any other form. Consequently, unless individual participants chose to discuss their experiences through their contributions (some did, and others did not), we cannot be entirely certain as listeners who is or has been 'homeless' and who is or has not. Within the recording, there are many different voices, and many different variations of tone, timbre, and rhythm born of class, gender, regional accent, and so on – and it feels natural to draw conclusions about those voices; whom they belong to; class status; wealth, relative privilege, etc. Doubt, however, remains – a doubt that to my mind invites curiosity, and which functions, perhaps, to suspend judgement for a moment long enough to seed the question: who or what is a homeless person; where do my assumptions lead me; can I be certain that they are valid?

A third and final observation here concerns the success of the project in creating a multi-perspectival portrayal of homelessness, and in invoking a sense of shared ownership of and responsibility for the issue. To begin with in this regard, an important point to stress about the sounds featured in *TNM* is their essential ubiquity and banality: wind, soup, silence, near silence, traffic, coffee, keys, locks, fire, television. With the exception of the fender sounds and other canal noises discussed at 10m00s, the auditory artefacts deposited by contributors to the MoH archive will be familiar to most listeners; a part of a shared everyday life and set of cultural references. When at the conclusion of the recording, for example, we hear the final speaker associate the peace and quiet of her newly acquired home with the freedom simply 'to be', or when we hear another contributor describe her experience of the sounds of soup bubbling in terms of collective care, what we hear are not the sounds of homelessness *per se* but rather the sounds of what it is to be human.

These are basic, I would venture to say universal, experiences and widely relatable aspirations that can underpin, and give tangible expression to, shared efforts to found a new everydayness. In Frances Dyson's *The Tone of Our Times* – a reflection on the sonorities of economic and ecological crisis in late capitalism – the author foregrounds the critical and pedagogical challenge of promoting and 'mov[ing] toward a shared sensibility' from which to build 'sense, the common, and common sense simultaneously'.[49] Taking on that task, one of the principal appeals *TNM* makes to its listeners is to attend to the world sensuously and in solidarity with others: to hear the wind, and to recall and share another's suffering as a precursor to action; to affix to one's own experience of peace, quiet, and rest, a shadow appreciation of another's lack thereof: indeed, to design and build the commons via the elaboration of a renewed common sense.

Extending this point, and considering further the question of social change, at 17m25s in the recording we hear a younger female participant in the workshop, her voice faltering and discomfort palpable – an 'outsider' in her own terms – offer a listening silence as her own contribution to the MoH archive. Without knowing more about this particular attendee's own history (and I do not), it is impossible to judge the candour of her statement. (Have 'life paths', as she suggests, truly sheltered her from homelessness? Is it possible to live in London and avoid the constant dreadful scene of people sleeping in doorways? What degree of evasion does this innocence involve?) What seems so valuable to me about this contribution, however, is, on the one hand, the attention it draws to 'outsiders' own vulnerability and uncertainty faced with others' suffering, and on the other, equally, the value it attributes to listening as a prerequisite for solidarity.

Returning to the notion of counter-mapping, a critical task Cohen designates as part of the practice is that of providing opportunities for dialogue: a space and framework 'for negotiating differences of standpoint and experience', a process which enables the charting of one's own self and experience, but which also 'explores dis/identifications with the Other – the other class, other ethnicity, other race, other generation.'[50] While to be sure, in this respect, *TNM* represents only a beginning, the effect of contributions like the one just discussed is both to document dialogue in action, and, more importantly perhaps, to reaffirm the point that solutions to issues like homelessness should be sought not only in the therapeutic rehabilitation of those affected most intimately by them, but also elsewhere, in the inclusion, education, and more broadly support of those outsiders who may care for and wish to act with, and on behalf of, marginalised constituencies (and here I include myself), but who do not necessarily possess an understanding of *how to act.* Faced with John Carman's above-cited comments on the seeming futility of 'outreach', it reminds us of the value of working with people who are like us, and who already agree with us, but who may feel powerless, to see what together we can achieve.

FINAL THOUGHTS

My purpose in unpacking *TNM* above has been primarily to reflect on the challenges facing the project of counter-mapping and to highlight the contribution that critical listening practices can make to its future. To the latter point, I hope that the work presented in this chapter prompts curiosity and experimentation. In saying that, however, I would repeat the point that I found my own studies of sound to be rather toothless as means of getting to the root of structural inequality. Perhaps, then, listening is best deployed as one tool among many. Turning to the prospects of counter-mapping, finally, I wish only to reiterate how valuable a concept noise has been to me in thinking through the (in)efficacy of my own efforts in the field and in problematising a term, community, and a process, communication, that previously I took too much for granted.

NOTES

1. This chapter reworks substantial parts of chapters 1, 5, and 7 of my doctoral thesis (Tourle 2020). The research it describes was funded by the Economic and Social Research Council. I remain enormously grateful to the team at the Museum of Homelessness, who back in 2018, worked with me to develop the project I present here.
2. Museum of Homelessness. 2019a. 'About Us'. *Museum of Homelessness* (blog). 2019. https://museumofhomelessness.org/about/.
3. Museum of Homelessness, 2019a.

4. Drever, John Levack. 2014. 'Silent Soundwalking: An Urban Pedestrian Soundscape Methodology'. 9 October 2014. https://www.academia.edu/4028850/Silent_Soundwalking_an_urban_pedestrian_soundscape_methodology; Westerkamp, Hildegard. 1974. 'Soundwalking'. Edited by WJ Langlois, Janet Caruthers, and Derek Reimer. *Sound Heritage Journal* 3 (4): 18–25.

5. British Library. 2011. 'UK Soundmap'. 2011. http://sounds.bl.uk/sound-maps/uk-soundmap. 2015. 'Sounds of Our Shores: Sounds of Our Shores'. 2015. http://www.bl.uk/sounds-of-our-shores.

6. Voegelin, Salomé. 2014. *Sonic Possible Worlds : Hearing the Continuum of Sound / Salomé Voegelin.* New York: Bloomsbury Academic, pp. 23.

7. Carlyle, Angus. 2014. 'The God's Eye and The Buffalo's Breath: Seeing and Hearing Web-Based Sound Maps'. In *Proceedings of Invisible Places / Sounding Cities. Sound Urbanism and Sense of Place*, edited by Raquel Castro and Miguel Carvalhais. Viseu, pp. 150.

8. Carlyle, Angus, 2014. pp. 150.

9. Tourle, Paul. 2016. 'White Noise: Sound, Materiality and the Crowd in Contemporary Heritage Practice'. *International Journal of Heritage Studies*, November, 1–14. https://doi.org/10.1080/13527258.2016.1261919;

Tourle, Paul. 2020. 'Sound, Heritage Homeless in the Age of Noise'. London: UCL (University College London).

10. See Holtorf, Cornelius. 2015. 'Averting Loss Aversion in Cultural Heritage'. *International Journal of Heritage Studies* 21 (4): 405–21. https://doi.org/10.1080/13527258.2014.938766.

11. Waldock. 2011. 'Soundmapping: Critiques and Reflections on This New Publicly Engaging Medium'. *Journal of Sonic Studies* 1 (1). http://journal.sonicstudies.org/vol01/nr01/a08; Waldock, Jacqueline. 2016. 'Crossing the Boundaries: Sonic Composition and the Anthropological Gaze'. *The Senses and Society* 11 (1): 60–67. https://doi.org/10.1080/17458927.2016.1164429.

12. Droumeva, Milena. 2017. 'Soundmapping as Critical Cartography: Engaging Publics in Listening to the Environment'. *Communication and the Public* 2 (4): 335–51. https://doi.org/10.1177/2057047317719469.

13. Droumeva, Milena. 2017. pp. 338.

14. For example, Tomas, Annabel, and Helga Dittmar. 1995. 'The Experience of Homeless Women: An Exploration of Housing Histories and the Meaning of Home'. *Housing Studies* 10 (4): 493–515.https://doi.org/10.1080/02673039508720834; Somerville, Peter. 1992. 'Homelessness and the Meaning of Home: Rooflessness or Rootlessness?' *International Journal of Urban and Regional Research* 16 (4): 529–39. https://doi.org/10.1111/j.1468-2427.1992.tb00194; Somerville, Peter. 2013. 'Understanding Homelessness'. *Housing, Theory and Society* 30 (4): 384–415. https://doi.org/10.1080/14036096.2012.756096; Ravenhill, Megan. 2008. *The Culture of Homelessness*. Aldershot: Ashgate.

15. Robinson, Catherine. 2009. 'Homelessness Felt'. *Cultural Studies Review* 15 (1): 167–72–167–72. https://doi.org/10.5130/csr.v15i1.2060; Robinson, Catherine. 2011. *Beside One's Self: Homelessness Felt and Lived.* 1st ed. Space, Place, and Society. Syracuse, NY: Syracuse University Press.

16. Ravenhill, Megan. 2008.

17. Lancione, Michele. 2013. 'How Is Homelessness?' *European Journal of Homelessness* 7 (2): 237–47.

18. Farrugia, David, and Jessica Gerrard. 2016. 'Academic Knowledge and Contemporary Poverty: The Politics of Homelessness Research'. *Sociology* 50 (2): 267–84. https://doi.org/10.1177/0038038514564436. pp. 279.

19. O'Neill, Bruce. 2017. *The Space of Boredom: Homelessness in the Slowing Global Order*. Duke University Press. https://doi.org/10.1215/9780822373278.

20. Williams, Jean Calterone. 1996. 'Geography of The Homeless Shelter: Staff Surveillance And Resident Resistance'. *Urban Anthropology and Studies of Cultural Systems and World Economic Development* 25 (1): 7–113; Conradson, David. 2003. 'Spaces of Care in the City: The Place of a Community Drop-in Centre'. *Social & Cultural Geography* 4 (4): 507–25. https://doi.org/10.1080/1464936032000137939; Datta, Ayona. 2005. '"Homed" in Arizona: The Architecture of Emergency Shelters'. *Urban Geography* 26 (6): 536–57. https://doi.org/10.2747/0272-3638.26.6.536; Johnsen, Sarah, Paul Cloke, and Jon May. 2005. 'Day Centres for Homeless People: Spaces of Care or Fear?' *Social & Cultural Geography* 6 (6): 787–811. https://doi.org/10.1080/14649360500353004.

21. Robinson, C. 2010. *Rough Living: Surviving Violence and Homelessness*. UTS e press. https://opus.lib.uts.edu.au/handle/10453/14227; Jordan, Lucinda. 2012. 'Spaces of Trauma: Young People, Homelessness and Violence'. *Youth Studies Australia* 31 (4): 11.

22. Farrugia, David. 2011. 'The Symbolic Burden of Homelessness Towards a Theory of Youth Homelessness as Embodied Subjectivity'. *Journal of Sociology* 47 (1): 71–87. https://doi.org/10.1177/1440783310380989.

23. Sibley, David. 1995. *Geographies of Exclusion: Society and Difference in the West*. London; New York: Routledge; Amster, Randall. 2003. 'Patterns of Exclusion: Sanitizing Space, Criminalizing Homelessness'. *Social Justice; San Francisco* 30 (1): 195; Tosi, Antonio. 2007. 'Homelessness and the Control of Public Space - Criminalising the Poor?' 1: 225–36; DeVerteuil, Geoffrey, Jon May, and Jürgen von Mahs. 2009. 'Complexity Not Collapse: Recasting the Geographies of Homelessness in a "Punitive" Age'. *Progress in Human Geography* 33 (5): 646–66. https://doi.org/10.1177/0309132508104995; Davis, Mike. 1991. 'Afterword - A Logic like Hell's: Being Homeless in Los Angeles'. *UCLA Law Review* 39: 325; Davis, Mike. 2006. *City of Quartz: Excavating the Future in Los Angeles*. New ed. London; New York: Verso.

24. Bretherton, Joanne, Caroline Hunter, and Sarah Johnsen. 2013. '"You Can Judge Them on How They Look…": Homelessness Officers, Medical Evidence and Decision-Making in England'. *European Journal of Homelessness* 7 (1): 69–92.

25. Bibbings, Jennie. 2018. 'So What Is Priority Need - and Why Do We Want to Get Rid of It?' *Shelter Cymru* (blog). 20 April 2018. https://sheltercymru.org.uk/blog-what-is-priority-need/.

26. Edmiston, Daniel. 2017. '"How the Other Half Live": Poor and Rich Citizenship in Austere Welfare Regimes'. *Social Policy and Society* 16 (02): 315–25. https://doi.org/10.1017/S1474746416000580; Woolford, Andrew, and Amanda

Nelund. 2013. 'The Responsibilities of the Poor: Performing Neoliberal Citizenship within the Bureaucratic Field'. *Social Service Review* 87 (2): 292–318. https://doi.org/10.1086/671072.

27. Robinson, Catherine. 2011. *Beside One's Self: Homelessness Felt and Lived.* 1st ed. Space, Place, and Society. Syracuse, NY: Syracuse University Press; Desjarlais, Robert R. 1997. *Shelter Blues: Sanity and Selfhood among the Homeless.* Contemporary Ethnography. Philadelphia, PA: University of Pennsylvania.

28. Robert R. 1997. pp. 222–225.

29. Robinson, Catherine. 2011. pp. 47.

30. Scarry, Elaine. 1985. *The Body in Pain: The Making and Unmaking of the World.* First issued as paperback. Oxford Paperback. New York, NY: Oxford Univ. Press.

31. Serres, Michel. 1995. *Genesis.* Ann Arbor: University of Michigan Press.

32. Lingis, Alphonso. 1994. *The Community of Those Who Have Nothing in Common.* Studies in Continental Thought. Bloomington: Indiana University Press. pp. 69–78.

33. Serres, Michel. 1995.

34. Terranova, Tiziana. 2004. 'Communication beyond Meaning on the Cultural Politics of Information'. *Social Text* 22 (3 80): 51–73. https://doi.org/10.1215/01642472-22-3. pp. 80–51; Terranova, Tiziana. 2012. 'Attention, Economy and the Brain'. *Culture Machine* 13: 1–19.

35. Bauman, Zygmunt. 1987. 'The Philosopher in the Age of Noise: A Reading of Richard J. Bernstein's Philosophical Profiles'. *Theory, Culture & Society* 4: 157–65; Bauman, Zygmunt. 2007. *Consuming Life.* Cambridge; Malden, MA: Polity Press.

36. Marcus, Anthony. 2003. 'Shelterization Revisited: Some Methodological Dangers of Institutional Studies of Homelessness'. *Human Organization* 62 (2): 134–41.

37. Dean, Jodi. 2005. 'Communicative Capitalism: Circulation and the Foreclosure of Politics'. *Cultural Politics* 1 (1): 51–74. https://doi.org/10.2752/174321905778054845.

38. Feld, Steven. 1987. 'Dialogic Editing: Interpreting How Kaluli Read Sound and Sentiment'. *Cultural Anthropology* 2 (2): 190–210. https://doi.org/10.2307/656355.

39. see Museum of Homelessness. 2018. 'Annual Review July 2017 - June 2018'. https://museumofhomelessness.org/wp-content/uploads/2017/02/MoH-Annual-Review-1718.pdf. pp. 8–10.

40. Museum of Homelessness. 2019b. 'Dying Homeless Project'. *Museum of Homelessness* (blog). 2019. https://museumofhomelessness.org/dying-homeless-project/.

41. Rose, Nikolas. 1999. *Powers of Freedom: Reframing Political Thought.* Cambridge: Cambridge University Press. Rose, Nikolas. 2000. 'Community, Citizenship, and the Third Way'. *American Behavioral Scientist* 43 (9): 1395–1411. https://doi.org/10.1177/00027640021955955; Tyler, Imogen. 2013. *Revolting Subjects Social Abjection and Resistance in Neoliberal Britain.* London; New York: Zed Books. http://site.ebrary.com/id/10685289; Gowan, Teresa. 2010. *Hobos,*

Hustlers, and Backsliders: Homeless in San Francisco. Minneapolis: University of Minnesota Press; Marcuse, Peter. 1988. 'Neutralizing Homelessness'. *Socialist Review* 18 (1): 69–96.

Marcuse, Peter. 2017. 'After Exposing the Roots of Homelessness – What?' *Urban Geography* 38 (3): 357–59. https://doi.org/10.1080/02723638.2016.1247601.

42. Carman, John. 2011. 'Stories We Tell: Myths at the Heart of "Community Archaeology"'. *Archaeologies* 7 (3): 490–501. https://doi.org/10.1007/s11759-011-9180-x. pp. 499–500.

43. Bauman, Zygmunt. 2007.

44. Lefebvre, Henri. 2008. *Critique of Everyday Life*. Special ed. Vol. 3. London; New York: Verso. pp. 29.

45. Cohen, Phil. 2017. 'Our Kind of Town (2)'. *Livingmaps Review* 0 (2). http://livingmaps.review/journal/index.php/LMR/article/view/44. pp. 9.

46. Perec, Georges. 2008. *Species of Spaces and Other Pieces*. Translated by John Sturrock. Penguin Classics. London: Penguin Books. pp. 11.

47. Perec, Georges. 2008. Pp. 11.

48. A pseudonym.

49. Dyson, Frances. 2014. *The Tone of Our Times: Sound, Sense, Economy, and Ecology*. Leonardo Book Series. Cambridge, MA: MIT Press. pp. 149.

50. Cohen, Phil. 2017. pp. 11.

BIBLIOGRAPHY

Amster, Randall. 2003. 'Patterns of Exclusion: Sanitizing Space, Criminalizing Homelessness'. *Social Justice; San Francisco* 30 (1): 195.

Bauman, Zygmunt. 1987. 'The Philosopher in the Age of Noise: A Reading of Richard J. Bernstein's Philosophical Profiles'. *Theory, Culture & Society* 4: 157–65.

———. 2007. *Consuming Life*. Cambridge; Malden, MA: Polity Press.

Bibbings, Jennie. 2018. 'So What Is Priority Need - and Why Do We Want to Get Rid of It?' *Shelter Cymru* (blog). 20 April 2018. https://sheltercymru.org.uk/blog-what-is-priority-need/.

Bretherton, Joanne, Caroline Hunter, and Sarah Johnsen. 2013. '"You Can Judge Them on How They Look…": Homelessness Officers, Medical Evidence and Decision-Making in England'. *European Journal of Homelessness* 7 (1): 69–92.

British Library. 2011. 'UK Soundmap'. 2011. http://sounds.bl.uk/sound-maps/uk-soundmap.

———. 2015. 'Sounds of Our Shores: Sounds of Our Shores'. 2015. http://www.bl.uk/sounds-of-our-shores.

Carlyle, Angus. 2014. 'The God's Eye and The Buffalo's Breath: Seeing and Hearing Web-Based Sound Maps'. In *Proceedings of Invisible Places / Sounding Cities. Sound Urbanism and Sense of Place*, edited by Raquel Castro and Miguel Carvalhais. Viseu.

Carman, John. 2011. 'Stories We Tell: Myths at the Heart of "Community Archaeology"'. *Archaeologies* 7 (3): 490–501. https://doi.org/10.1007/s11759-011-9180-x.

Cohen, Phil. 2017. 'Our Kind of Town (2)'. *Livingmaps Review* 0 (2). http://livingmaps.review/journal/index.php/LMR/article/view/44.

Conradson, David. 2003. 'Spaces of Care in the City: The Place of a Community Drop-in Centre'. *Social & Cultural Geography* 4 (4): 507–25. https://doi.org/10.1080/1464936032000137939.

Datta, Ayona. 2005. '"Homed" in Arizona: The Architecture of Emergency Shelters'. *Urban Geography* 26 (6): 536–57. https://doi.org/10.2747/0272-3638.26.6.536.

Davis, Mike. 1991. 'Afterword - A Logic like Hell's: Being Homeless in Los Angeles'. *UCLA Law Review* 39: 325.

———. 2006. *City of Quartz: Excavating the Future in Los Angeles*. New ed. London; New York: Verso.

Dean, Jodi. 2005. 'Communicative Capitalism: Circulation and the Foreclosure of Politics'. *Cultural Politics* 1 (1): 51–74. https://doi.org/10.2752/174321905778054845.

Desjarlais, Robert R. 1997. *Shelter Blues: Sanity and Selfhood among the Homeless*. Contemporary Ethnography. Philadelphia, PA: University of Pennsylvania.

DeVerteuil, Geoffrey, Jon May, and Jürgen von Mahs. 2009. 'Complexity Not Collapse: Recasting the Geographies of Homelessness in a "Punitive" Age'. *Progress in Human Geography* 33 (5): 646–66. https://doi.org/10.1177/0309132508104995.

Drever, John Levack. 2014. 'Silent Soundwalking: An Urban Pedestrian Soundscape Methodology'. 9 October 2014. https://www.academia.edu/4028850/Silent_Soundwalking_an_urban_pedestrian_soundscape_methodology.

Droumeva, Milena. 2017. 'Soundmapping as Critical Cartography: Engaging Publics in Listening to the Environment'. *Communication and the Public* 2 (4): 335–51. https://doi.org/10.1177/2057047317719469.

Dyson, Frances. 2014. *The Tone of Our Times: Sound, Sense, Economy, and Ecology*. Leonardo Book Series. Cambridge, MA: MIT Press.

Edmiston, Daniel. 2017. '"How the Other Half Live": Poor and Rich Citizenship in Austere Welfare Regimes'. *Social Policy and Society* 16 (02): 315–25. https://doi.org/10.1017/S1474746416000580.

Farrugia, David. 2011. 'The Symbolic Burden of Homelessness Towards a Theory of Youth Homelessness as Embodied Subjectivity'. *Journal of Sociology* 47 (1): 71–87. https://doi.org/10.1177/1440783310380989.

Farrugia, David, and Jessica Gerrard. 2016. 'Academic Knowledge and Contemporary Poverty: The Politics of Homelessness Research'. *Sociology* 50 (2): 267–84. https://doi.org/10.1177/0038038514564436.

Feld, Steven. 1987. 'Dialogic Editing: Interpreting How Kaluli Read Sound and Sentiment'. *Cultural Anthropology* 2 (2): 190–210. https://doi.org/10.2307/656355.

Gowan, Teresa. 2010. *Hobos, Hustlers, and Backsliders: Homeless in San Francisco*. Minneapolis: University of Minnesota Press.

Holtorf, Cornelius. 2015. 'Averting Loss Aversion in Cultural Heritage'. *International Journal of Heritage Studies* 21 (4): 405–21. https://doi.org/10.1080/13527258.2014.938766.

Johnsen, Sarah, Paul Cloke, and Jon May. 2005. 'Day Centres for Homeless People: Spaces of Care or Fear?' *Social & Cultural Geography* 6 (6): 787–811. https://doi.org/10.1080/14649360500353004.

Jordan, Lucinda. 2012. 'Spaces of Trauma: Young People, Homelessness and Violence'. *Youth Studies Australia* 31 (4): 11.

Lancione, Michele. 2013. 'How Is Homelessness?' *European Journal of Homelessness* 7 (2): 237–47.

Lefebvre, Henri. 2008. *Critique of Everyday Life*. Special ed. Vol. 3. London; New York: Verso.

Lingis, Alphonso. 1994. *The Community of Those Who Have Nothing in Common*. Studies in Continental Thought. Bloomington: Indiana University Press.

Marcus, Anthony. 2003. 'Shelterization Revisited: Some Methodological Dangers of Institutional Studies of Homelessness'. *Human Organization* 62 (2): 134–41.

Marcuse, Peter. 1988. 'Neutralizing Homelessness'. *Socialist Review* 18 (1): 69-96.

———. 2017. 'After Exposing the Roots of Homelessness – What?' *Urban Geography* 38 (3): 357–59. https://doi.org/10.1080/02723638.2016.1247601.

Museum of Homelessness. 2018. 'Annual Review July 2017 - June 2018'. https://museumofhomelessness.org/wp-content/uploads/2017/02/MoH-Annual-Review-1718.pdf.

———. 2019a. 'About Us'. *Museum of Homelessness* (blog). 2019. https://museumofhomelessness.org/about/.

———. 2019b. 'Dying Homeless Project'. *Museum of Homelessness* (blog). 2019. https://museumofhomelessness.org/dying-homeless-project/.

O'Neill, Bruce. 2017. *The Space of Boredom: Homelessness in the Slowing Global Order*. Duke University Press. https://doi.org/10.1215/9780822373278.

Perec, Georges. 2008. *Species of Spaces and Other Pieces*. Translated by John Sturrock. Penguin Classics. London: Penguin Books.

Ravenhill, Megan. 2008. *The Culture of Homelessness*. Aldershot: Ashgate.

Robinson, C. 2010. *Rough Living: Surviving Violence and Homelessness*. UTS e press. https://opus.lib.uts.edu.au/handle/10453/14227.

Robinson, Catherine. 2009. 'Homelessness Felt'. *Cultural Studies Review* 15 (1): 167–72. https://doi.org/10.5130/csr.v15i1.2060.

———. 2011. *Beside One's Self: Homelessness Felt and Lived*. 1st ed. Space, Place, and Society. Syracuse, NY: Syracuse University Press.

Rose, Nikolas. 1999. *Powers of Freedom: Reframing Political Thought*. Cambridge: Cambridge University Press.

———. 2000. 'Community, Citizenship, and the Third Way'. *American Behavioral Scientist* 43 (9): 1395–1411. https://doi.org/10.1177/00027640021955955.

Scarry, Elaine. 1985. *The Body in Pain: The Making and Unmaking of the World*. First issued as paperback. Oxford Paperback. New York, NY: Oxford Univ. Press.

Serres, Michel. 1995. *Genesis*. Ann Arbor: University of Michigan Press.

Sibley, David. 1995. *Geographies of Exclusion: Society and Difference in the West*. London; New York: Routledge.

Somerville, Peter. 1992. 'Homelessness and the Meaning of Home: Rooflessness or Rootlessness?' *International Journal of Urban and Regional Research* 16 (4): 529–39. https://doi.org/10.1111/j.1468-2427.1992.tb00194.x.

———. 2013. 'Understanding Homelessness'. *Housing, Theory and Society* 30 (4): 384–415. https://doi.org/10.1080/14036096.2012.756096.

Terranova, Tiziana. 2004. 'Communication beyond Meaning ON THE CULTURAL POLITICS OF INFORMATION'. *Social Text* 22 (3 80): 51–73. https://doi.org/10.1215/01642472-22-3_80-51.

———. 2012. 'Attention, Economy and the Brain'. *Culture Machine* 13: 1–19.

Tomas, Annabel, and Helga Dittmar. 1995. 'The Experience of Homeless Women: An Exploration of Housing Histories and the Meaning of Home'. *Housing Studies* 10 (4): 493–515. https://doi.org/10.1080/02673039508720834.

Tosi, Antonio. 2007. 'Homelessness and the Control of Public Space - Criminalising the Poor?' 1: 225–36.

Tourle, Paul. 2016. 'White Noise: Sound, Materiality and the Crowd in Contemporary Heritage Practice'. *International Journal of Heritage Studies*, November, 1–14. https://doi.org/10.1080/13527258.2016.1261919.

———. 2020. 'Sound, Heritage Homeless in the Age of Noise'. London: UCL (University College London).

Tyler, Imogen. 2013. *Revolting Subjects Social Abjection and Resistance in Neoliberal Britain*. London; New York: Zed Books. http://site.ebrary.com/id/10685289.

Voegelin, Salomé. 2014. *Sonic Possible Worlds: Hearing the Continuum of Sound / Salomé Voegelin*. New York: Bloomsbury Academic.

Waldock. 2011. 'SOUNDMAPPING: Critiques and Reflections on This New Publicly Engaging Medium'. *Journal of Sonic Studies* 1 (1). http://journal.sonicstudies.org/vol01/nr01/a08.

Waldock, Jacqueline. 2016. 'Crossing the Boundaries: Sonic Composition and the Anthropological Gaze'. *The Senses and Society* 11 (1): 60–67. https://doi.org/10.1080/17458927.2016.1164429.

Westerkamp, Hildegard. 1974. 'Soundwalking'. Edited by WJ Langlois, Janet Caruthers, and Derek Reimer. *Sound Heritage Journal* 3 (4): 18–25.

Williams, Jean Calterone. 1996. 'Geography of the Homeless Shelter: Staff Surveillance And Resident Resistance'. *Urban Anthropology and Studies of Cultural Systems and World Economic Development* 25 (1): 75–113.

Woolford, Andrew, and Amanda Nelund. 2013. 'The Responsibilities of the Poor: Performing Neoliberal Citizenship within the Bureaucratic Field'. *Social Service Review* 87 (2): 292–318. https://doi.org/10.1086/671072.

Chapter 3

Mapping the Right to the City

City Perception as a Shaping Force

Giulia Carones

> *What is the heart of a city? And its soul?. . . How do you know a city? How do you get to know your city?*
>
> *Method: one should either give up talking about the city or force oneself to talk about it in the plainest, most obvious, most familiar terms. Getting rid of pre-conceived ideas. Quitting thinking in ready-made terms, forgetting what urban planners and sociologists said. We shall never be able to explain or justify the city. The city is here, it is our space and it is our only one.*[1]
>
> George Perec, *Espèces d'espaces* 1974

This chapter engages with activists' perception of the urban environment and practice of the urban commons via their political engagement. More specifically, the research featured here has sought to address how participating in an urban social movement which advocates the grassroot, citizen-led creation and management of the urban commons can shape individuals' perception of the city.

To explore this, I have devised and relied on a cognitive-mapping tool as a component of semi-structured interviews with members of the activist collective Laboratorio Universitario Metropolitano (LUMe) in Milan, Italy. The resulting maps have proved conducive to meaningful conversations and reflections on city perceptions; from this evidence, it has become clear that activism does affect city perception through the investigation and assimilation of new landmarks in the urban fabric; the maps conveyed participants' creative and operational gaze on the city, as well as a positive notion of political confrontation[2] as a generative and necessary rupture of the status quo. Drawing on these materials, the work described here sets out to conceptualise city perception as an urban shaping force, capable of impacting how city dwellers *make* a city's time and space.

In what follows, I first introduce the context where the research was conducted, briefly situating LUMe within the political landscape of Milanese social movements. I go on to present the rationale for proposing cognitive mapping as a tool for participatory research. Then, I describe the methodological choices which led me to adapt Kevin Lynch's analytical categories for cognitive mapping. I discuss some of the material which emerged using this approach and its significance within the overarching research frame, especially in relation to the urban commons and the generative potential of political confrontation. In conclusion, I present a few reflections on the reception of my research work by participants and its medium-to-long-term implications.

MAPPING AS VOICING ALTERNATIVE PERSPECTIVES ON THE CITY

Conducting ethnographic research in Milan on how political activists perceive the city through the lenses of their explicit advocacy for the urban commons has prompted me to devise a participatory methodology to explore the nexus between political engagement, the Lefebvrian framework of the right to the city, and the urban commons.[3] More specifically, I was interested in how the purposeful long-term squatting practices of activists within the urban fabric could shape participants' perceptions of the city and the city itself. This is in the context of an increasingly regulated and neoliberal globalising urban environment, characterised by privatisation, gentrification, and uneven spatial development.[4] These issues led me to consider the extent to which contemporary citizens can effectively reclaim their right to the city as the right to make the city, to engage in a creative endeavour giving rise to cities as (in Lefebvre's terms) an *oeuvre*.

Completed between February and May 2018, my fieldwork research consisted of an ethnography of a *Centro Sociale Occupato Autogestito* (CSOA) (literally, Self-managed, Occupied Social Centre), the Laboratorio Universitario Metropolitano (LUMe) in Milan. LUMe brings activists, university students, and recent graduates together with young artists and musicians from a variety of backgrounds. They share a political commitment against fascism, racism, sexism and, more contentiously internally, against capitalism. LUMe has been active in Milan since April 2015 and has occupied three locations in the city. Its current site is shown by the dot on the map in figure 3.1. The movement's explicit references to notions of re-appropriation and de-alienation of urban space and to the right to the city

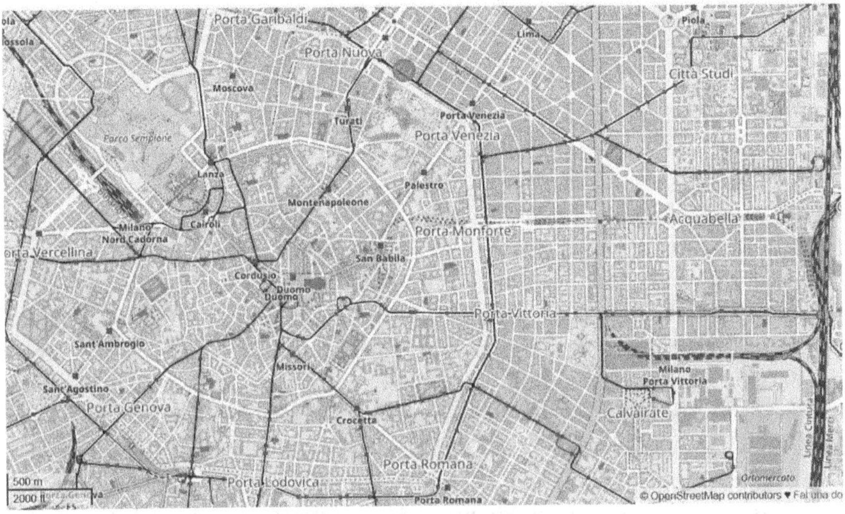

Figure 3.1 Map of the City Centre of Milan Showing LUMe at 24 Via Vittorio Veneto. OpenStreetMap 2018 (CC).

render LUMe especially significant for the issues outlined above, conveying a research aware, politically charged commitment to reshape the city.[5]

Milan currently hosts several kinds of occupied cultural spaces, ranging from the long established, to the ones mostly lived through the experiences of high-school students, LUMe is the only social centre explicitly responding to the desire for mobilisation of university students, specifically. In terms of the purposive case selection for the research design, the rationale for selecting LUMe as case study has been to acknowledge and work on a self-proclaimed *deviant* case amongst urban social movements practicing occupation in the Milanese environment. According to several participants to LUMe, the latter represents specific traits which set it apart from other urban social movements experiences in Milan, such as its participants' pool, its strongly characterised cultural offer, as well as its political subjectivation.

LUMe emerged out of Università degli Studi's political collective in 2015. Most of its participants are university students, although enrolled in other institutions, too, as well as young theatre actors, artists, and musicians. LUMe relies on a pool of members between sixty and eighty people – including less-assiduous participants. The horizontal decision-making body, the assembly, whose membership is voluntary, spontaneous, and fluid, at the time of fieldwork numbered between twenty and thirty members. Attendance to the management assembly is open to all; it is based on the will to be involved in decision-making processes concerning LUMe and acts on the principle of consensus.

In the mission statement released on LUMe's blog, the chief aim of the movement is to awaken citizens and the administration to the potential cultural value of unused and abandoned spaces within the urban landscape.[6] The occupations enacted by LUMe have the proclaimed objective of claiming the municipality's attention and to denounce the devaluation of sections of the city. In an on-going dialectic of occupation and eviction, LUMe has moved three times. In November 2017, LUMe moved to its third site of occupation, a former maintenance shed in the public gardens around Porta Venezia, abandoned for years and used by heroin dealers and addicts, as hundreds of syringes cleaned up from the ground during the first days of their occupation testified. Tales of the conditions in which LUMe found the space when they first entered it convey incredulity and anger at an historic site's decay. 'You really don't want to know what it looked like – syringes, rubble and garbage amassed with tools from when the public gardens' shed was still in use, more than 100 years old' (Giano, Conversation 9th February 2018).

As I was organising my fieldwork, I had gotten in touch with Giano, an acquaintance of mine from my teenage years and one of LUMe's co-founders, hinting at the possibility to collaborate on a research project focussing on the urban commons in Milan. Aware of my interest in politics, Giano had invited me to visit the various social movements he had got involved with, although up to that moment study commitments abroad had always prevented me from taking up his offer. During my first visit to LUMe, on 9 February 2018, Giano was enthusiastic to speak to me about what LUMe considers good practice in the creation of the urban commons, pointing at Napoli's legislation as exemplary. In contrast, LUMe looked with suspicion at Bologna and Torino's experiences, as they require associations to have some formal recognition to be assigned the use of public space. As far as Milan City Council is concerned, subsequent administrations have showed varied mindsets and approaches towards negotiating with self-labelled alternatives to private property. Amongst others, one of the main obstacles is that Milanese alternative arrangements rarely formally constitute themselves as registered associations, thus remaining in the realm of the 'informal', refusing to enter the jurisdictional domain.

THE PROJECT

Given my dual commitment to participatory research and activist anthropology, the results of the present work must be understood within the framework of engaged research, building on shared values and political stances which are between the researcher and participants.[7]

Importantly, the methodological link between walking in cities as a research method and cognitive mapping to prompt and enhance empathy and mutual understanding[8] is increasingly being explored and exploited. In this light, cognitive mapping is meant to explore the possibility to represent alternative claims to the right to the city and contribute to the negotiation of such claims.

Cognitive mapping was chosen as a tool because of its potential to generate meaningful exchanges, constituting a gateway to political conversation and reflection. As a tool increasingly adopted in the field of Urban Anthropology,[9] cognitive mapping by LUMe participants has the potential to intersect counter-mapping[10] practices and participatory mapping processes, to represent and make visible the ephemeral realities of urban occupied spaces and their cultural scene.

My aim was twofold: to capture and describe city perception from the often side-lined standpoint of political activists and to explore notions of the urban commons as a practice within the city and a way to assert a comprehensive right to the city. As such, the cognitive-mapping tool lies at the intersection of practices of counter-mapping – which give voice to alternative, grassroots perspectives on the city and advances their visual representation – and *care-tographies* as suggested by Hind (chapter 16, this volume), which emphasise that representation entails a form of responsibility and the practice of power which can be understood in terms of a relationship of care for a place. This latter consideration also bridged the practice of squatting and the act of representing the city, precisely via the attitude of care towards the city, of taking care of a place by taking responsibility for it.

CREATING A TOOL FOR PARTICIPATORY EXPLORATORY RESEARCH

In order to achieve these two aims and to build a replicable, consistent methodological framework, I have relied mostly on Lynch's seminal work on city images and imageability, specifically targeting individuals' perception of the urban environment and experience.[11] I have also drawn on the localisation and actualisation of Lynch's work carried out by Pezzoni in her use of cognitive maps to investigate the perception of the city of migrants passing through Milan. I have made adaptations as necessary to create a tool which will work in an activist context with participants who live in Milan. Figure 3.2 illustrates how I have adapted Lynch's categories for this work.[12]

Lynchean category (Lynch 1960)	Research-adapted category	Methodological Notes
Landmarks / *Reference points people cannot enter*	Landmarks / *Quali sono i tuoi punti di riferimento a Milano?* What are your reference points in Milan?	The Lynchean restriction of non-access to Landmarks was foregone.
Paths / *Travel channels*	Mobility / *Quali sono i tuoi spostamenti abituali e come ti muovi all'interno della città?* What are your habitual paths and how do you usually move around the city?	Mobility was open to feeling-based judgments on preferred modes of transport according to destinations.
Districts / *Medium-large, two-dimensional areas*	Urban commons / *Cosa sono per te i beni comuni urbani e li collocheresti a Milano? Se sì, dove?* What are the urban commons for you and would you say they are present in Milan? If yes, where?	The category Urban commons represents the element of novelty within the framework. It was purposefully left to participants to construct personal meaning for said category
Nodes / *Focus of large areas*	Confrontation[49]/ *Cosa è per te il conflitto all'interno della città e dove lo collocheresti?* What is confrontation/struggle for you within a city environment and where would you place it?	The participatory approach to research has proven especially insightful for the category Confrontation, as the latter has been the most questioned, criticised and shaped by participants, as well as being debated amongst participants themselves following the cognitive mapping interviews.
Edges / *Real or perceived boundaries*	Boundaries / *Dove finisce la città? Pensi che la tua percezione corrisponda ai confini geografici?* Where does the city end? Do you think your perception differs from the city's geographical boundaries?	The second question was used to prompt further discussion.

Figure 3.2 Methodological Framework for Cognitive Mapping. The table shows the questions in Italian, focused on perceptions of the city, which were used in the nineteen semi-structured interviews (fifteen with LUMe members and four with Milan residents not involved in urban social movements) together with their English translations. They maintain a dialectic relationship with the Lynchean genealogy.

CAPTURING PERCEPTIONS OF THE CITY: EMERGING TRENDS

Interviews featuring cognitive mapping were conducted between 10 February and 15 May 2018, and the data shown in the figures below refers exclusively to these semi-structured interviews. Amongst the varied range of findings suggested by the data, two key observations have emerged from the cognitive mapping with LUMe members which relate to perceptions of the city. The first is about the emergence of a creative, operational gaze on the city and the making of new landmarks and places of belonging. The second reveals a normatively positive phenomenology of confrontation, confirming Prujit's remarks[13] on the centripetal role which it plays in political urban social movements which practice occupation and which perceive confrontation as a positive, creative, generative rupture within the city.

MAKING AND INTEGRATING NEW LANDMARKS

The making of new landmarks, usually in the form of CSOA (Centro Sociale Leoncavallo – a self-managed social centre), and their steadfast adoption and integration in individuals' perceptual landscape of the city, is corroborated by analysis of the data collected during the cognitive-mapping interviews. More specifically, it is noteworthy that all the LUMe participants involved in the cognitive-mapping interviews reveal that making new landmarks within the city is perceived as a creative gesture, and that fourteen out of fifteen participants confirm that new landmarks are straightforwardly integrated into individual city perceptions. Thus, the triangulation of data collected via cognitive-mapping interviews, participant observation, LUMe's research output, and extant literature suggests that participation in social movements is associated with a creative gaze on the city aimed at disrupting its established order, allowing for spontaneous encounters and for an alternative way to live in time and space. One of the strategies for creating such an alternative is by the making and assimilation of new landmarks.

To explore the visual relevance of both trends in the perception of the city, I now consider two of the maps from LUMe participants, both produced by women under 25 years old.

The map in figure 3.3 is representative of the tendency to assimilate new and recent landmarks within the city. It shows how squatting practices may be framed as a creative endeavour within the urban fabric which delivers genuine change in the perception of abandoned spaces. In the participants' account, former sites of occupation are not abandoned places and have not ceased to be reference points for an emotional navigation of the city. On the

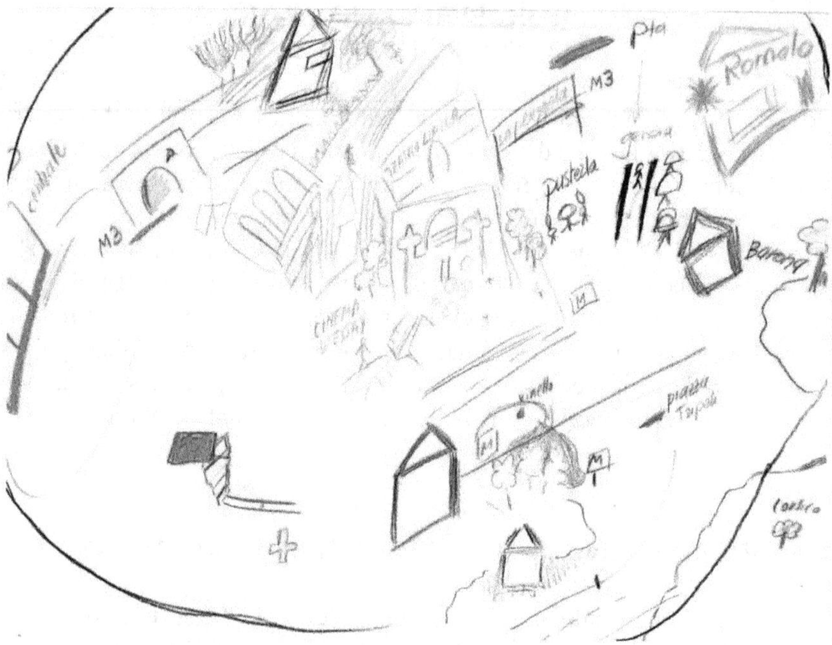

Figure 3.3 Participant Cognitive Map of Milan. Author's Image.

contrary, LUMe's former sites of occupation are present on the map, and participants attach to them the creative potential which their activism aims to unlock. Drawn in red, in the imaginary bottom right quadrant and signposted with a hand-drawn cross, is LUMe's foundational, first occupation site. The map also features sites of Temporary Autonomous Zones (TAZ). Despite the ephemeral, transient nature of these experiences, which are self-contained or subject to eviction, they have produced long-lasting change in participants' perception of the city perception as they are progressively integrated as landmarks and reference points despite their formal disappearance. At the other end of the spectrum, the second cognitive map supports the observed trend of looking at the city with a creative, operational gaze, in a proactive search for new landmarks which is coupled with the desire to *make* the city. Understood in this way, activists' creative urban gaze stretches both into the past and into the future, spanning time with the possibility of crafting the city.

In the imaginary bottom right quadrant, the participant has referenced a *villetta abbandonata*, an abandoned house, as a landmark, written in green and circled. The rationale behind this choice of representation was to express the potential which the participant saw in the unused and decaying place. She

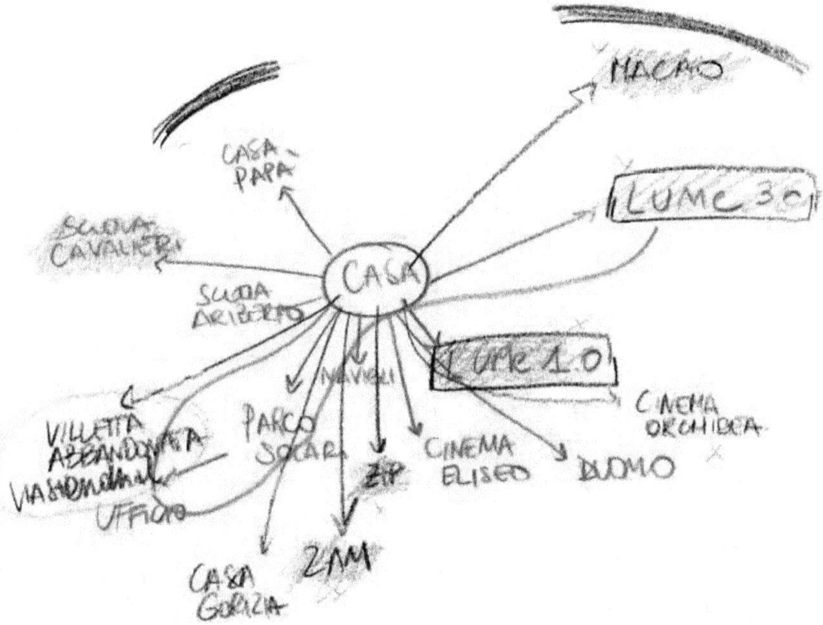

Figure 3.4 Participant Cognitive Map of Milan. Author's Image.

told me she wished to turn the abandoned house into something useful for the neighbourhood, something alive, and continued saying that on her daily commute to work she could not help looking at it and wondering what LUMe could make out of it.

THE PHENOMENOLOGY OF CONFRONTATION

The second theme to emerge from the cognitive-mapping interviews concerns the phenomenology of confrontation. When asked to define confrontation and to place it in Milan, if they thought it was present, participants' responses fell into one of the three phenomenological categories identified by the diagram represented in figure 3.5. More specifically, responses could be traced back to:

a) A descriptive judgement – regarding the presence or absence of confrontation and its relative occurrences.
b) A normative negative judgement identifying confrontation as undesirable.
c) A normative positive judgement, deeming confrontation as a necessary, generative moment of rupture which is to be hoped for and sought after in the city.

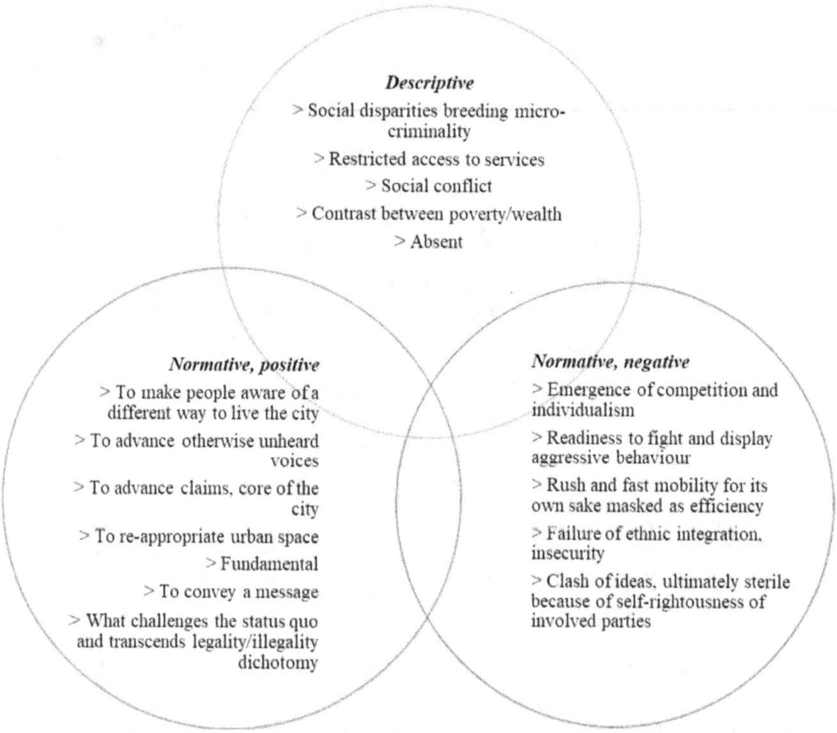

Figure 3.5 Groups Participants' Responses, including Features and Definitions of Confrontation within Milan and Cities more Generally, and Refers Them to One of the Phenomenological Categories.Author's Image.

PHENOMENOLOGY OF CONFRONTATION

Two of the longest-standing LUMe members who participated in the cognitive mapping suggested that the three responses which fell into the normative negative phenomenological description of confrontation were attributable to having joined LUMe only very recently (around one year or less before the interviews were conducted). In their opinion, and, as far as the data was concerned, this was corroborated by research. More sustained participation in militant activism proved transformational for how LUMe members viewed confrontation, changing the mainstream, confrontation-avoiding stance which was consistently characterised by non-militant views. Indeed, the very narrow sample of non-militant participants involved in the cognitive-mapping interviews compromises the reliability of the data for the latter category. However, as far as activist participants are concerned, the positive view of confrontation is consistent with Prujit's tenet that such a view is central to political occupations. One participant's response offered an insightful mediation between the descriptive and the normative positive judgement on confrontation, bridging the gap amongst LUMe participants, by suggesting that confrontation had two possible connotations, a passive one and an active one. In the passive sense, confrontation occurred in the tangible differences in wealth amongst citizens, whilst the active sense of confrontation entailed the occupation and liberation of urban places to make them accessible, free, and inclusive.

EXPLORING NOTIONS OF THE URBAN COMMONS AS A PRACTICE: EMERGING TRENDS

The diagram in figure 3.6 collects the main notions of the commons gathered from LUMe members by means of the interviews and the cognitive-mapping task. It is worth bearing in mind that although not perfectly homogenous

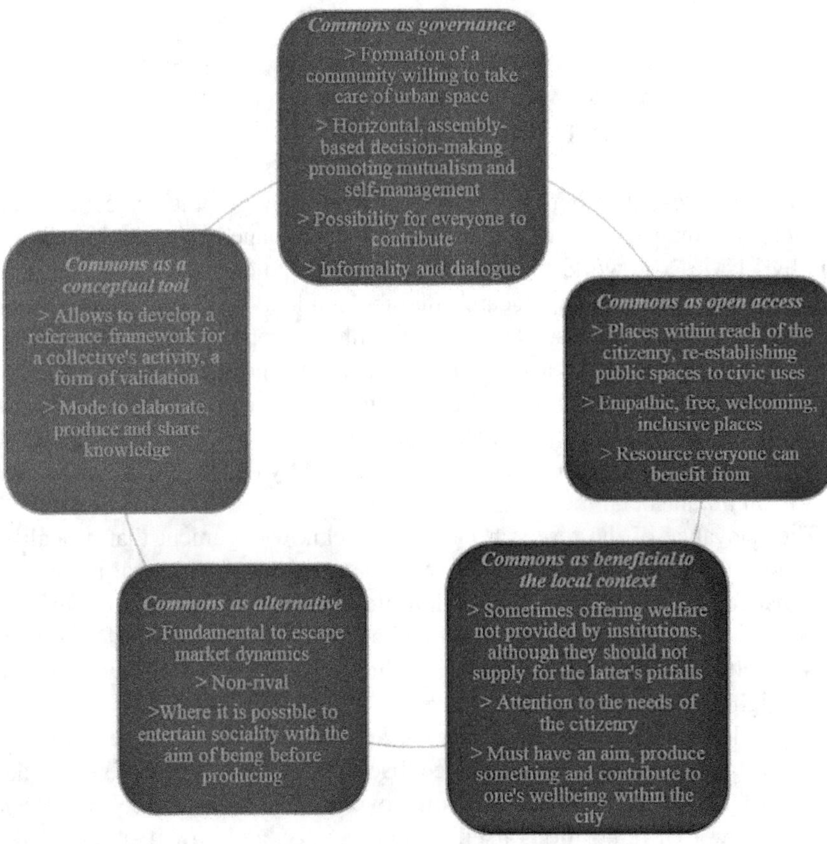

Figure 3.6 Polysemy of the Commons. Author's Image.

across LUMe membership and attendance, these notions are not mutually exclusive and reinforce the federal, inclusive nature of the movement, allowing for coexistence of different schools of thought, often influenced by individuals' educational backgrounds.

These findings suggest that the preliminary and all-encompassing conception of the urban commons as a mode of production of space, comprising governance, local empowerment, open access, conceptual innovation and there being an alternative to the *status quo*, informs and influences LUMe participants' perception of the city.[14] It does this in relation to research of new reference points and how confrontation manifests itself in the city.

As well as illuminating alternatives to the neoliberal paradigm through occupation, these results, coupled with the counter-mapping potential of the cognitive-mapping interviews, hint at the possibility of a novel way to look at the city, at the intersection of precarity and neoliberalism, and strategies to engage with the latter.[15]

WHERE TO NOW? COGNITIVE-MAPPING AND PARTICIPATORY METHODS BEYOND EXPLORATORY RESEARCH

The research described here is a part of a wider ethnographic project carried out through three months of fieldwork in Milan, shaped by participation in and by LUMe. The work discussed here has tried to convey how cognitive-mapping and participatory research may be conducive to more comprehensive and inclusive visions of the city we inhabit, representing and voicing otherwise unheard claims, and co-creating a tool for political discussion. The study has been achieved through fifteen semi-structured interviews based on a cognitive-mapping exercise which explored activists' perception of the city and their notion of the urban commons as shaped by their political participation in an urban social movement.

The relevance of the research lies in the acknowledgement that globalising cities are increasingly characterised by encroaching neoliberal urbanism and that fora for citizens' participation are standardised and top-down. By contrast, my research has focused on claims of self-legitimacy by looking at how squatting conveys and expresses claims to the right to the city in terms of the right to make the city.

When discussing the value of this research and of the cognitive-mapping tool with participants, most of the feedback I received was positive: it had prompted new ways of reflecting on how people perceive the city they inhabit, how LUMe positions itself within the city, and how it supports the on-going process of political subjectivation and self-awareness. However, some participants also expressed considerable scepticism about the extent to which the research described here would positively impact LUMe itself, beyond self-knowledge and exploration of differing notions of confrontation in the city and the urban commons. For the engagement, I have described to be truly meaningful, longer, closer participatory research is needed. Further study could envision how to transform exploratory grassroots research into a genuinely political tool for negotiation towards more inclusive cities. Visualising voices and claims which are often side-lined through cognitive-mapping has the potential to become a tool for such negotiation.

NOTES

1. All translations from French and Italian are the author's unless stated otherwise.

2. During interviews, 'confrontation' has been referred to as 'conflitto', because it is a polyvalent term, assuming different values within leftist, antagonist social movements. However, the questions aimed at exploring first, instinctual definition and localisation of confrontation within the city, precisely to investigate possible divergences within its phenomenology, namely, the way it is perceived. The rationale for the choice to explore perceptions of confrontation lies in Prujit's observation that for urban political occupation configurations, confrontation constitutes a non-renounceable component of activism. Interestingly, there was a tendency amongst LUMe's most recent acquisitions and participants external to LUMe to identify confrontation as a negative phenomenon. Prujit, H. 2013. The Logic of Urban Squatting, *International Journal of Urban and Regional Research*, 37(1), 19–45.

3. MacDonald, C. 2012. Understanding Participatory Action Research, *Canadian Journal of Action Research*, 13(2), 34–50; Lefebvre, H. 1996. The Right to the City, in *Kofman, Eleonore; Lebas, Elizabeth, Writings on Cities*. Cambridge, MA: Wiley-Blackwell.

4. Mayer, M. 2013. First World Urban Activism, *City*, 17(1), 5–19. Available: http://dx.doi.org/10.1080/13604813.2013.757417; Brenner, N. and C. Schmid. 2015. Towards a New Epistemology of the Urban? *City*, 19(2–3), 151–182, DOI: 10.1080/13604813.2015.1014712.

5. LUMe. 2017. *Spaziolibero*. Issuu. Available: https://issuu.com/lume2/docs/spaziolibero.

6. *Ibid*.

7. Sheper-Hughes, N. 1995. The Primacy of the Ethical: Propositions for a Militant Anthropology. *Current Anthropology*, 26(3), 409–440. Available: http://links.jstor.org/sici?sici=0011-3204%28199506%2936%3A3%3C409%3ATPOTEP%3E2.0.CO%3B2-C.

8. Myers, M. 2017. Way from home: Share a walk elsewhere. Livingmaps Review, 2. Available: http://livingmaps.review/journal/index.php/LMR/article/view/

61/87; Pezzoni, N. 2013. *La città sradicata*. Milan: O Barra O Edizioni; Pezzoni, N. 2017. I Migranti Mappano l'Europa. *Planum: The Journal of Urbanism*. Available: http://www.planum.net/i-migranti-mappano-l-europa-migrants-mapping-europe; Plichon, J.; Baralic, S.; Sabino, L.; Pramanik, D.; Mazzuco, S. 2017. Empathy Walks: amplifying layers of the city. *Livingmaps Review* (3). Available: http://livingmaps.review/journal/index.php/LMR/article/view/88/148.

9. Genz, C.; Lucas-Drogan, D. 2017. Mapping, in Urban Ethnography Lab – Methods. Available: https://urban-ethnography.com/methods/mappings/.

10. Peluso, N. L. 1995. Whose Woods are These? Counter-mapping Forests in Kalimantan, Indonesia. *Antipode*, 27(4), 383–406.

11. Lynch, K. 1960. *The Image of the City*. Cambridge, MA: MIT Press.

12. Pezzoni, N. 2013.

13. Prujit, H. 2013. The Logic of Urban Squatting. *International Journal of Urban and Regional Research*, 37(1), 19–45.

14. Fumagalli, A. 2017. *L'economia politica del comune. Sfruttamento e sussunzione nel capitalismo bio-cognitivo*. Roma: DeriveApprodi; Vercellone, C.; Brancaccio, F.; Giuliani, A.; Vattimo, P. 2017. *Il Comune come modo di produrre*. Verona: Ombrecorte.

15. Fish, M. 2016. Contested Spaces/Radical Places, in D. B. Shaw and M. Humm, eds. *Radical Space*. London: Rowman & Littlefield, pp. 109–127.

BIBLIOGRAPHY

Brenner, N. and C. Schmid. 2015. Towards a New Epistemology of the Urban? *City*, 19(2–3), 151–182, DOI: 10.1080/13604813.2015.1014712.

Fish, M. 2016. Contested Spaces/Radical Places. In D. B. Shaw and M. Humm, eds. *Radical Space*. London: Rowman & Littlefield, pp. 109–127.

Fumagalli, A. 2017. *L'economia politica del comune. Sfruttamento e sussunzione nel capitalismo bio-cognitivo*. Roma: DeriveApprodi.

Genz, C. and Lucas-Drogan, D. 2017. Mapping, in Urban Ethnography Lab – Methods. Available: https://urban-ethnography.com/methods/mappings/.

Lefebvre, H. 1996. The Right to the City. In *Kofman, Eleonore; Lebas, Elizabeth, Writings on Cities*. Cambridge, MA: Wiley-Blackwell.

LUMe. 2017. *Spaziolibero*. Issuu. Available: https://issuu.com/lume2/docs/spaziolibero.

Lynch, K. 1960. *The Image of the City*. Cambridge, MA: MIT Press.

MacDonald, C. 2012. Understanding Participatory Action Research. *Canadian Journal of Action Research*, 13(2), 34–50.

Mayer, M. 2013. First World Urban Activism. *City*, 17(1): 5–19. Available: http://dx.doi.org/10.1080/13604813.2013.757417.

Myers, M. 2017. Way from home: Share a walk elsewhere. *Livingmaps Review*, 2. Available: http://livingmaps.review/journal/index.php/LMR/article/view/61/87.

Peluso, N. L. 1995. Whose Woods are These? Counter-mapping Forests in Kalimantan, Indonesia. *Antipode* 27(4): 383–406.

Pezzoni, N. 2013. *La città sradicata*. Milan: O Barra O Edizioni.

Pezzoni, N. 2017. I Migranti Mappano l'Europa. *Planum: The Journal of Urbanism*. Available: http://www.planum.net/i-migranti-mappano-l-europa-migrants-mapping-europe.

Plichon, J., Baralic, S., Sabino, L., Pramanik, D., Mazzuco, S. 2017. Empathy Walks: amplifying layers of the city. *Livingmaps Review* (3). Available: http://livingmaps.review/journal/index.php/LMR/article/view/88/148.

Prujit, H. 2013. The Logic of Urban Squatting. *International Journal of Urban and Regional Research*, 37(1), 19–45.

Sheper-Hughes, N. 1995. The Primacy of the Ethical: Propositions for a Militant Anthropology. *Current Anthropology*, 26(3), 409–440. Available: http://links.jstor.org/sici?sici=0011-3204%28199506%2936%3A3%3C409%3ATPOTEP%3E2.0.CO%3B2-C.

Vercellone, C., Brancaccio, F., Giuliani, A., Vattimo, P. 2017. *Il Comune come modo di produrre*. Verona: Ombrecorte.

Part 2

RECLAIMING THE TERRITORY

Chapter 4

TalkingMaps and Diasporic Community

Jina Lee

As an artist and researcher, it has long been my interest to bring together the disparate areas of critical cartography and drawing practice. Since 2011, when I started running workshops with the Korean Joseonjok community in southwest London, I have been particularly concerned with socially displaced and overlooked women within this group. Through an organised workshop, I came up with a particular method that I thought of as a *TalkingMap*. This is a means of using drawing to show the thinking processes going on between participants in my workshops, and that can be incorporated into the practice of critical cartography. As for the title *TalkingMap*, this is a map that emerges from a conversation between the participants in my project, and it is also a map that is itself in conversation with other maps. The writing in this chapter, like my practice, will take a conversational form, one that is informal, free-wheeling, and in the style of my own storytelling.

MAPPING THE SUBALTERN: SEN AS A DIASPORIC COMMUNITY

Before I move on to present my practice, I would like to bring to the reader's attention the term 'subaltern'. This was specifically defined by Gayatri Chakravorty Spivak in her essay 'Can the Subaltern Speak?' (1988), a text I have used in order to give perspective to my research subject, namely the diasporic community. I first read Spivak's essay in early 2007, during my third year at university in South Korea, when it was introduced by my lecturer Jang-Un Kim.[1] Back then, during my early twenties, I found Spivak's writings staggering, because that was a period when most students were inspired by French intellectuals, such as Deleuze, Derrida, and Foucault, taught by professors who had studied in Europe, especially in Germany and France, in the 1970s and 1980s. I first encountered Spivak's work through her conversation-based essay 'Who Sings the Nation-state? Language, Politics, Belonging'. This opened up a new way of thinking and of seeing the world. Spivak's rigorous questioning of Western writers gratified my curiosity and encouraged me to further investigate these writers with my own critical questions. What was most revelatory for me, and what still impresses me today, is the sharp approach with which Spivak exposes Foucault's and Deleuze's accounts of power, in which she argues they are ideologically blind to 'the subalterns'.

What then does 'subaltern' mean? Subaltern is a term coined by the Italian Marxist philosopher Antonio Gramsci, now used in postcolonial studies in order to identify specific people, or social groups, that are socially, politically, and geographically excluded and displaced from within the hierarchies of power. However, Spivak critiques this deployment of the term by Western writers, suggesting that the 'subaltern is not just any excluded or oppressed people, but everything that has restricted or no access to the cultural imperialism'.[2] Nearly ten years have passed since I first read Spivak's text, which means it has been nearly thirty years since her essay was written. Why does this matter to me now, and how much has changed since the initial construction of 'Can the Subaltern Speak?' Unfortunately, ignorance towards the subaltern continues today, and the need to recognise and act on this issue remains as urgent today, especially as it remains a global problem. Spivak's argument thus became a key concept as I was researching the displaced Korean community of Joseonjok women living in north London.

The Joseonjok are people who live in the Yanbian Autonomous Prefecture in China's Jilin Province, who used to be regarded as Korean. There are many Koreans living in this area because the north-eastern part of China (above the current North Korean border) frequently used to be Korean territory. In 1964, at the end of the war, a secret treaty was made between China and Korea on the national border, as is shown on contemporary maps. In other words, the Joseonjok are legally Chinese people living in China, although they are culturally Korean, speaking the Korean language, the *hamkyung* dialect, which is very similar to North Korean. Joseonjok can be seen as a unique group of migrants within a global migration. Jin Young Lee (2012) states that amongst 193 million Joseonjok, 40 per cent are living in foreign countries. Most of these people live in South Korea, but there are sizeable populations in Japan and the United States too, as well as smaller groups all around the world including in the United Kingdom.

The reason I began to work with the Joseonjok women is down to an accident of personal history. When I came to London from South Korea in 2011 to continue my studies, I didn't know much about other London Koreans (including South/North Koreans and Joseonjok) as I didn't have any other social relationships besides my immediate student network. When I had my first child in 2014, I had a back injury, which meant that I couldn't properly look after the newborn baby crying beside me. My husband and I urgently searched for a nanny, preferably Korean, who could look after me and my little child, and also help with housework such as cooking. By chance, a friend introduced me to her Joseonjok nanny, and this is when I first got to know about the Joseonjok people in London. I soon discovered that global political and economic processes were deeply linked to the trajectory of the Joseonjok diaspora and to the fact that these people were an overlooked and suppressed minority.

The question I want to address in this chapter is, 'Can the Joseonjok women's community in London, que the subaltern, speak?' In Spivak's definition, 'subaltern' is the structured place from which the capacity to access power is radically obstructed'.[3] In other words, 'subaltern' is less an identity than a predicament. The empirical focus of my research with Joseonjok women and their predicament of subalternity was based on an interrogation of the historical immigration records related to their movement to south-west London and the insertion of fragmentary and speculative accounts of this process generated by Joseonjok women's voices. As an example of how the global and local intersect in these accounts, a Joseonjok who ran a restaurant in New Malden High Street told me how micro-level changes in the composition of the high street were entangled with a huge project of roadworks linked to the expansion of the local economy as a result of world trade.

From what I have found out, there are few ways for Joseonjok to enter the United Kingdom – I would say 'were' because it is currently nearly impossible for them to come into the country. If they do come, it is either through a legal work permit, claiming to be Chinese, or through illegal South Korean passport forgery. The majority use false passports, which worsens their status, first because they have to repay huge debts to brokers and, second, because they claim to be North Korean defectors seeking asylum. According to the Embassy of the Republic of Korea in London, there were no Joseonjok immigrants to the United Kingdom until 1997.[4] Before 1997, Joseonjok tended to enter the United Kingdom for short-term visits rather than to immigrate. Since then, the numbers of Joseonjok immigrants has tended to grow rapidly, reaching more than 2,000 by 2004, many with genuine work permits, although one has to bear in mind that all the data is approximate.

The Joseonjok people's migration route to the United Kingdom is unclear, and it is difficult to find accurate numbers.[5] Dramatic changes occurred from 2004, when the British government's approach to immigration changed. Prime Minister Tony Blair announced a political 'tipping point' towards asylum seekers in 2004,[6] but economically the situation worsened as the global financial crisis in 2008 made it harder for Joseonjok to find jobs in the United Kingdom. Now, there are only 400 or less as a result of stricter immigration controls imposed by the British government.[7] Of course, many of these rules do not apply to expatriate professionals in high-value sectors such as financial departments and high technology sectors whom the government wishes to attract and retain as permanent settlers. Whenever there is a crisis, minorities such as the Joseonjok are the first ones to be prohibited and find themselves substituted by other temporary migrants.

Within this context, what can I do as an artist? The *TalkingMap* can be thought of as a 'new map' that can re-position mapping as a social activity. It is designed as a tool to tell narratives that can open up new worlds and communities for people who can't speak for themselves.

My latest research has been specifically with Joseonjok women residing in London. These impoverished women, who mostly engage in domestic work, chose to leave their country due to lack of employment opportunities in their homeland and to suffer family separation by living abroad. Those who live in London are mostly based in Kingston-upon-Thames, specifically in New Malden, where many South Koreans were first established in 1990s. As regards to their national identity it is utterly impossible for them to become embedded anywhere for, politically, having Chinese nationality while socially engaging with Korean culture and geographically living in the United Kingdom. The Joseonjok people clearly cannot locate themselves within any conventional political geography of map centred on nation-states. Consequently, there is need for a new map for the Joseonjok, one which opens up inter-relationships between identities, territories, and borders, thus revealing the Joseonjok as a diasporic society within the world map.

The *TalkingMap* is intended to encourage a contemporary rethinking of the meaning of territory, exploring ways in which fresh boundaries are constantly recreated. Through this practice, I have been able to produce mental constructions of what is missing from those representations in the hegemonic accounts of conventional maps and, furthermore, in this respect, the *TalkingMap* functions as a mediatory practice, a way of narrating personal journeys and stories.

While I was conducting interviews within the Joseonjok community, I chose to make drawings as tools for recording, a process I called the *TalkingMap*. No camera, no video, but only pen and paper helped me to build trust between the participants and myself. This was especially useful method because the Joseonjok were unwilling to share their personal stories in other ways, feeling that their personal data might be given away.[8] The act of drawing on paper enabled me to make them feel more secure in terms of privacy and trust. Furthermore, as time went by and after several interviews, deeper relationships began to develop between the Joseonjok and myself and, most important of

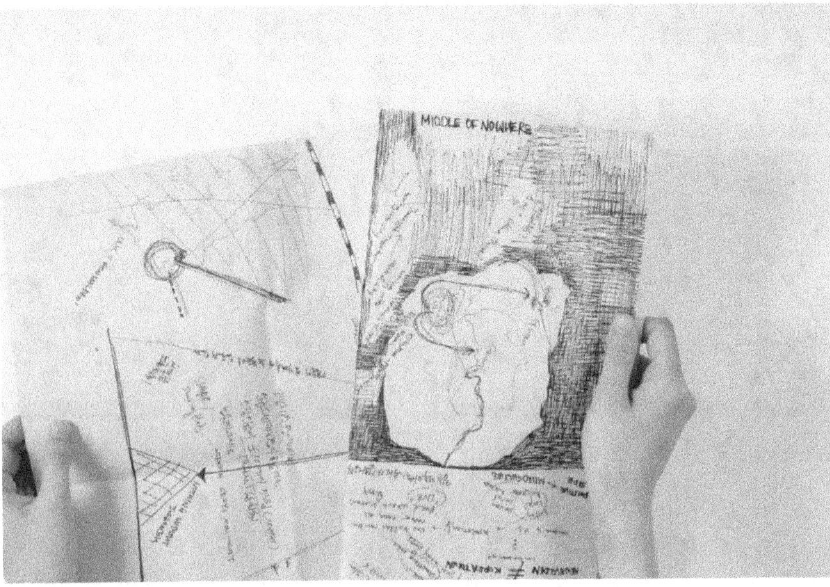

Figure 4.1 Jina Lee, *TalkingMap #1* (detailed), pencil on paper, 2016. Author's Image.

all, they made the transition from being observers to becoming active participants in a collaborative map-drawing process. This is precisely the outcome I wanted to see through my project: the transformation from passive subjects to actively speaking independent people. As part of a natural progression, the drawing paper becomes an interactive platform on which each participant could join in the process of mapping while the interview was in progress.

As you can see, this map is not an objective representation of reality, but represents a partial and necessarily biased perspective. A participant once told me that, even if she could tell her families or friends back in China that she is living in the United Kingdom, she is not sure where and what the United Kingdom is. She regards New Malden as the United Kingdom, as it is where her actual life is. Moreover, even though what I had drawn was her idea of a world map, it only consisted of her homeland (including Korea), the United Kingdom, the United States, and somewhere ideal that she dreamed of going. Ignoring geographical locations or scale, these nations are situated according to what, in her imagination, the 'world' is. However, this does not mean that less educated people have an inadequate view of what the 'world' is. We have to understand the many differences and perspectives involved in an understanding of what the 'world' is and the tactical importance of situated knowledge.[9]

The interviewee mentioned that she had not had a single moment to Skype her family in the four years since she came to the United Kingdom

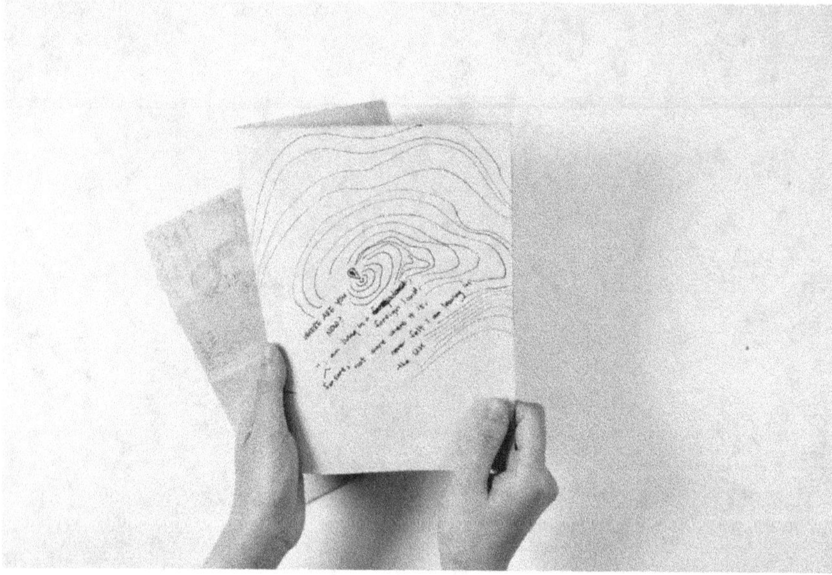

Figure 4.2 Jina Lee, *TalkingMap #3* (detailed), pen and pencil on paper, 2016. Author's Image.

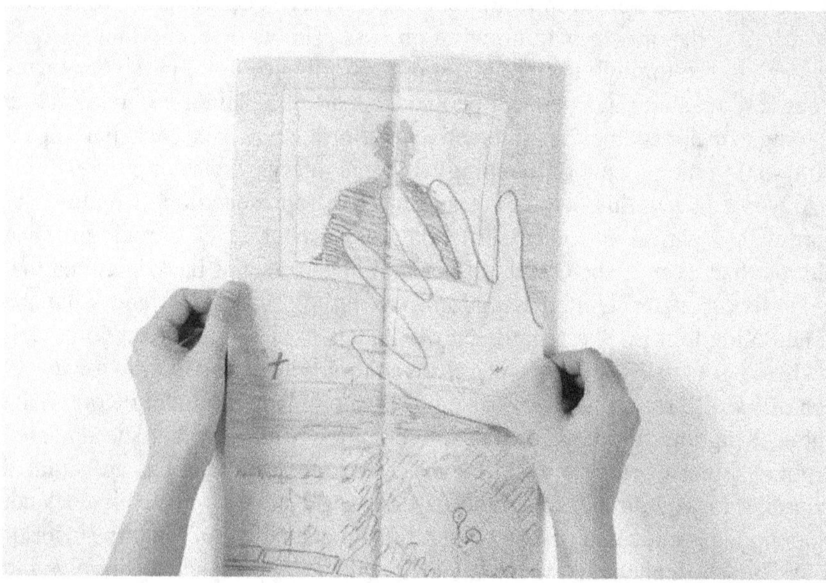

Figure 4.3 Jina Lee, *TalkingMap #1* (detailed), pencil on paper, 2016. Author's Image.

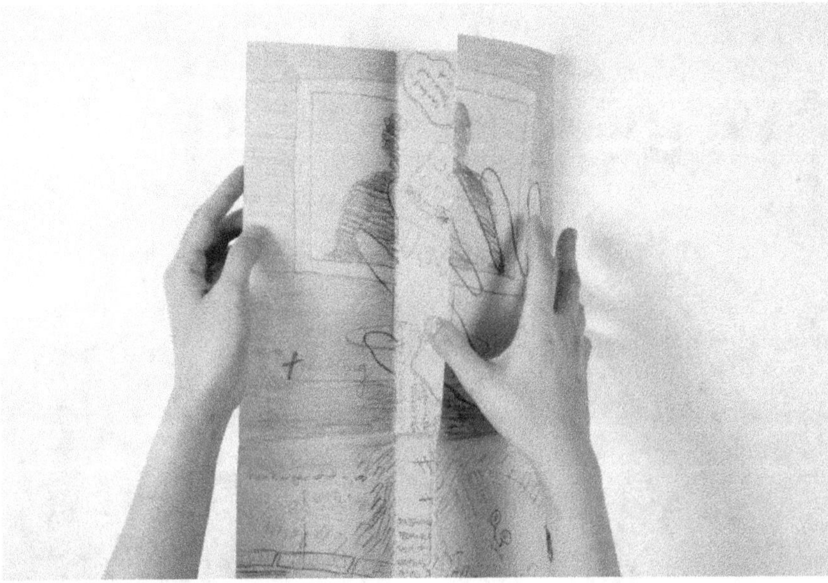

Figure 4.4 Jina Lee, *TalkingMap #1* (detailed), pencil on paper, 2016. Author's Image.

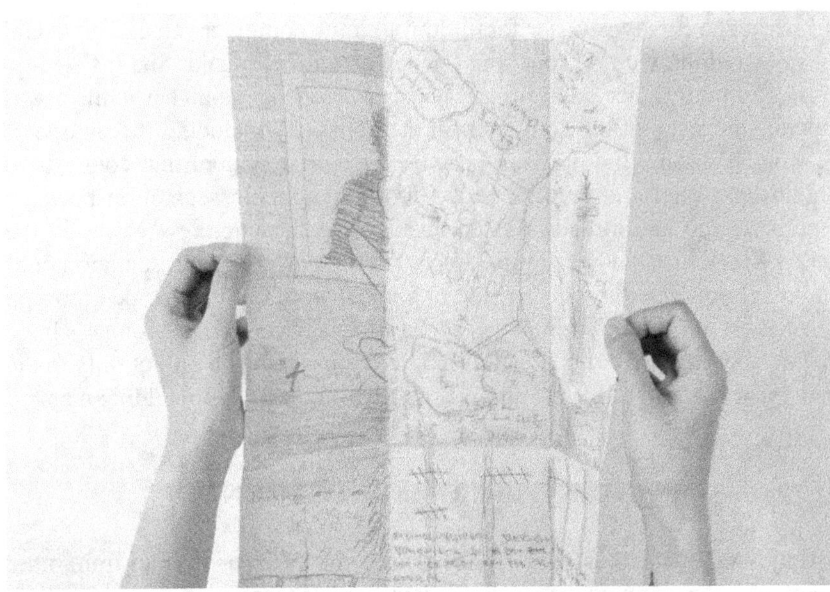

Figure 4.5 Jina Lee, *TalkingMap #1* (detailed), pencil on paper, 2016. Author's Image.

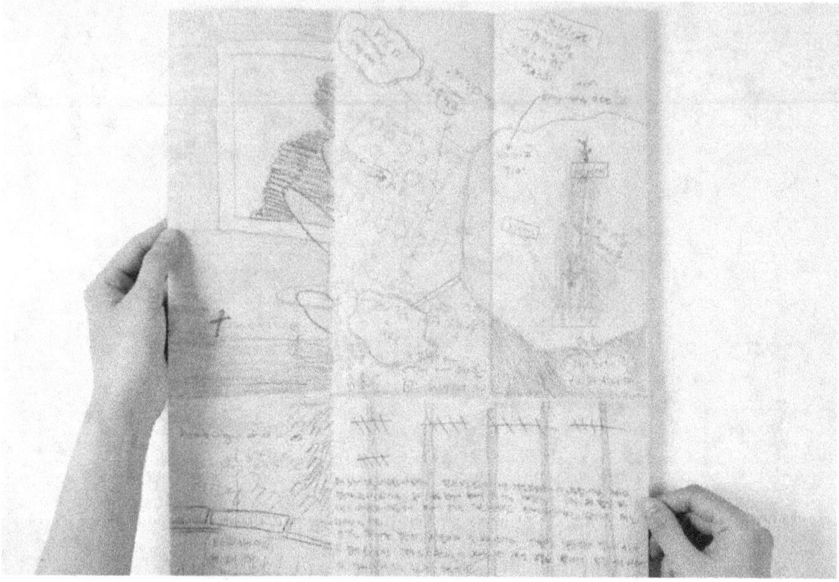

Figure 4.6 Jina Lee, *TalkingMap #1* (detailed), pencil on paper, 2016. Author's Image.

and when she finally saw her family through the computer screen, she could not say anything but cry, touching the screen with her hand. She insisted on physically putting her hand on the paper and tracing around it while I was drawing an image of the computer on the paper. By including the outline of the woman's hand, the drawing became a material symbol that connects to a highly personal and emotive story about geographic separation between a mother (who had already passed away while the interviewee was in the United Kingdom) and her family. Access to the power of representation does not have to operate through written or spoken words; it can happen through the smallest of gestures, which can act as a form of voice. The woman's hand on the paper did exactly that as this single gesture meant a lot not only to the participant but also to me as, and potentially to others who read this map.

FINDING UNCOMMON GROUND

Joseonjok migrants to the United Kingdom can be seen as distinct from other Joseonjok people around the world. That is because there is no place other than New Malden – not even in South Korea – that the Joseonjok, South Korean, and North Korean people can negotiate the shared space of their transnationalism. Joseonjok are politically Chinese, however, interestingly they have hardly any relationship with the Chinese community or Chinese government in the

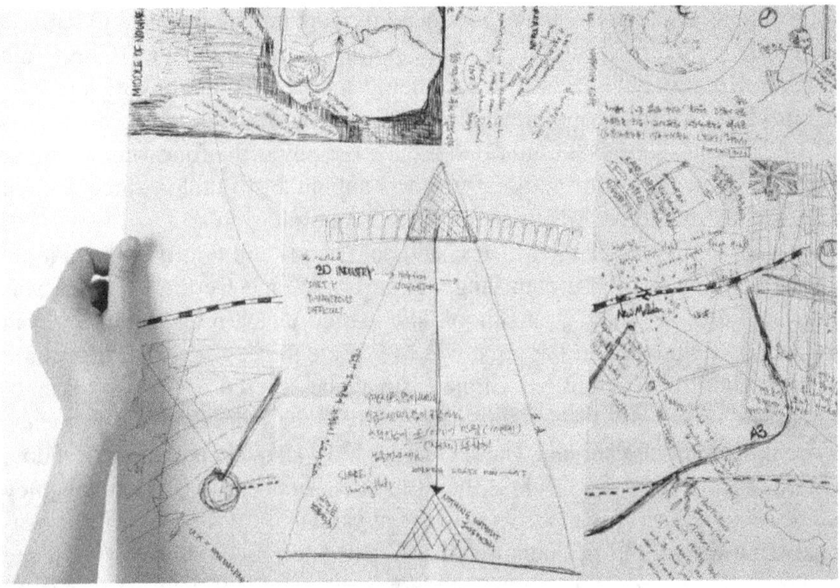

Figure 4.7 Jina Lee, *TalkingMap #2* (detailed), pen and pencil on paper, 2016. Author's Image.

United Kingdom . Rather, trying not to be obtrusive to others, they have created their own community around the South Koreans in New Malden, taking jobs in the service sector, for example, in Korean restaurants or the repair industry. Consequently, this has established another hierarchy within the wider hierarchy. Such a scenario should lead us to further complicate our understanding of subalternity (here with particular regard to Joseonjok women) as a structural space of discrimination that is always subject to exclusion by hegemonic regimes of power. The Joseonjok can neither be included in South Korean society nor in North Korean society. However, in the labour market, the Joseonjok have become an indispensable substitute for Koreans, because they can speak the language, understand Korean culture, and are prepared to work in migrant jobs characterised by the three Ds – dirty, dangerous, and demanding.

The Joseonjok normally speak Korean, and because of this I always thought that Korean was their mother tongue. However, through *TalkingMap*, I discovered that they have their own language, which is a combination of Chinese, Japanese, and Korean. Nonetheless, their language is bound to be 'wrong', or itself displaced, for they have been required to obliterate their own language and learn Korean words in order to work for Korean people. One of my oldest participants, who is in her seventies, reported that she was able to speak and understand Korean, but the words that were used in the restaurant kitchen were unfamiliar, such as ingredient names and kitchen appliances.

For example, a carrot is called *dang-geun* in Korean, but she used to call it a *ninzi* (which is very similar to Japanese *ninjin*), back in Yanbian. Chives are *buchu* in Korean, but she used to call them *yeom-ji*. Without having any time to learn these words properly, she had to figure out Korean words on site, as managers or head chefs would often reduce her pay if the food was wrong or late. However, things are looking up for Yanbian's in London since she got here, mainly due to improvements in their economic status.

Another participant, in her fifties, said her friends and family spoke a more correct version of the Korean language at home, and by the time she was 8–9 years old, in Grade 3 at school, she started to learn Chinese. An even younger participant, who I assume was in her late thirties or early forties, told me her school had taught her Chinese since she entered it (Joseonjok enter school at 6–7 years old) and there were many other schools that didn't teach or speak Korean in Yanbian. These changes have also taken place in London. About twenty years ago, when the Joseonjok first came to London, they often suffered discrimination, working in abysmal circumstances with ridiculously low wages and frequent dismissals without notice. However, now the Joseonjok have grown in stature economically and politically by owning their own shops and obtaining permanent visas within the United Kingdom. Adding up these little stories one by one, the *TalkingMap* has created a space for such counter-narratives, often revealing unofficial and self-made paths, virtual lines of desire articulating the maps of their journeying.

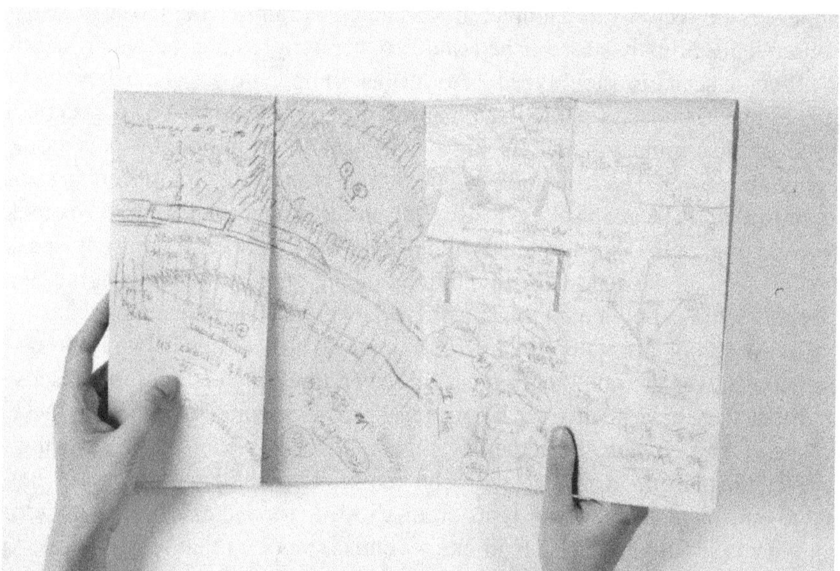

Figure 4.8 Jina Lee, *TalkingMap #1* (detailed), pencil on paper, 2016. Author's Image.

COUNTER-MAPPING

There are a growing number of cartographers and artists who claim that maps can be used to make counter-claims, to express competing interests, to make visible otherwise marginal experiences and hidden histories, to make practical plans for social change or to imagine utopian worlds.[10] *TalkingMap* is about discovering and mapping hidden stories that lie around us that can be realised through different perspectives and socio-political views. Each map contains a unique dialogue, but it also creates a shared agenda for the exchanging of differences in position and experience. For the people who think their stories are worthless, or shouldn't be told, it isn't an easy job enabling them to speak. Although it took me a while, it was worth it, to draw out their exclusive knowledge and to facilitate it into my mapping project. Thus, it is not only the final production of a map that is important, but the process itself that is the key factor in my project. In fact, there is no finishing point on the map. The map can always be restarted from wherever it stopped and another story added when there are more stories to incorporate. Within this framework, the holistic process of meeting people and exploring new ways of sharing each other's stories became a process of learning together, producing new knowledge by bringing multiple perspectives together through mapping.

NOTES

1. Jang-Un Kim is a Korean art critic and curator who majored in art theory and cultural theory. He recently worked as an associate professor at Kewon Art College and is the director of the National Museum of Modern and Contemporary Art in Seoul.

2. Morris, Rosalind C., and Gayatri Chakravorty Spivak. *Can the subaltern speak? Reflections on the history of an idea*. New York: Columbia University, 2010, p. 38.

3. Morris, Rosalind C., and Gayatri Chakravorty Spivak. *Can the subaltern speak? Reflections on the history of an idea*. New York: Columbia University, 2010, p. 8.

4. Lee, Jean-Young. 'Korean-Chinese (Chosunjok) in U.K.: Migration Settlement and Social Relations', *Journal of Northeast Asian Studies*, 62nd edition, 2012.

5. Hyun-Mee Kim claims that approximately 20,000–30,000 South Koreans, 1,500–2,000 Joseonjok immigrants and 100 North Korean asylum seekers were funded in 2008. Yet in the same year, Jean-Young Lee claims that there are 850 North Koreans living in the UK. He then states that there are 46,829 South Koreans, approximately 1,000 Joseonjok and approximately 850 North Korean asylum seekers.

6. More information on the immigration legacy of Tony Blair can be found at http://www.migrationpolicy.org/article/immigration-legacy-tony-blair/, accessed on 15 April 2016.

7. Shin, H. R. *Koreans in London (South Korean, Joseonjok, North Korean): London Korean immigrants' conflict and assimilation in conjuncted area (New Malden),* translated by Jina Lee, Available from <*community.snu.ac.kr/bbs/servlet/DownLoad?SEQ=88213*> [accessed 20/11/2015], 2014.

8. Jobertová, Daniela. and Alice Koubová. Ed., *Artistic research: is there some method?* Prague: Academy of Performing Arts, 2017, p. 157.

9. Lee, Jina. Drawing Life Maps of New Malden: Ethnographical enquiry into the Joseonjok people in New Malden by means of drawing practice, *Livingmaps Review*, 2nd issue, 2017.

10. Firth, R. Critical Cartography. *The Occupied Times of London*, p. 9. Available from <http://theoccupiedtimes.org/PDFs/OT27.pdf>, [accessed 20.04.16], 2015.

BIBILIOGRAPHY

Coles, Alex. 2000. *Site-specificity: the ethnographic turn.* Black Dog Publishing Ltd.

Jobertová, Daniela, and Alice Koubová. Ed., *Artistic research: is there some method?* Prague: Academy of Performing Arts, 2017, p. 157.

Lee, Jean-Young. Korean-Chinese (Chosunjok) in U.K.: Migration Settlement and Social Relations. *Journal of Northeast Asian Studies*, 62nd edition, 2012.

Morris, Rosalind C., and Gayatri Chakravorty Spivak. 2010. *Can the subaltern speak?: reflections on the history of an idea.* New York: Columbia University.

O'Reilly, Karen. 2009. *Key concepts in ethnography.* SAGE.

Spivak, Gayatri Chakravorty. 2012. *An aesthetic education in the era of globalization.* Harvard University Press.

Spivak, Gayatri Chakravorty. 2015. *Nationalism and the Imagination.* Seagull Books.

Chapter 5

Stories of the Unmappable

Marija Biljan

When, drawing a sketch map for a friend, I take my line for a walk, I retrace in gesture the walk that I made in the countryside and that was originally traced out as a trail along the ground. Telling the story of the journey as I draw, I weave a narrative thread that wanders from topic to topic, just as in my walk I wandered from place to place. This story recounts just one chapter in the never-ending journey that is life itself, and it is through this journey – with all its twists and turns – that we grow into a knowledge of the world about us.

Tim Ingold[1]

Watching a map of a city, reading its streets, squares, landmarks, a game of associations and memories occurs. Even when living in a city all our lives, our knowledge is always a fragment of the whole picture. It's always partial, local. When we are mentally strolling through streets of that city, our route is full of inscribed experiences. Names of streets or monuments might stay the same, but the strict cartographic compositions that map makers have annotated, almost melts away. If we would transfer our inner navigation onto a paper next to it, it would almost look as if unmapping the city that is so objectively presented on the 'real' map in front of us. Even if the city in urbanistic terms does change over time (and so do their maps), experiences and histories that are being inscribed and held there stay in some other domain, so often untold.

I have recently started exploring the concept of maps and how its main characteristic of (re)presentation of physical places can be also reversed in order to describe experiences that are lived there, on the ground, between walls, along the shortcuts, revealing the hidden dimension of concrete, solid spaces. If all map makers are humans, and all maps are subjected to the decisions of those who make them, how would the city look if it was mapped according to each individual's perception? Where would be their limits, if there would be any? In the following text, I will describe how I used mapping as a storytelling tool that transformed the London map into an artwork full of colours and stories, and storytellers into map makers.

From November to December of 2017, I conducted two community intergenerational mapping workshops with BAME women activists in London. These sessions were organised as a part of the Ubele organisation's project called PatHERways[2] that aims to empower young BAME women on their paths to becoming changing agents in their communities.

The first workshop was organised with a group of young women, while in the second workshop, older generations of women joined in. Even though the main objective of the sessions was to create and/or strengthen these intergenerational bonds, while conducting the workshops, other potentials of community-mapping methodology emerged.

Playing with the concept of mapping (assuming and measuring distances, positioning ourselves, creating our own maps) but also using 'real' maps to locate certain memories, we explored different ways of perceiving ourselves in space and place (see figure 5.1).

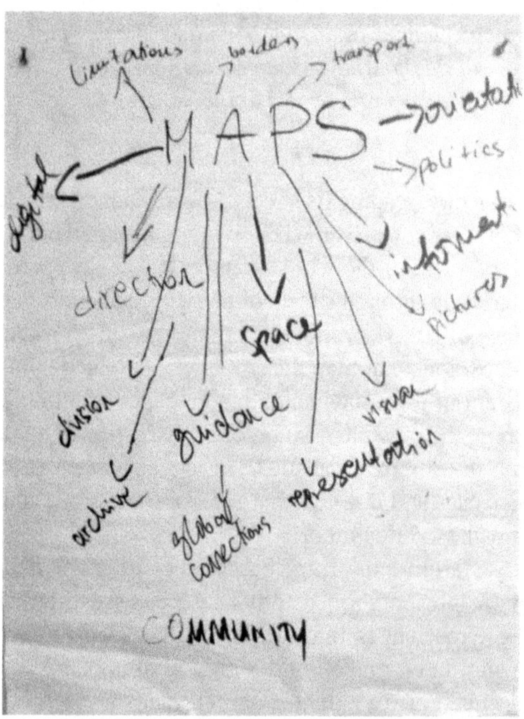

Figure 5.1 Author's Image.

There were three types of mapping activities: in the first one, I used a coloured rope to mark river themes on an empty floor.

The participants had to take a position according to the river that would correspond to where their home is in London. One by one participants would stand on 'their' place and while doing it, hold another coloured rope that at the end created a web showing their connections, physical proximities/distances (see figure 5.2).

The goal of this activity is to create group cohesion and start reflecting on the city and personal experiences in it. However, even though the floor was an imagined map of London and participants just had to stand on a position that vaguely represented their physical home, this activity turned out to be more than just a warming up exercise. In their twenties, most of these young women have already changed apartments and houses. They moved out of parents' house, they travelled, and they are still in a transitional phase of life. Moreover, having their family origins stretching sometimes from Africa/Asia, over Jamaica, all the way to United Kingdom, contributed to their immediate identity reflection when I asked to 'find their home' on the map of London. Some of them went 'back' to their parents' place, some found their

Figure 5.2 Author's Image.

feet at the place they live at the moment, but some had a really difficult time deciding. A simple action of standing in an improvised space representing a city where most of their lives are taking place is actually a strong demonstration of what it means *to locate*. To locate our home is more complex than just marking a physical house, it can seem defining and definitive, while so often our personal maps and locations stretch beyond the physical, 'real and objective' limits of maps, that is, of one city.

The second exercise was taken to another room where participants had to think of their journeys from home to their favourite place (in the city, if possible) and then present it by creating their own maps. They were free to use papers, magazines, colours, stickers, and so on. This was a very stimulating exercise that allowed them freedom of interpretation and expression.

While some maps were made of words, others of lines, images, and words, most of them were collages with images that represented their cultural and personal identities. The favourite place that I imagined to be found in London, finally appeared in abstract expressions, or in physically distant and sometimes even not yet existent places. There were images showing African landscapes, professional paths, or sceneries that are still to be arrived to in the future (see figure 5.3).

In the third activity, we used printed maps of London and stickers in different colours. One of the goals was to collect data that would afterwards be used for creating a digital map showing their internal connections. For that reason, participants were asked to write and mark the place where they live (again), where they went to school, and a place signifying for their PatHERways journies. They were also asked to add personal information to it, a memory or what they like about that place.

This, as well as the first exercise, was repeated almost equally when conducting the workshop in a mixed group where both young and 'older' BAME women activists participated (see figures 5.4–5.6).

Unlike the first time, when I mostly just focused on collecting their data, the second time when the activity was held in a bigger and age more diverse group, mapping places and evoking memories or impressions resulted as a very inspirational storytelling method. Participants were separated into two age mixed groups and had to share between them memories of school, their first community activism places and events, who was an important person for them and why, or thoughts and memories on their first jobs. Each participant chose one topic, marked it on the map of London that was lying in front of them, and shared their story with the rest of the group. Sometimes similar, sometimes completely different contexts and stories created a storytelling flow that soon started transforming that row map of London into a stage where real life happened and happens. Both groups were asked to create a new map showing a common path where their stories meet. They used various

Figure 5.3 Author's Image.

materials with which they finally spiced the initial mono-story, neutral map of the city. In this activity, in which participants had to share something personal but just to create something new together, the group connections and activist spirit arose. Through talking over different topics, these women not only created new bonds, but they also supported each other. Having different generations sitting by the same table and talking about the same topics, but different contexts, created a real exchange. The London map suddenly had another layer, more abstract and palpable at the same time (figures 5.6 through 5.8). One participant, showing and explaining the map of their connections over London that they created, said:

> *We used the tree as the root that binds the community together, and we used different animal. It is the heart, the heartbeat of all people at this table and how they feel about their environment. We used a lot of white colours, we feel that the future of our community as activists is bright. The aura of this human being is telling us about the aura that he carries as a community activist, he unites and ejects light into the people's lives. Activists in this all started off where there was something they wanted to do. We all thought it was coincidental, but it isn't. We are all here to pass on the baton. We have the duty that we carry. We allow the light that it carries, to regenerate and ignite the light. All the beautiful colours*

Figure 5.4 Author's Image.

Figure 5.5 Author's Image.

Figure 5.6 Author's Image.

Figure 5.7 Author's Image.

Figure 5.8 Author's Image.

that are here. Colours represent our aura, our life, and because the organisation is international this is where the tree comes in. Birds feed from the trees and take the seed and plant it all over the world. And the root of the tree is to say that we are rooted. So through the struggles or challenges that we may have, we are not going to be turned over, we may lose a branch or two but we are solid beneath the ground. The birds carry the seed all over the world.

The data that was collected is going to be presented in a digital map that will show intergenerational connections across London between the BAME women. As we are talking about activists that have been bringing changes into their communities since the 1970s, marking those places on a map will also show places where values were and are able to live. If we go back to concept of maps and mapping to the idea of utility and practicality of regular maps, with the examples presented, we can also see how, when making personal maps that are based on individual experiences, instead of lines that represent roads and directions, some more subtle connections appear: personal stories that make graphical lines perceived as places of human connections. They happen, and they are gathered in physical places that through these experiences acquire values. As they hold our stories, we are also choosing them to represent parts of who we are in a place.

Therefore, using maps and mapping activities in working with communities and different groups can be a way to tackle questions of belonging and identity. Even if they are not directly placed and identified with the physical space (house, city, country), our experiences or lack of them always have their loci. Acknowledging and linking them to some other places, events, or even some abstract ideas can bring a stronger feeling of appropriation and ownership of something that we are already part of (but not aware of). Instead of looking at the map from 'above' and seeing it as something already defined and imposed, using it and transforming it according to our experiences and needs might also lead to a more active and aware participation in different aspects of our life: personal, in a community, or even wider.

NOTES

1. Ingold, Tim. 2007. Up across and along. *Life of lines*. London: Routledge, pp. 49.
2. https://www.ecos.pt/patherways.html.

BIBLIOGRAPHY

Tim Ingold. *Life of Lines*. London: Routledge, 2007.

Chapter 6

Former Fresnans

Mapping Home through a Memory Palace

Blake Morris

I was born and raised in Fresno, California, 'the best little city in the U.S.A', as the entrance sign on Van Ness Ave. declares (see figure 6.1). I moved away twenty years ago: university in Santa Cruz; a stint in Seattle; long periods in New York City and London; a couple of years in Northampton. I haven't lived in Fresno for over two decades, and while I still have a strong connection to my Central Valley birthplace, it doesn't feel like *home*. Or at least not *my* home. It is where I am from, where my parents live, where I was raised.

In 2012, I was living in New York City and thinking about leaving. As I contemplated moving abroad, I wondered where home was. I had been in New York for six years; I hadn't passed the mythical ten-year marker that confirms one as a New Yorker,[1] but the city felt like home. I wondered how other Fresnans in the city felt. Why did they leave? What was their relationship to Fresno? To New York City? to home? I decided to find former Fresnans who might help me explore these questions.

I started with friends and expanded my network through social media and word of mouth, seeking out old friends, and friends of friends. I asked them to take me on a walk between two places they considered home and help me document our experience through the creation of an image for a memory palace devoted to former Fresnans. This initial push resulted in fifteen walks with former Fresnans. I walked with some of my best friends and total strangers; with people I once knew well, but didn't any longer; and with those who were only ever acquaintances. What we all had in common was the place we lived – New York City – and the place we no longer lived – Fresno. Since starting the project, I have walked with twenty-one former Fresnans in New York, London, Oakland, and San Francisco. The resulting project, 'Former Fresnans' (2012–ongoing), is a memory palace sited in Fresno's Tower

Figure 6.1 Fresno CA Van Ness Portal. David Jordan (CC BY-SA 3.0 <https://creativecommons.org/licenses/by-sa/3.0>, via Wikimedia Commons).

District that brings to life the stories of those walks through a sculpture park activated by the imagination.

THE MEMORY PALACE

Fittingly, a former Fresnan introduced me to the concept of the memory palace. On a walk from my apartment in Astoria, Queens to Socrates Sculpture Park on New York City's East River,[2] Heather Gardner told me its origin story:

> The memory palace, or method of loci, is attributed to Simonides of Ceos, Ancient Greece's first for-profit poet, in a story that is likely apocryphal.[3] At a banquet for Scopas, Simonides recited a poem that praised Castor and Pollux, two newly appointed gods. Scopas, affronted, refused to pay Simonides in full and referred him to the twin gods to collect the other half of his payment. Shortly after, a messenger summoned Simonides and told him two young men were waiting for him outside. He left the building, but the young men were nowhere to be found. Upon his exit, the banquet hall's roof collapsed and those inside were killed and disfigured beyond recognition. Simonides, the only surviving member of the party, remembered the table and the positions of each person at the banquet, and was able to supply family names to the dead.

Figure 6.2 The Tower Theatre at Night. Richard Harrison (CC BY-SA 4.0 <https://creativecommons.org/licenses/by-sa/4.0>, via Wikimedia Commons).

> Through this action he realised the power of spatial memory and the memory palace was born.

Memory palaces link symbolic images to specific locations. To create a memory palace, one chooses a specific space and imagines vivid symbolic images throughout it. The more absurd the image, the easier it is to recall. To retrieve the memories, one imagines walking through the space and looking at the different images. Since that conversation with Gardner, I have been using them to document my walks.[4]

WALK #8 - HARLEM TO CENTRAL PARK WITH CECE

The Tower District is named for the historic Tower Theatre, which dominates the skyscape of the low-rise neighbourhood (see figure 6.2). In a memory image created with CeCe Olisa, the hills of Marcus Garvey Park in Harlem are nestled beneath the awnings of the theatre. They are surrounded by green hedgerows from Central Park, which mark the perimeter of the theatre's entrance. Atop the hill is a large, sparkling, pink C. The C is, of course, for CeCe, who walked with me from her apartment in Harlem, up and over the hills of Marcus Garvey Park, into the quiet, hedge-rowed gardens of Central Park.

As we walked, she explained the freedom she felt when she first moved to Harlem. Fresno is a diverse city with a large Latinx and East-Asian demographic; Black and African American residents, however, only comprise around 5 per cent of the population.[5] Though the Black and African American population of Harlem has lowered in recent years, it was still 62 per cent of the population in 2008;[6] CeCe described the experience of being surrounded by people that looked like her as profound. Suddenly she was aware of an experience that she didn't even realise she was lacking. She took me through her Harlem and explained that she had just quit her corporate job to become a full-time blogger focussed on plus-size fashion and body positivity. The C we put in the Tower was aspirational. A vision of the life she wanted to build for herself. Since walking with me, she has founded CurvyCon, modelled for Nike, and become a successful influencer across social media platforms.[7] Her Buzzfeed video linking Black Lives Matters to her work with body positivity went viral,[8] and her TEDx Talk at California State University, Fresno has been viewed over 800,000 times.[9]

THE TOWER DISTRICT

Fresno is characterised by suburban sprawl,[10] which made it a challenge to think of a walkable, centrally located neighbourhood to locate the memory palace. The Tower District, or the Tower as it's known colloquially, satisfied both of these requirements; additionally, its position as Fresno's key cultural district meant it was likely to be familiar to the former Fresnans with whom I would be walking.

Locally, the Tower's reputation is bohemian, though that exists in tension with a national perception of Fresno as impoverished, both culturally and economically.[11] The neighbourhood has a relatively dense collection of galleries, performing spaces, independent restaurants, bars and shops, and provides the city with its most active cultural scene. In an article for *BOOM: The Journal of California*, Aris Janigian 'guess[es] that the stretch from downtown to the Tower District has more artists' studios (as opposed to art galleries) per capita than anywhere in California'.[12] As a teenager, I spent countless days and nights in Fresno's 'bohemia'. This also turned out to be true for many of the former Fresnans with whom I would walk. The Tower was a cultural hub.

THE WALKS IN NEW YORK CITY

My instructions were simple: *take me on a walk between two places you consider home*. They always decided the route, and thus the length, of the walk.

Generally, we started where they were currently living, though the instruction offered enough flexibility that we didn't have to start where they currently lived. At the end of our walk, we sat down and discussed what we wanted to remember. Together we transformed that memory into an image and placed it somewhere in Fresno's Tower District.

The walks varied wildly. Carla led me on a ten-mile trek over the Triborough Bridge. We started inside her current apartment and ended inside her old one, with a large order of pancakes at a diner in Queens in between. Grace took me for a ten-minute stroll through Bedford-Stuyvesant in Brooklyn, which ended with Bloody Marys in a wood-panelled bar. Patrick and I walked from Williamsburg to Hell's Kitchen, in an encounter that ended up stretching on until the next morning. Brett led me along the Brooklyn waterfront from his apartment in Red Hook to the small loft by the Navy Yard that we shared for four years when we first arrived in New York City together.

As we walked, we shared stories and memories and what we were up to now. They told me the significance of the places they had chosen; why they considered them *home*. We engaged in what Misha Myers has termed 'conversive wayfinding':

> A spatial practice that conducts percipients' attention to landscapes through mediated/live aural performance; perceptual and dialogic strategies of interacting and knowing place – shared viewpoints, earpoints, conversational conviviality and critical witnessing; the use of different paces, paths and places of narrative; and performance as a way of knowing.[13]

For Myers, this mode of walking, 'where a percipient becomes more a wayfarer than a map reader', activates and invites 'modes of participation that generate places and knowledge of places' through a conversational, social and convivial interaction between people and the landscapes they inhabit.[14] In 'Former Fresnans,' participants drew on their cognitive maps of the city to lead me to places they associated with home.[15] In doing so, they shared with me their relationship to New York City, as well as its associations with Fresno (and I did the same).

WALK #10 - MIDTOWN MANHATTAN TO THE WEST VILLAGE

In a back alley in the Tower District, a single cherry tree blossoms. It is in between the dumpsters and the stage door of Good Company Players' 2nd Space Theatre (see figure 6.3). If you look closely, you can see a knothole, similar to the one where Boo Radley placed his gifts for the children in *To*

Figure 6.3 The Stage Entrance to Good Company Players' Second Space Theatre. Author's Image.

Kill a Mockingbird (1960). Inside the knothole is a chessboard with pieces made by the drama students of Roosevelt School of Performing Arts' class of 2000. This image remembers my walk with Justin on a crisp spring morning. We started in midtown Manhattan near his apartment, and he led me to the West Village's Cherry Lane Theatre, where he received his first professional theatre credit. Justin and I had once been close friends. We had not, however, spoken in years. There was some tension between us – the specifics of which were long forgotten, replaced by a vague sense of clashing personalities.

As we walked, we reminisced about high school and what our classmates might be doing. He took me to see the chessboards on Thompson Street, a favourite pastime of his since moving to the city. He bought me a crepe at one of his favourite eateries, before taking me to the Cherry Lane Theatre. When we walked together he was embarking on a new adventure: producing Broadway plays. He has since been nominated for two Tony awards, fulfilling a life-long dream. The walk, and our mutual love of New York City, quickly diffused any lingering tension. As we headed downtown, we discussed the different journeys that had brought us to the city. Our paths had diverged, but we could trace them back to a formative location: Fresno's Good Company Players. Its most famous alumna is Audra McDonald, and countless Fresnans have had their artistic futures shaped by the work done in GCP's theatres. Justin and I shared the role of Jem in a production of *To Kill a Mockingbird* (1997). With this common bond, we decided to place our image at the stage entrance to the theatre, in a location where we had spent much time rehearsing together (and performing apart).

THE WALKS IN FRESNO

Every time I return to Fresno I share the memory palace publicly. I choose a selection of images from the palace and guide audiences through it, describing the images and relating the stories that led to them. Certain locations stood out in the memory of this small sample of former Fresnans, what I have termed hubs of resonance: the area around Good Company Players' performance and rehearsal spaces; the parking lot outside Livingstones where kids would hang out away from prying parental eyes; or the ever-changing coffee shop (always Café Revue in my mind), which served the same social function despite the various names under which it operated.

These spaces also contained memories for those who participated in tours of the palace. Megan Bohigian, a poet based in the Tower District (whose daughter contributed one of the memory images), spoke of the rush of memories that flooded back to her when walking through the parking lot of the 2nd Space Theatre. Others agreed, mentioning the time spent parked while waiting for shows to be over or rehearsals to be done. The tour of the palace made Bohigian consider how the 'kinetic experience of walking [. . .] stimulates the 5 senses, and creates memory.'[16] It evoked the 'sense memory' of past experiences, while also creating a new experience of imagining the works in the memory palace.

The tour highlights the memories of my walks with former Fresnans, but for participants in Fresno it also overlaps with their own memories of the spaces through which we are walking. For Bohigian, the stories and images brought back from afar highlighted 'the way we carry our various life "geographies" with us, always'.[17] The iconic images that punctuate the memory palace, such as the hedgerows or the Bethesda Fountain at Central Park, San Francisco's Golden Gate Bridge or the Brooklyn Bridge, and the stories of my walks through these locations, engage the audience's own memory of them, either through their personal encounters or images filtered down through television, films, or other forms of visual media.

As I have written elsewhere, through memory palaces, I link memory to place and engage audiences in creative re-imaginings of the landscape. Unlike an object created in response to a walk, a memory palace has to be imagined, and each participant manifests the images differently depending on their individual interpretations.[18]

As Bohigian notes, the memory palace methodology puts 'imagination at the root of collaboration'.[19] It is a shared experience of imagination in both its creation with former Fresnans after the walk and its presentation to the audience who must imagine the sculpture. It connects 'a real landscape, route, map, landmarks and so on--with inspiration, imagination and creation (and collaboration)'.[20] In this way, it creates an intersection for

Figure 6.4 Map of Works in the Former Fresnans' Memory Palace (2020). Author's Image.

multiple experiences of space, both local and further afield, and makes visible what Doreen Massey has referred to as 'space as a simultaneity of stories-so-far.'[21]

WALK #13 - WILLIAMSBURG TO THE LOWER EAST SIDE WITH RICHARD

Sometimes when I visit my aunt she mentions the 'dancing tomato' in front of what used to be the local record shop. The image to which she refers is actually a large, anthropomorphic red apple standing in the snow. The apple, who sports a pink mohawk and has Mickey Mouse hands, holds a compact disc of Talking Heads: 77 in front of his shrivelled apple genitals.

The image records a walk with Richard, who showed me nearly a dozen places he had lived in Brooklyn and Manhattan. One of them didn't have an address – it was just the place with the apple painted on it. As we walked, he told me stories of his time in New York: performing in Stomp and going on national and international tours (sometimes sporting said pink mohawk); the challenges of sleepwalking in the winter when one also sleeps nude (that particular story included a Free AOL CD-ROM he found on the street and used to retain some modesty); the music he got to make and the musicians with whom he worked. We ended in a bar on the Lower East Side for an afternoon drink. Or perhaps it was the West Village. I think the bar was long and narrow, but it might have had big open tables. My aunt misremembers the image. I likely misremember the walk, and I certainly misremember Richard's stories.

HOME REIMAGINED

I started 'Former Fresnans' with a question: What is my relationship to home? Through the project I have embraced a dispersed definition: home was New York City, but it was also Fresno. More recently, it has been London and Northampton. As I prepare to move back to the United States, I find myself wondering where home will be next. Fresno to begin. It's where my family is. Indeed, my family was a key point of inspiration for the project. While I now work through methods of digital exchange that foreground walking together at a distance,[22] at the time my art required physical presence, and they had not been able to participate. Former Fresnans was a chance to make work they could see. Though the project encompasses more than my family, its engagement with them is essential. After all, they are the reason I go back to Fresno in the first place.[23]

Everyone I walked with left Fresno for a reason, but we all appreciated the role it played in our formation. In places like Fresno, which are dominated by conservative Christian values, marginal voices often move to cities more welcoming of various lifestyles. As a little queer kid growing up in a homophobic town, I had few role models.[24] As an adult, I removed my voice from the community. The people who moved away mean Fresno has fewer Black Lives Matter activists, fewer advocates for the homeless, fewer writers, performers, singers, dancers, and chefs. Some have moved back, adding their perspectives away to the tapestry of life back home; others visit, adding their voices to the valley when they can. Former Fresnans was my contribution: an imaginary sculpture park that shares the stories of those who have left. It is an invisible addition to Fresno's stretch of studios, galleries, and performance spaces. For those who can imagine it, Fresno's Tower District is transformed.

NOTES

1. See, for example, the opening of Jonathan Wolfe's article "New York Today: Becoming a New Yorker", in *The New York Times*, January 5, 2017, sec. New York (available at: https://www.nytimes.com/2017/01/05/ nyregion/new-york-today-becoming-a-new-yorker.html.

2. Socrates Sculpture Park is a public, open-air exhibition park and studio, that artist Mark di Suvero and a group of community activists founded on a former dump site on New York City's East River in 1986. Socrates Sculpture Park. "Mission & History." (available at: https://socratessculpturepark.org/about-us/mission-and-history/).

3. The story of Simonides and the memory palace was first recorded in the ancient Roman text Rhetorica ad Herennium, traditionally attributed to Cicero, though its authorship has been disputed. Gardner was drawing on Joshua Foer's book *Moonwalking with Einstein: The Art and Science of Remembering Everything* (London: Penguin, 2006). For a more thorough discussion of the history of the memory palace, see Francis Yates, *The Art of Memory* (London: Pimlico, 1966). For information on Simonides as the first Greek poet to accept coins as payment for his poems, see Anne Carson, *Economy of the Unlost* (Princeton University Press, 2009).

4. For a more detailed discussion of this see the preamble to my book, *Walking Networks: The Development of an Artistic Medium* (London: Rowman and Littlefield International, 2020).

5. The most recent census data from the government is not yet available and the demographic data cited here was provided by Claritas Segmentation & Market Solutions, a data-driven marketing company. See Healthy Fresno County. "Healthy Fresno County :: Demographics :: County :: Fresno :: Race," 2020 (http://www.healthyfresnocountydata.org/demographicdata?id=247&sectionId=940).

6. Sam Roberts, 'No Longer Majority Black, Harlem is in Transition', *New York Times*, sec. New York, 5 Jan. 2010.

7. See https://ceceolisa.com/.

8. Nora Whelan, 'This Woman Nails How Race Fits into Body Positivity', Buzzfeed, 14 July 2016 (Available at: https://www.buzzfeed.com/norawhelan/race-body-positivity.)

9. CeCe Olisa How to Build Self Confidence. Accessed October 13, 2020 (Available at: https://www.ted.com/talks/cece_olisa_how_to_build_self_confidence.)

10. Mark Arax wrote about this for the *Los Angeles Times* in 1994: "Sprawl Threatens Central Valley, Study Says: Agriculture: State Could Lose $5.3 Billion a Year If Farm Belt Cities Don't Stem Urban Expansion, Researchers Say. They Call for Controlled Growth to Preserve Farmland." *Los Angeles Times*. October 26, 1995. https://www.latimes.com/archives/la-xpm-1995-10-26-mn-61251-story.html. Compare this to the time-lapse map put together by *The Fresno Bee* in 2016, which demonstrates the residential development of the period: Jody Murray. "Watch Three Decades of Urban Sprawl in Fresno and Clovis, Squeezed into a Few Seconds." *The Fresno Bee*. August 17, 2017, sec. Local (available at https://www.fresnobee.com/news/local/article167829902.html).

11. See, for example, James Fallows' article for *The Atlantic*, "Why the Least Hip Town in California Thinks It Can Be a Center of Coolness", April 6, 2015 (available at: https://www.theatlantic.com/national/archive/2015/04/creating-californias-new-bohemiain-an-unexpected-locale/389691/); or, also in *The Atlantic*, a series on Fresno's high poverty rates: Gillian B White, 'Fresno's Ugly Divide', August 25, 2018, (available at: https://www.theatlantic.com/entertainment/archive/2018/08/unequal-fresno/568558/).

12. Aris Janigian, 'Come See California's Future', in *BOOM: The Journal of California*, 6(1), 2016, 102–107.

13. Misha Myers, 'Walk with Me, Talk with Me': The Art of Conversive Wayfinding', *Visual Studies*, 25(1), 2010, 59–68, DOI: 10.1080/14725861003606894, p. 67.

14. Ibid.

15. Kevin Lynch, *The Image of the City* (Cambridge, MA: The MIT Press, 1960).

16. Megan Bohigian, 'Thoughts about your memory Palace Project', unpublished, 2013.

17. Bohigian, op. cit.

18. Morris, *Walking Networks,* p. xix.

19. Bohigian, op. cit.

20. Bohigian, op. cit.

21. Doreen Massey, *For Space* (London: Sage, 2005), p. 9.

22. See, for example, my work with Morag Rose, 'Pedestrian Provocations: Manifesting an Accessible Future', *Global Performance Studies*, 2(2), 2019, DOI:10.33303/gpsv2n2a3.

23. I enlisted my niece to 'install' the images in the palace. Installation of memory palaces is a curious experience. We would walk to the spots together and I would describe the images to her. She would help me position them precisely (up until then they only existed in my memory of the space) and ensure I was able to describe the

image in a vivid way to someone who didn't know where and what they were. She is the person who has seen the images the most.

24. This is, of course, relative. For example, Fresno native Eric Aoki has written that, while Fresno is less LGBTQ friendly than a city such as San Francisco, it has more spaces of LGBTQ visibility than Fort Collins, Colorado, where he lived at the time of publication. See Eric Aoki, 'Strategic Liminality and Trans-regional Mobility: Engaging Diverse City Spaces to Constitute and Negotiate Intersectional Identities of Newfound Class Privilege, Repressed Ethnic Anger, and the (In) visibility of Gay (Male) Life', in eds. Ahmet Atay and Jay Brower, *Communication, Culture, and Making Meaning in the City: Ethnographic Engagement in Urban Environments* (Lanham: Lexington Books, 2017), p. 214 n. 13.

BIBLIOGRAPHY

Anne Carson. *Economy of the Unlost*. Princeton: Princeton University Press, 2009.

Aris Janigian. 'Come See California's Future'. *BOOM: The Journal of California* 6, no. 1 (2016): 102–107.

Blake Morris. *Walking Networks: The Development of an Artistic Medium*. London: Rowman and Littlefield International, 2020.

Blake Morris and Morag Rose. 'Pedestrian Provocations: Manifesting an Accessible Future'. *Global Performance Studies* 2, no. 2 (2019).

CeCe Olisa. *How to Build Self Confidence*, 2018. https://www.ted.com/talks/cece_olisa_how_to_build_self_confidence.

Doreen Massey. *For Space*. London: Sage, 2005.

Eric Aoki. 'Strategic Liminality and Trans-Regional Mobility: Engaging Diverse City Spaces to Constitute and Negotiate Intersectional Identities of Newfound Class Privilege, Repressed Ethnic Anger, and the (In) Visibility of Gay (Male) Life'. In *Communication, Culture, and Making Meaning in the City: Ethnographic Engagement in Urban Environments*. Lanham: Lexington Books, 2017.

Francis Yates. *The Art of Memory*. London: Pimlico, 1966.

Gillian B White. 'Fresno's Ugly Divide'. *The Atlantic*, 25 August 2018. https://www.theatlantic.com/entertainment/archive/2018/08/unequal-fresno/568558/.

James Fallows. 'Why the Least Hip Town in California Thinks It Can Be a Center of Coolness'. *The Atlantic*, 6 April 2015. https://www.theatlantic.com/national/archive/2015/04/creating-californias-new-bohemiain-an-unexpected-locale/389691/.

Joshua Foer. *Moonwalking with Einstein: The Art and Science of Remembering Everything*. London: Penguin, 2006.

Kevin Lynch. *The Image of the City*. Cambridge, MA: MIT Press, 1960.

Mark Arax. 'Sprawl Threatens Central Valley, Study Says: Agriculture: State Could Lose $5.3 Billion a Year If Farm Belt Cities Don't Stem Urban Expansion, Researchers Say. They Call for Controlled Growth to Preserve Farmland'. *Los Angeles Times*, 26 October 1995. https://www.latimes.com/archives/la-xpm-1995-10-26-mn-61251-story.html.

Misha Myers. ''Walk with Me, Talk with Me': The Art of Conversive Wayfinding'. *Visual Studies* 25, no. 1 (n.d.): 59–68.

Nora Whelan. 'This Woman Nails How Race Fits into Body Positivity'. *Buzzfeed*, 14 July 2016. https://www.buzzfeed.com/norawhelan/race-body-positivity.

Sam Roberts. 'No Longer Majority Black, Harlem Is in Transition'. *New York Times*, 5 January 2010.

Chapter 7

The Busyness of Button-Mapping

Exploring Children's Everyday Politics in Belfast

Amy Mulvenna

INTRODUCTION

Inspired by feminist materialist thought that takes seriously issues of the material, socioeconomic, and crucially, geopolitical narratives within and beyond children's geographies,[1] my PhD project engaged with creative mapping approaches with the aim to give visibility to ways in which geography matters in less overt and less obvious ways for children living in Belfast, Northern Ireland.[2] This chapter outlines one of those mapping approaches – button-mapping – co-developed with participants aged 7–11 across four field sites in Belfast. 'Button-mapping' refers to a more than visual and intra-active mode of mapping that represents one's city using tiny things like buttons, shells, paperclips, marbles, bits of straws, cocktail sticks, beads, or anything else to hand. In this chapter, I present an empirical example of this approach in order to focus on the complex and open-ended ways that participants practiced everyday minor politics through mapping.[3] In doing so, I seek to highlight researcher responsibilities when enrolling 'creative' and 'playful' approaches to place-mapping, which can include sensitive questions around identity, belonging, and security.

CONTEXT

At the time of my fieldwork in 2018, two decades had passed since signing the Belfast (Good Friday) Agreement. Nine years since the introduction of Austerity measures in the United Kingdom. Almost three years without a Northern Ireland power-sharing Executive. And, in the words of one of my ten-year-old participants, over three years since 'Brexshit happened'. These political moments resonate with Peter Shirlow's assertion that the 'fundamental problem affecting Belfast is that Geography matters in ways that are overt and obvious'.[4] Where Shirlow talks about geography mattering in 'overt and obvious' ways, I ask: whose geographies are we talking about – overt and obvious to whom? Framed by whom? In this context, he refers to normative geographies of division that persist in Northern Ireland today; geographies of division and ethno-sectarianism that arise primarily from conflicting, macro-scale, 'big P' Political understandings of 'the nation', nationhood, and national identity. It is true that the material realities of geographies of division and everyday conflict remain a reality for many children and young people in Belfast: in the twenty-six peace walls, street murals and flag displays; in family stories, cultural traditions, and segregated schooling; by barriers both visible and invisible. And yet, along with these monoliths of political and cultural expression, children and young people's everyday lives and geographies are complexly and obliquely interwoven across various sites and flows of the social, material, cultural, and political. So, what of those everyday matterings? How might we as researchers reach at those less obvious, less overt, and unpredictable geographies of the everyday without foreclosing engagement with 'big P' Political issues of the day?

Within this context, I developed an approach to mapping children's geographies that tried to approach these complexities, centring on what matters to children within their local geographies in Belfast today. Specific focus was paid to the human and more-than-human encounters and intra-actions that emerged through four creative mapping approaches over an eight-month period from 2017 to 2018, including button-mapping. Button-mapping involves participants playing, experimenting, and thinking with everyday, typically throwaway or overlooked things, including assorted beads, buttons, paperclips, threads, shells, and more – anything that could be retrieved from the bottom of a bag or from deep inside a pocket! In practice, it was about engaging participants to reflect on processes of re/ordering, organisation, and categorisation – of different spaces, peoples, and things therein. Button-mapping further involved mobilising playful exploration of political and cultural identity that had already begun to emerge during our earlier mapping activities: in other words, giving visibility to children's critical geopolitics. The notion of the minor came into play here insofar as my intent was to trace, and possibly produce, 'alternative subjectivities, spatialities, and temporalities', as I now explain.[5]

MAPPING AND THE MINOR

A significant methodological grounding underpinning the project was the idea that participants' creativity (and my own) emerged from on-going, experimental, experiential processes between makers (the children and myself), materials, tools, and other non-human things. From this starting point, I engaged with the theoretical notions of processuality, performativity, and the materiality of maps and mapping, as well as the notion of intra-active 'thing play'[6] as acts of potentially radical and *minor* subjectivation.[7] Importantly, maps and mapping involve more than physical and geographical spaces and places: ideologies, cultures, politics (and more) are crucial touchstones to be plotted. The 'minor' comes into play as a mode of producing theory that 'reworks the major' (i.e. 'big P' Political discourse and theory, which could be said to include ethno-sectarian demographies in NI) from within. Accordingly, the researcher is called to situate themselves not at the margins of the field but rather within the field in order to decompose those forms and practices that sustain hegemonic systems.[8] In this regard, the minor can be said to 'refus[e] mastery in both the academy and research practices . . . striv[ing] to change theory and practice simultaneously'.[9]

Transformation is therefore a key aspect of the minor. Cindi Katz posits an interesting way that transformative politics and crucially a positioned sense of 'becoming minor' can work through what she calls 'renegade cartographies at once situated, fluid and incorporative'.[10] That is, she appeals for 'renegade cartographers' to plot their place(s) somewhere between particular and 'idiosyncratically centered' mental maps and (allegedly) objective maps in order to

> produce maps of becoming that recognise how sociosymbolic and politicaleconomic positions are at once individual and shared; mobile and situated.[11]

I argue that button maps are 'maps of becoming', which can be used to trouble ideas around the politics of positionality (e.g. class, ethnicity, nationality, age, and race), understanding that while these politics are not homologous, they are not separate worlds, geographically speaking. Moreover, Katz' calls for a processual approach to mapping (also championed in the past decade by scholars such as Martin Dodge, Rob Kitchin, and Chris Perkins)[12] that challenges the politics of representation:

> Maps of and for a kind of informed politics that refuse . . . the pigeonholes of identity but embrace position, difference, and the multidimensionality of change.[13]

I take Katz' advocacy of 'maps-of-becoming' as a clarion call for researchers to enact different ways of mapping and doing children's everyday geographies, particularly those with a focus on minor geo/politics that seek to give visibility to the complex lifeworlds of children, which are frequently enmeshed within unequal power relations. I call for 'more-than mappings', which are an active mode of 'staying with the trouble'[14] of children's geographies in Northern Ireland today – geographies that are entangled with ethno-sectarian demographies, geographical divisions, and a legacy of protracted period of conflict referred to as the 'Troubles'. In response to partial and essentialist constructions of what maps can be and do in Northern Ireland, I argue for modes of mapping in the midst of and with the trouble – bearing in mind Noora Pyyry's words, which state that knowing is 'more than just a human issue' – including cartographic knowing.[15] This vitalist, inherently feminist idea behoves researchers to be reflexive regarding the technologies and tools of map-making, including the materialities of maps and map data. Using an agential realist lens inspired by Karen Barad, my project was about exploring the insights that creative mappings can offer to the more-than-rational, affective, and embodied nature of participants' lifeworld that goes beyond narrow discussions of sectarian difference and conflict.[16]

In developing a conceptual framework, I was inspired by recent developments around 'post-representational cartographies' that follows on from Dodge *et al.*'s rethinking of maps, eschewing singular approaches to mapping, and indeed the 'illusion of cartographic objectivity' advocated by Brian Harley.[17] Moving beyond Harley's work, 'post-representational cartographies' can be seen to respond to the tensions between representationalism and non-representational geographies while still affirming the transient and fleeting nature of being, understanding the interrelatedness of the following: mapping; materiality; performance; processuality; everydayness; affect, embodiment, and playfulness.[18] What 'post-representational cartographies' do differently, as it were, is to recognise maps as being both representations and practices simultaneously, maintaining that maps have political value which should be variously interrogated, resisted, and re-interpreted, recognising the nuances of attendant power structures that we cannot ethically disavow.[19] Much work in this field is increasingly informed by creative, experimental, processual, and praxis-oriented approaches to mapping that entwine arts-based practice with process-oriented philosophies and methodologies, as well as feminist new materialisms – although this is an avenue that remains as yet under-explored – with some notable exceptions.[20] My intervention into this field comes through a focus on 'intra-action', minor politics and playfulness within creative mapping practices, conceiving of children's everyday cartographies as a critical mode of unsettling and troubling children's geographies in a post-conflict society.

BUTTON-MAPPING: THE 'THINGNESS' OF INTRA-ACTIVE MAPPING ENCOUNTERS

Originally developed by Sue Moffat, director at the New Vic Theatre, alongside researchers involved with the Community Animation and Social Innovation Centre (CASIC) at Keele University, button-mapping recognises ordinary little everyday things including (but not limited to): buttons, shells, nuts, bolts, stamps, beads, and pebbles as having the potential to crosscut boundaries; to 'spark new thinking, connect people together and translate between them.'[21] In my iteration of button-mapping with children in Belfast, small groups of two-three participants were presented with a jumbled array of mixed 'loose things', spread before them on a table surface. Without any other explanation, I invited them to do three things in order, giving them only a few minutes for each stage:

- First sort the buttons (by colour, material, use, decoration, or other means.)
- Next, organise and label the buttons, based on how they thought Belfast was organised.
- Lastly, organise and label the buttons, based on how they thought Belfast should be organised in the future.

In this case, button-mapping operated in a rhizomatic sense: it was an open and connectable mode of mapping, yielding multiple entryways, as well as having the potential to be constantly modified and reworked as participants (both individually and in small groups) handled, sorted, and discussed the assorted tiny things before them.[22] The maps that materialised were provisional and iterative re/orderings of everyday lived spacings. Two examples are shown in figures 7.1a and b, below.

These maps were never completed. Sites and scales, keys and borders, and more were never fixed but rather were fluid and shifting works-in-progress, open to constant interruption and disruption – both from myself and other

Figure 7.1a and b Button-Mapping with P5 at Cavehill PS, North Belfast. Author's Image.

participants as well as the things themselves! More than anything, the children enjoyed the tactility of these familiar tiny items. At times, I noted tensions between participants' decision-making and desires to simply play with stuff. Above all, I was frequently struck by how the cornucopia of small things played or acted back with individuals – that is, where everything was an entanglement; where everything occurred as an 'imaginative ongoing moment'.[23] This is where the concept of 'intra-action' (a neologism from Karen Barad) comes into play.[24]

Interacting bodies are understood to exist independently from one another, both prior to and after the moment of exchange. Distinguishable from interaction, intra-action refers to a kind of action coming from within. Bodies can 'intra-act', as Barad puts it, in 'co-constitutive ways' – that is, where individuals materialise through intra-actions. In this respect, the ability to act emerges from within the relationship – not outside of it. Intra-action is, in other words, the comingling of people and things, where 'stuff' can act. Correspondingly, agency is understood here, not as something or someone has but rather as the ways in which it emerges and is enacted between bodies, including the more-than-human.[25]

In line with agential realism thinking, Barad reworks the dichotomy of subject/object (where the Cartesian cut between subject and object is an onto-epistemological given). Differences and distinctions are neither fixed nor assumed. Rather, 'subject' and 'object' are only distinguished within the phenomenon itself. What is being troubled here is 'the metaphysics of individualism (the belief that there are individually constituted agents or entities, as well as times and places)'.[26] Against this thrust of individualism sits the argument that all is relational. All is processual. Regarding questions of mattering, when we recognise that those foundational philosophical constructs (namely: causality, agency, space, time, matter, meaning, knowing, being, responsibility, accountability, and justice) do commingle and are ontologically inseparable, we must interrogate phenomena as entangled, intra-acting agencies; as 'spacetimematterings'.[27] Barad explains:

> Agential realism . . . makes inquiries into how differences are made and remade, stabilized and destabilized, as well as their materializing effects and constitutive exclusions. Since cuts are understood to be enacted rather than given (it is the cut that makes the individual and not the other way around), all manner of questions regarding the nature of mattering come together here – that is, questions of matter in the multiple senses of meaning, being, and valuing.

Reflecting on our map-making endeavours in Belfast, relationally speaking, I suggest that the child, who maps, exists only within the materialising relations in the on-going, iteratively reconfiguring process of button-mapping.

Our agency did not exist separately but emerged from the relationships in intra-action. 'Mattering' came about where participants entered relationships with the materials to hand. Their ability to do emerged, changed and transformed. Child mappers, intimately and imaginatively engrossed in the doing of mapping, entered into entanglement with the tools of mapping (the array of tiny items) which were not a mass of 'inanimate given-ness' but rather substantial in their 'iterative, intra-active becoming' – becoming-with-child as it were, insofar as the maps created could be regarded as a 'congealing of agenc[ies]'.[28] As I now detail, ideas of classification, juxtaposition, scale, relationality, distance, and peripherality emerged in fascinating ways, underscored by matter that had the ability to act and do.

BUSILY MAPPING THE 'BONEY': ON THE RELATIONALITY AND TEMPORALITY OF BUTTON-MAPPING ENCOUNTERS

During a workshop at one primary school in a predominantly Unionist/Protestant area of North Belfast, I engaged with eleven-year-old Corey. Corey generally preferred to create independently during our workshops. He was quiet and well respected by his peers but engaged with them in a gruff and sometimes curt manner. Midway through the session, I noticed him meticulously collecting small cocktail sticks and larger, uniformly sized black buttons, then watched on as he set to work, carefully engineering with quiet intent a structure that grew upwards, and up, and up, and up. I approached him. He told me that he was making a 'boney' – a model 'landmark' bonfire, commemoratively marking the Protestant/Unionist cultural celebrations that were soon approaching. Held annually on the evening of 11 July, bonfires and associated material markers of Unionism were familiar to these children. I asked him to tell me more, explaining that I had never before attended the celebration. He gave detailed accounts of where he went to watch the celebrations and with whom, as well as the preparations that he was involved within the weeks leading up to the event: the materials they collected (or purloined) to construct the 'boney' and how local children would be enlisted to guard the site during day and night shifts. He confided some rather alarming details, too – encounters he and his brother's friends had had when challenged by rival local gangs, and the intimidation tactics used by older men onto younger children and youths. All the while, he continued to build his little edifice. He was happy for me to watch and make photos but would not be distracted from his undertaking (see figure 7.2).

A week later, during our last button-mapping workshop together, the children were using the assorted materials to showcase and discuss their best

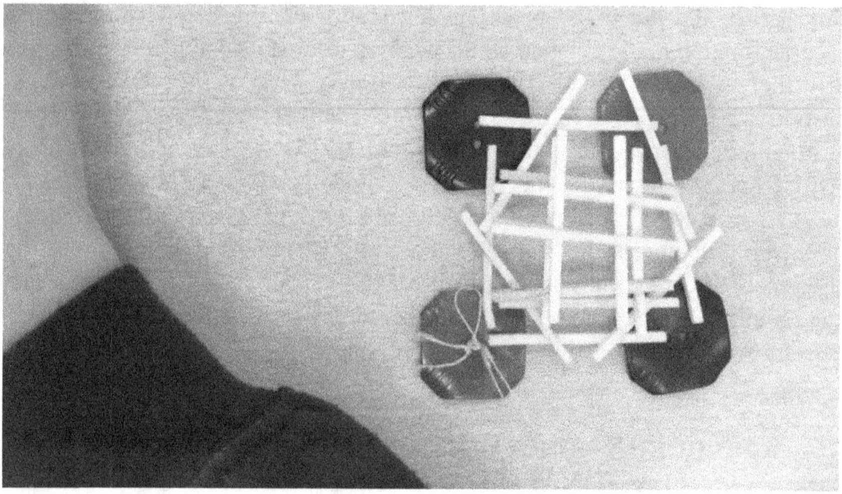

Figure 7.2 'Mapping the Boney', North Belfast. Author's Image.

Figure 7.3 'Knife for Amy'. Author's Image.

'future Belfast(s)'. Corey approached me and presented me with a little pretend hand-held knife that he told me he had whittled using a small, wooden lollipop stick and a pair of scissors (see figure 7.3).

Corey had meticulously fashioned a handle from soft wool wrapped around and around and a brightly coloured name label. A personalised knife, he told me, to keep me safe during the next stage of my research. 'That's what you need in Belfast. Like at the boney, ya'remember'. Whereas the other participants were gamely engaging in future-thinking exercises, Corey was fixed on our conversation two weeks prior. Clearly, my lack of local knowledge had stayed with him; it set me apart from other adults that he knew, and that mattered to him, moving him in that moment to engage with the matter around him to make for me a protection of sorts, being concerned for my wellbeing as a researcher in the field. There is an irony here, insofar as Corey presented me with a weapon – a tool of destruction, and arguably tribalism – in order to invite me in, to build a relationship with me, and invite me into his sense of being-there, within his community.

However, Corey did not wish to discuss the knife for long. He did not disclose if and where he had seen 'protection' like this before. Nor did he want the other participants to witness his gifting. He seemed slightly abashed, but spoke with conviction, nonetheless. However, in the minutes immediately after this, I observed other participants in the room noticing and acting on the exchange of the handmade knife, entering busily, laughingly, and playfully into relationships with the available materials, fixing together, whittling, scratching, stringing, and gluing. The individual/group binary can be called into question here, as can the seeming distinction between bodies entangled in the phenomenon of the Twelfth of July celebrations that was being performatively enacted here. Through intra-action, we human actors (the participants and myself) were brought together noisily with a messy amalgam of things: cutting things and sticky things, knotty things and loose, trailing things, sharp things, and splintered things. The 'comingtogetherness' inherent in the makings and doings in the rooms that day was concomitant with the emergent phenomenon of the Twelfth celebrations, which included the children's bodies, discourses on Belfast, cultural celebrations around parading and bonfires, the role of politics, local community leaders, news channels, fear and protection. And yet, this intra-action also separated us into new, co-constitutive subject positions by virtue of what was created, namely the significance of the weapons. Participants knew that, at the end of the session that day, they were not allowed to take their little knives into the playground (although some tried to escape nonetheless!). In contrast, the contained space of the workshop provided a freer space wherein linkages could be performatively developed and enacted, that is, linkages between affective interior worlds and the exterior world of the city. Being-with the city through button-mapping, we became (temporarily) protected and non-protected, involved and the not involved, at risk and not at risk. And further, our discussions emerging from this intra-action revealed how questions of mattering come to be, in the multiple senses of matter as 'meaning, being, and valuing'.[29] Participants and myself were responsible for matter produced in this intra-action, that is, discourses around place detail and use, spatial control and surveillance, as well as our subject positions concerning in/security and belonging.

REFLECTIONS

In this chapter, I have argued that thinking with intra-action is a mode of engaging with the transformative aspect of the minor through the practice of button-mapping. I have built on the notion of 'becoming-minor' through mapping, insofar as we give up representationalist designations of subject–object binaries, as well as notions of individual agency. By entangling the minor with intra-active practices such as button-mapping, we can rethink the potential of what minoritarian-mapping can be and do in a post-representational sense. Mapping in this sense can be a way to queer the familiar by opening possibilities for troubling or unsettling the given ways that are used to know and understand cities. Exploring the interconnectedness (and contestations) of power, knowledge, and spatiality through an empirical case study of button-mapping with children in Belfast, I contend that intra-active mapping can critically bring into view, if only temporarily, the spaces of mattering in children's lives. Button-mapping *is* a form of everyday praxis, not simply a concept; a processual mode of 'playing with place' that takes place through encounters of difference and diversities which foreground more-than-rational, affective, and embodied nature of participants' lifeworlds. This, I argue, can potentially lead to new ways of theorising children's geopolitical praxis in Belfast insofar as these encounters prompt us to ask different questions concerning the 'withness' of 'otherness' – what Haraway (2008) would call 'another worldliness' – that actively decompose and rework narrow ethno-sectarian political framings and normative (typically adult) notions of difference and conflict.

NOTES

1. Benwell, M. C. and Hopkins, P. (2017). *Children, young people and critical geopolitics*. London: Taylor & Francis; Horton, J. and Kraftl, P. (2006a). What else? Some more ways of thinking and doing 'Children's Geographies', *Children's Geographies*, 4: 69–96; Horton, J. and Kraftl, P. (2006b). 'Not just growing up, but going on: Materials, spacings, bodies, situations', *Children's Geographies*, 4(3): 259–276; Skelton, T. (2013). 'Young people, children, politics and space: A decade of youthful political geography scholarship 2003–13', *Space and Polity*, 17(1): 123–136.

2. Shirlow, P. (2006). 'Belfast: the "post-conflict" city', *Space and Polity*, 10(2): 99–107.

3. Katz, C. (1996). 'Towards minor theory', *Environment and Planning D: Society and Space*, 14: 487–499.
4. Shirlow, P. (2006). 'Belfast: the "post-conflict" city', *Space and Polity,* 10(2): 99–107.
5. Katz, C. (1996). 'Towards minor theory', *Environment and Planning D: Society and Space*, 14: 487–499 (p. 490).
6. Barad, K. (2003). 'Posthumanist performativity: Toward an understanding of how matter comes to matter', *Signs: Journal of Women in Culture and Society*, 28(3): 801–831; Barad, K. (2007). *Meeting the universe halfway: quantum physics and the entanglement of matter and meaning*. Durham, NC: Duke University Press.
7. Deleuze, G. and Guattari, F. (1986). *Kafka: Towards a minor literature*. Trans Dana Polan. Minneapolis and London: University of Minnesota Press; Katz, C. (1996). 'Towards Minor Theory', *Environment and Planning D: Society and Space*, 14: 487–499.
8. Ibid (p. 491).
9. Ibid (p. 491).
10. Ibid (p. 495).
11. Ibid (pp. 495–496).
12. Dodge, M., Kitchin, R. and Perkins, C. (2009). *Rethinking maps: New frontiers in cartographic theory*. Abingdon, Oxon: Routledge.
13. Katz, C. (1996). 'Towards Minor Theory', *Environment and Planning D: Society and Space*, 14: 487–499 (pp. 495–496).
14. Haraway, D. (2016). *Staying with the trouble: Making kin in the Chthulucene*. Durham, NC: Duke University Press.
15. Pyyry, N. (2017). 'Thinking with broken glass: Making pedagogical spaces of enchantment in the city', *Environmental Education Research*, 23(10): 1391–1401.
16. Barad, K. (2003). 'Posthumanist performativity: Toward an understanding of how matter comes to matter', *Signs: Journal of Women in Culture and Society*, 28(3): 801–831.
17. Harley, B. (1989). 'Deconstructing the map', *Cartographica*, 27: 1–23; Kitchin, R. and Dodge, M. (2007). 'Rethinking maps', *Progress in Human Geography*, 31(3): 331–344.
18. Rossetto, T. (2015). 'Semantic ruminations on 'post-representational cartography', *International Journal of Cartography*, 1(2): 151–167; Caquard, S. (2015). 'A post-representational perspective on cognitive cartography', *Progress in Human Geography*, 39(2): 225–235; Gerlach, J. (2014). 'Lines, contours and legends: coordinates for vernacular mapping', *Progress in Human Geography,* 38(1): 22–39.
19. Rossetto, T. (2015). 'Semantic ruminations on 'post-representational cartography', *International Journal of Cartography*, 1(2): 151–167 (p. 160); Del Casino, V. J., and Hanna, S. P. (2006). 'Beyond the "binaries": Methodological intervention for interrogating maps as representational practices', *ACME*, 4(1): 34–56.
20. Rossetto, T. (2019). 'The skin of the map: Viewing cartography through tactile empathy', *Environment and Planning D: Society and Space*, 37(1): 83–103.
21. Steele, J. (2015). 'Boat Tree Game Button: How ordinary objects can spark extraordinary change'. Cultural Animation and Community Organising Practice

Report. Available at: www.keele.ac.uk/media/keeleuniversity/ri/risocsci/thelegacyo fconnectedcommunities/Boat_Tree_Ga me_Buttons_Practice_Report_single_page_p rint_updated%20(3)%20(1).pdf. Accessed 22/10/2019.

22. Deleuze, G. and Guattari, F. (1980). *A Thousand Plateaus*. Trans. Brian Massumi. London and New York: Continuum, 2004. Vol. 2 of Capitalism and Schizophrenia. 2 vols. 1972–1980. Trans. of Mille Plateaux. Les Editions de Minuit: Paris.

23. Ainsworth, P. (2019). 'Map, terrain, string and paper.' *Living Maps Review*, 6. Available at: http://livingmaps.review/journal/index.php/LMR/article/view/160/295. Accessed 13/10/2019.

24. Barad, K (2010). 'Posthumanist performativity: Toward an understanding of how matter comes to matter', *Signs*, 28(3): 801–831 23; Barad, K. (2012). 'Intra-Actions', *Mousse*, 34: 77.

25. Ibid (p. 77).

26. Ibid (p. 80).

27. Ibid (p. 77).

28. Ibid (p. 77).

29. Ibid (p. 77).

BIBLIOGRAPHY

Ainsworth, P. (2019). 'Map, terrain, string and paper', *Living Maps Review*, 6. Available at: http://livingmaps.review/journal/index.php/LMR/article/view/160/2 95. Accessed 13.10.19.

Barad, K. (2003). 'Posthumanist performativity: Toward an understanding of how matter comes to matter', *Signs: Journal of Women in Culture and Society*, 28(3): 801–831.

Barad, K. (2007). *Meeting the universe halfway: quantum physics and the entanglement of matter and meaning*. Durham, NC: Duke University Press.

Barad, K (2010). 'Posthumanist performativity: Toward an understanding of how matter comes to matter', *Signs*, 28(3): 801–831.

Barad, K. (2012). 'Intra-Actions', *Mousse*, 34: 77.

Benwell, M. C. and Hopkins, P. (2017). *Children, young people and critical geopolitics*. London: Taylor & Francis.

Caquard, S. (2015). 'A post-representational perspective on cognitive cartography', *Progress in Human Geography*, 39(2): 225–235.

Del Casino, V. J., and Hanna, S. P. (2006). 'Beyond the 'binaries': Methodological intervention for interrogating maps as representational practices', *ACME*, 4(1): 34–56.

Deleuze, G. and Guattari, F. (1986). *Kafka: Towards a minor literature*. Trans. Dana Polan. Minneapolis and London: University of Minnesota Press.

Deleuze, G. and Guattari, F. (1980). *A thousand plateaus*. Trans. Brian Massumi. London and New York: Continuum, 2004. Vol. 2 of Capitalism and Schizophrenia. 2 vols. 1972–1980. Trans. of Mille Plateaux. Les Editions de Minuit: Paris.

Dodge, M, Kitchin, R and Perkins, C. (2009). *Rethinking maps: New frontiers in cartographic theory*. Abingdon, Oxon: Routledge.

Gerlach, J. (2014). 'Lines, contours and legends: co-ordinates for vernacular mapping', *Progress in Human Geography,* 38(1): 22–39.

Haraway, D. (2016). *Staying with the trouble: Making kin in the Chthulucene*. Durham, NC: Duke University Press.

Harley, B. (1989). 'Deconstructing the map', *Cartographica,* 27: 1–23; Kitchin, R. and Dodge, M. (2007). 'Rethinking maps', *Progress in Human Geography*, 31(3): 331–344.

Horton, J. and Kraftl, P. (2006a). 'What else? Some more ways of thinking and doing 'Children's Geographies', *Children's Geographies*, 4: 69–96.

Horton, J. and Kraftl, P. (2006b). 'Not just growing up, but going on: Materials, spacings, bodies, situations', *Children's Geographies*, 4(3): 259–276.

Katz, C. (1996). 'Towards minor theory', *Environment and Planning D: Society and Space*, 14: 487–499.

Pyyry, N. (2017). 'Thinking with broken glass: Making pedagogical spaces of enchantment in the city,' *Environmental Education Research*, 23(10): 1391–1401.

Rossetto, T. (2015). 'Semantic ruminations on "post-representational cartography", *International Journal of Cartography*', 1(2): 151–167.

Rossetto, T. (2019). 'The skin of the map: Viewing cartography through tactile empathy', *Environment and Planning D: Society and Space*, 37(1): 83–103.

Shirlow, P. (2006). 'Belfast: the "post-conflict" city', *Space and Polity*, 10(2): 99–107.

Skelton, T. (2013). 'Young people, children, politics and space: A decade of youthful political geography scholarship 2003–13', *Space and Polity*, 17(1): 123–136.

Steele, J. (2015). 'Boat Tree Game Button: How ordinary objects can spark extraordinary change.' Cultural Animation and Community Organising Practice Report. Available at: www.keele.ac.uk/media/keeleuniversity/ri/risocsci/thelegacyofconnectedcommunities/Boat_Tree_Ga me_Buttons_Practice_Report_single_page_print_updated%20(3)%20(1).pdf. Accessed 22/10/2019.

Chapter 8

Mapping the Overlaid Life of Places of Play

Joel Seath and Kelda Lyons

OUR WORKING UNDERSTANDING OF PLAY, AND PLAY IN THE BUILT ENVIRONMENT

There are many ways to view and understand play. It is generally considered that there might be many potential benefits to playing, such as therapeutic, physical and mental health, skills development, and others.[1] Without discounting these potential benefits, for this chapter, we are viewing and discussing play as something that children (and adults) do for fun, for its own sake and because they want to: we are talking about play for the sake of playing. *Play for a Change* presents a perspective on play from Brian Sutton-Smith, describing how children's worlds can be made more exciting and amusing through play.[2] The idea is put forward that 'what play prepares you for is more play, and what that gives you is more satisfaction in being alive'.[3]

Play is something that comes from people's own actions rather than being something that happens to them. The fact that play comes from people themselves may mean that there is an inherent freedom in play, or in the act of playing. Many children and adults can initiate playful states of being by moving into their own imaginary worlds, pretending that they are a zombie or that they can fly. A child in a buggy might imagine that they are in a spaceship, then create a whole narrative around this. Sutton-Smith said that 'play is typically a primary place for the expression of anything that is humanly imaginable'.[4] If we relate this expression of imagination to the built environment, thinking about these environments in terms of something that children should be able to enjoy and use freely, and the fact that all children play (in some way, everyone may play differently to everyone else), then built environments that support children's play could be viewed as environments that could or should support the freedom, that is, play itself.

In this chapter, we argue that if the built environment were to facilitate more playfulness, then adults knowing how to listen to and interpret children's thoughts on the environments they use could contribute to informing their planning, design, development, and maintenance, whether these are *destination* places for children, such as large, fenced playgrounds, or *in-between* places, such as streets, alcoves, or along or on garden walls. A child's default state of being is play; however, for all that play is a fundamental aspect of who children are, their voices regarding that play (and therefore about themselves and their experiences) are not always well heard or understood. Mapping play, by various means described in this chapter, gives voice to the children who are, after all, also valid members of society and its built environments. Children deserve better respect and recognition in all that we routinely see about us.

In the first two sections of this chapter, we provide a brief overview of adventure playgrounds in the United Kingdom (research for this chapter's play mapping was based at one such site in London) and the general purpose of playwork (being the work which is undertaken by skilled adults in service of children's play). In the next section, we introduce the idea of 'gentle and subtle listening' with regard to carrying out research with children. This is followed by information that provides an overview of, and methodology for, play-mapping work undertaken in London (UK) between April 2015 and January 2016. Two preparatory sketch maps made during the course of this research are also included here. In the following section, we draw on the writing of the cultural anthropologist Margaret Mead in exploration of adults' understandings of the meanings of children's artefacts (e.g. their drawings), with regard to those children's expressions about their play. The next section highlights the map-territory thinking of Alfred Korzybski, in the initial stages of investigating how places develop as the overlaid experiences of multiple players. Subsequently, we develop a deeper thinking on the idea of *place* and how its form is created by way of various given names and stories. We relate some of the names and stories that have been gathered during research with both children and older generations in a particular area of West London. The next section further investigates the stories told by older generations, and observations made, of *destinations* (such as parks), *in-betweens* (such as streets and alleys), various *edges*, *zones*, and *routes*. We highlight the everydayness of play: it can happen anywhere and everywhere. In the penultimate section, we present a brief written account of an overlaid map created using many of the stories, drawings, and observations from the previous sections. We explain an embracement of the concept of 'What-If?' in the mapping process: that there was a playing with the maps in an act, like children's play itself, of finding out. In the final section of the chapter, we present our conclusions, including a short investigation of the state of affairs in contemporary UK town planning as regards children, their play, and the built environment.

A BRIEF OVERVIEW OF ADVENTURE PLAYGROUNDS

Children play everywhere and anywhere; however, the research project that this chapter is based on began on a London adventure playground. Adventure playgrounds tend to be located in and around cities. Many of them were created from pieces of land that were previously bomb sites and which were used for play by children. Some of these sites were developed into playgrounds after adults perceived a need for safe places for children to play. From these sites and others, adventure playgrounds evolved where children could come and play, swing, build, climb, make fires, cook, make things, relax, and contribute, with the support of adult volunteers or staff. Adventure playgrounds are usually open after school, on weekends and in school holidays, with children aged six to fifteen years old (typically) being able to attend. They normally have indoors and outdoors areas, are staffed by playworkers (who support children to play in ways that they, the children, want and need to), and are open access, meaning that children can go in and out of the playground during play sessions as they please. Playworkers are always present in play sessions. Some adventure playgrounds welcome children with disabilities in their sessions. There is a network of adventure playgrounds in the United Kingdom that are specifically for children with disabilities and their siblings (some of these playgrounds do not have an open access policy, for safety reasons). There is normally no payment to attend, with adventure playgrounds being free at point of entry. Some adventure playgrounds welcome parents during play sessions; in others, parents can only drop off and collect their children and mostly have to stay in the gate or entrance area. Most adventure playgrounds are run by local authorities, charities, or play associations.

Adventure playgrounds are often very much part of their local area, by way of personal and community connections. They are not isolated facilities. On some playgrounds, parents spend time, volunteer, socialise, share advice, tell playworkers local news, host events, and make new friends with other parents. Adventure playgrounds tend to have the feel of a community hub and sometimes function as this rather than having the feel of a public service offer.

Playworkers sometimes see more of children's everyday lives than what goes on inside the playground, for instance in the action and interactions that happen around playground entrance gate areas. These gate areas can be placed on busy roads, within parks, in cul-de-sacs, or in other types of locations. Some adventure playground gates are kept closed (not locked) to stop dogs running in, or for other reasons, such as to discourage the public from wandering in. Some playground gates are kept wide open while play sessions are running.

Incidental observations made by playworkers around the thresholds of gate areas can give playworkers a sense of who children know and are comfortable to talk with, how children interact with their wider community, and how safe children feel to leave the playground. As these kinds of observations are incidental, they are typically non-intrusive for children and can happen regularly and take place over very long periods of time (if the same playworkers and children are present on a playground over a number of months or years). A playworker's extended general awareness of children's everyday lives can contribute valuable knowledge to play-mapping exercises such as the one described in this research. Playworkers might use this general knowledge to help create mapping exercises that will have relevance and be understandable for children in this area, or it may just give playworkers some reference points and context for the information that children might convey about their play and the perceptions of the areas that they spend time in.

THE PURPOSE OF PLAYWORK

Playwork is what playworkers practice and do on adventure playgrounds and other places, to run them as such. Our current understanding of playwork is that it exists to offer children time, space, and support to play.[5] Playworkers may work on adventure playgrounds, after school clubs, in prisons, or doing outreach work in public places. Playworkers on adventure playgrounds prepare materials, are ready to interact with and support children, and are there to help organise resources or more things to do if children ask for this. Playwork is not about running play sessions with a time-fixed programme, where children are directed through a series of activities. Ideally, there are lots of resources available for play sessions, including food, toys, tools, musical instruments, and many other small and large objects and materials which, in accordance with the Playwork Principles Steering Group, children can choose to use how they want. Adventure playgrounds normally have interesting outdoors environments that may include features such as swings, sandpits, fire pits, chill out areas, ponds, wooden play structures, gardens, trees, playhouses, sculptures, plants, sports pitches, and others. Playworkers tend not to direct children's play, but rather facilitate play and playful atmospheres, and join in with children's play if invited to do so by children. It should be up to the children to decide if they want playworkers to join in with their play or not. Playworkers normally aim to establish and maintain non-authoritarian relationships with children, while offering children whatever level of psychological, physical, and emotional support that they need to be able to play freely.

GENTLE AND SUBTLE LISTENING

Subtle and very attentive listening is an important part of carrying out research with children. Conversations and commentary that children have and make can benefit research by contributing verbal information to the body of knowledge about the artefacts that children create, even though not all these conversations may be recorded or recorded in detail. Gentle and subtle listening may be required to interpret, or try to interpret some of what children say, and to indicate to children that they are being taken seriously and that it is worth them talking further. The more conversations and commentary that children make on the artefacts they create, and the more that this is listened to and understood by adult researchers, the clearer adult understanding will be of children's ideas. Ideally, the least possible meaning-making is needed at the end of the process.

There are some ways in which playworkers (who are known to some of the children in the area) doing the research may benefit the information-gathering: children sometimes feel more confident to share more personal details of their play and routines, and their thoughts and feelings on this play, with adults who they already know and trust. This can also be a person who is new to the children; however, children tend to be aware that the adults who know the children's routines, friends, places of play, schools, family, and social networks will be able to put the information that the child gives into context more easily. Children may feel that playworkers will fully understand the importance of the small details of play.

Playworkers are skilled practitioners likely to understand children if the children talk about things in a non-literal way and/or move into the realm of talking about or with the imaginary; children are likely to understand that they are talking with an adult who can *tune into* or follow what they are saying. These comments do not discount any existing research methods or approaches but simply express our ideas on the subtleties of the way that research is done with children, and particularly in a playwork way or setting.

MAPPING PLAY

In the process of what we term *gentle and subtle listening*, which is considered consultation by observation and other means into the how and where of play, action research was undertaken, and on-going mappings were made, so as to further support children's play in White City, West London (UK). Qualitative data for this study was collected between April 2015 and January 2016. Ultimately, the aims of this study were two-fold: to increase children's opportunities to play and to remove barriers to that play (i.e. raising adult awareness and tolerance of what children do). Mapping the play (of past and present) was integral to the process of finding out about the *how* and *where*; however, it also proved beneficial in developing a further means of representing a long-held understanding of the *places* where children play (for instance, the overlaid life of those places: all the play that they have seen).

METHODOLOGY

Play was observed on the adventure playground in the middle of the estate and off-playground at local events, at outreach sessions in the streets and parks and in the everydayness of children's lives around the estate, as appropriate.

Three Year 2 (6–7 year olds) and school council groups also worked with one of the researchers, Joel Seath, at a local school, linking new mapping exercises with the children's existing study on the United Nations Convention

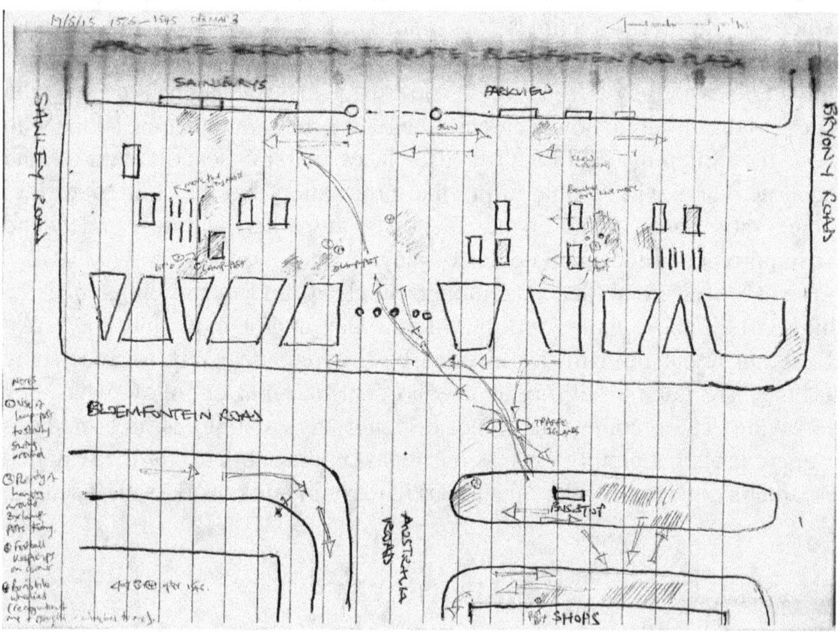

Figure 8.1 Example of an Observation of Preferred Play Places and Routes in the Public Arena. Author's Image.

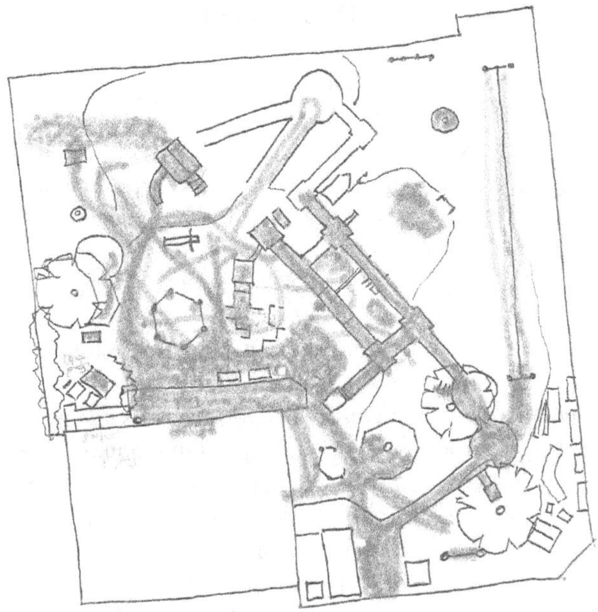

Figure 8.2 Example of Amalgamated Observations of Preferred Play Places and Routes on the Adventure Playground. Author's Image.

on the Rights of the Child, Article 31 – children's right to play.[6] This mapping aimed to ascertain children's favoured places to play (by drawing them). Local working-age and elderly adults' play stories of their own childhoods were also gathered by way of either online communications, short written accounts at local events, or in individual or small group audio-recorded discussions.

In the process of mapping, playworker-researcher thinking and practice was influenced by a small collection of key ideas from authors of various fields, embracing but not limited to the following:

- The content and intent of the play belongs to the child[7]
- The time and space needed for children to exercise their right to play[8]
- The four registers of the Good City, being *repair*, *relatedness*, *rights,* and *re-enchantment*[9]
- Five kinds of urban settings where play seems to happen most: paths, intersections, boundaries, thresholds, and props[10]
- The notion of a 'collective wisdom', that individual communities have a range of unique local knowledges, children and adults alike[11]
- A what-if approach to researching, just as play is a what-if approach to action[12]

MEANING-MAKING CONSIDERATIONS

As playworkers, we thought how adults understood the meanings of children's artefacts deserved more exploration. Part of the mapping process was asking the children to create artefacts. In this case, artefacts were drawings; however, our understanding of a playful artefact is any leftover representation of play that is or has been. An important consideration for us when presenting the children's artefacts at a Livingmaps seminar was to be aware of individual children's subjective experiences of places, their own play experiences, and of different kinds of play. The cultural anthropologist Margaret Mead took the perspective that 'it's a good thing to think about "the child" as long as you remember that "The Child" doesn't exist. Only *children* exist [as individuals] . . . every time we lump them together we lose something.'[13] This means that using the concept of *the child* is not helpful when trying to understand individual children's experiences. What individual children may get out of a place or experience may differ greatly from person to person. For example, a competitive game of jumping off high objects may be fun for children who are competitive and like heights but an unpleasant experience for children who are scared of heights and who enjoy more co-operative play or games. A stairwell in a block of flats might offer some children a private place that they enjoy away from adults; other children may feel uncomfortable with no known or *safe* adults present and may want to know that there are adults within sight, even if these adults are at a distance and not involved in the children's play. We believe that interpreting or attempting to make meaning from children's artefacts about places of play needs to be done extremely sensitively so that this process results in the clearest possible understanding of what children might have been thinking about, feeling, experiencing, or trying to convey when they were creating the artefacts.

The considered observation of children at play in the urban environment – in this case, with playworkers making the observations on an adventure playground, at local events, outreach sessions, and in everyday local life – can also be seen to go some way towards shifting the adult mind-set away from just thinking about objects (such as toys, play structures, and street furniture items that children play with) and the more visible physical activity that are part of playable moments, and towards a deeper understanding of what children are doing, where, and possibly why there.

PLACES AS LIVED EXPERIENCES AND TEMPORAL MULTIPLICITY

Whereas maps are flat representations, the city proper is a lived experience: it is 'territory' and we are a part of it, in our many navigations through its streets (like toothpaste) and other public areas and arenas, and as we navigate around its many monuments and other greater or lesser objects or props of the built environment. Play is intrinsic to culture and, as such, children (the play experts) are essential elements of city cultures. We might airbrush out the lived experiences when devising maps, but the physical traces of lived experiences of play can be seen to linger in the city proper, in the territory. With the idea of 'what we believe is true' in mind, what if we were to conceive of a map, which is not the territory itself, that we may move through, sense, and feel? What if, to the three dimensions we usually and can most easily perceive ourselves to operate within, we also mapped the added layer of time? It is not too great a leap to jump from the two-dimensional flat representation of a map to the four-dimensional map representation as we navigate within it and as we perceive it. The task of re-interpreting that multi-dimensional lived experience though, back into a two-dimensional map representation, is the greater challenge, to which we shall return.

The map is not the territory,[14] but equally so, the territory is not the map: by which we mean that the nowness of territory is not replete, to the casual eye, with the fullness of the memory maps carried around in current and past players' heads. If the territory is 'now', then the maps in our heads can and do hold a temporal multiplicity. They are replete with an abundance of what was and, perhaps, what could be. Jay Griffiths writes (in terms of the rural but the sentiment can just as easily apply to the urban) that 'a damp hollow by a riverbank is not an occasional place to visit but a permanent part of who you are'.[15]

No map is 'true', in the sense of being totally comprehensive; all maps are selective and serve to orientate their holders not necessarily just with where they are spatially but, potentially, also with where they are cognitively, emotionally, developmentally, temporally: what we believe is true. If the territory is the 'now', then this 'now', and then this 'now', a rapid fluidity, then the map, *this* map, is a tracing of the selected nows that were. It is this memory map that is carried in the head (made accessible to others by pressing into two dimensions on paper), and it is a four-dimensional map. In this sense, the memory map is far richer than the simple territory.

PERCEPTION OF PLACE

How is it that we might perceive? There can be a stance, a co-ordinate position (due to experience in the privileged proximity to play), a way of seeing, that perceives of *place*, for example, rather than *space*: *place* being full of all manner of past and present and possible future, emotional attachment, sensory engagement, and potential returns. In short, places are weighed with short or long stories: what has happened here, is happening, or what will happen, all fizz in the multi-dimensional map that we can walk through. Play is replete with stories, and they stick to the surfaces and sink deep into the whole. That which has been played, or is being played, is in some fashion indelibly marked within the fabric of any part of the city.

PLACE NAMES AND STORIES

For current and past players, places often have names to accompany their stories. These names are sometimes prosaic, or esoteric, or direct markers of the stories themselves. In conversation with children, and with older generations who either still currently live or used to live and play in the area around White City and Shepherd's Bush, London, places were variously named other than their two-dimensional map representations: BBC Park or The New Park or The Chinese Jungle (being all one and the same), The Cage (a roughly triangular wedge where football was played), The Wormholt Alleys (which were haunted), The Luxury Flats (because they had sliding doors), The Frontline (the shops where the local children gather), The Rubber (a sports pitch).

Some of these places have since been wiped from the conventional map, but in the perception, by way of the stories, they're still there. The Rubber, a dilapidated hard court on South Africa Road, used by community members, was built over and made into pay-to-play football pitches. Beneath the Rubber though, literally and in this multi-dimensional map perception, still lie remnants of the old Franco-British White City Exhibition buildings of 1908. Beneath this are the fields before London expanded this way. In similar ways, and by way of stories of the past and observation of the present, here are the air raid shelters between the tenement blocks where music was played; where later the balconies rang out with mischievous children running along them, being chased by officers of the law; where much later, chalk was left overnight and in the morning the place was covered in hopscotch grids and snakes and the tags of names and various other offerings.

Here is the park where the adventure-playground-constructed trolley full of things to play with was emptied and the summer-full of children and their parents played with it all; beneath this are the ghost remains of the prefabs (whose gardens everyone was jealous of because they were the only gardens around); beneath this is the waste land, traversed over to climb the fence to the *Adventure* when it was largely unbuilt on; beneath this are the traces of the neat white Edwardian edifice of the White City which lends its name to the estate.

Here is the modern plaza on Bloemfontein Road (or Blom Road in the local colloquial) where the local children gather around the lamp-posts and the benches, riding their bikes through on regular routes; beneath which is the trace of the old Blom Road swimming baths, where you could distract the lifeguard and jump over the wall to get in for free; this is just a short hop in time from when the dead were laid out during the war bombings (the bodies taken away before the afternoon pool session started), all a long time before they built a supermarket on the site.

DESTINATIONS, IN-BETWEENS, AND THE EVERYDAYNESS OF PLAY

Play is an everydayness of being. It can happen anywhere and everywhere that children have the will and opportunity to play. Whether play was a focus of the architects of the White City Estate or not, in the 1930s when it was officially opened, is unknown but some modern expectations of corralled play in the green strips between tenement blocks are not met: from the child's perspective, why *would* they wish to play in such narrowly overlooked and confined quarters? These are not the local destination places of choice (past generations had, for example, the old White City Stadium, which is now the site of BBC Media Village, another supermarket, and other developments). Instead, now, the children have the two main green parks, the various other designated areas for play (the adventure playground, the *pitches* beyond, the smaller fixed play equipment areas between tenement blocks), and so on. Rather than just happening in the destination places though, play is an everydayness: it happens on the streets themselves, in the myriad nooks and crannies of lobbies, kerbsides, pathways, in and around trees and bushes, and so on – play, we can see, also happens in these in-between places.

If we acknowledge play as being part of historical local memories and of the fabric of smaller geographic areas, while looking at the significance for play of seemingly insignificant in-between places, we might consider an aspect of built environment design that could speak to children (and others) in a subtle and playful way: we might factor local play histories into urban design or development in much the same way as heritage consultants do. Features and aspects such as artwork, symbols, signs, steps, and parts of walking routes could subtly acknowledge or reference local play histories: a possibility for the design of the many parts of the public realm that children regularly use and experience.

IN THE IN-BETWEENS: ARCHAEOLOGICAL LAYERS OF PLACE – INTERGENERATIONAL, INTER-TEMPORAL, GHOSTS OF PLAY

Where children once offered car owners the *peace of mind*, protection as it were, for their parked cars, for a fee on New Zealand Way, while the local team, Queens Park Rangers Football Club, played at the nearby stadium, now children pull others crammed three to the back of a makeshift trolley; old furniture was thrown from the balconies for the community bonfire and now those paved or tarmac surfaces below are ad-hoc football pitches; beneath the balconies where children would play *Knocking Dollies Out of Bed* (you knock on a door and run away) or tying string to multiple door knockers so they all go off at once, others now balance on the walls, run out of the adventure playground throwing water balloons or chalk on the ground.

EDGES, LIVE AND DEAD ZONES, URBAN PROPS AND ROUTES

On the adventure playground and nearby, in an unfenced green area nominally for the purpose of being an outside *adult exercise gym*, the site edges are noted in terms of play: on the adventure playground, the children by and large stayed clear of the very periphery (perhaps because this was perceived by them to be too dead a zone); in the adult gym area, the bushes right up against the nearby housing office wall proved conducive for hiding in and for looking out from, for stockpiling away the play equipment, hung on the branches as if looted treasure. In the in-betweens of the estate, in the destination places of the parks, and on the adventure playground alike, there are unplayed in zones that are *too dead* or *too live* (the very centres of the parks, for example, where the potential player is, perhaps, too visible) and in each there is the played in zone, the *Goldilocks Zone*, as it were: an *in-between* in itself.

Children's play will happen in conducive circumstances (physical environments and human environments, where *time and space* for play are acknowledged). Their routes to destination places are often dictated in the modern world by enforced subservience to the needs of car drivers, but children's mappable and favoured routes are also play opportunities in themselves. In the observation, children will go out of their way to splash in the puddles or to examine a door of sensory importance. They will trail their fingers along the railings or the flowers; they will take detours up and over walls or into and out of courtyards, just because they are there. The urban furniture, other fixed objects or *props* are significant. (On the South Bank, on the River Thames, there are a series of installations: orange benches, nominally, some of which are shaped in such ways or have pieces missing so as to render them functionally deficient – the children play on and over them, nonetheless.) In White City, this is where the tree that was the goal post was; now, here is the wall which is balanced along, around the modern SuDS (sustainable drainage system) in Bridget Joyce Square; there were the Canning House sheds, around which biking was played; here is the tree that the children like to shake the spring blossom from, near what the older generation call the Thru'penny Bit Flats.

Play has happened, does happen, and will happen, anywhere and everywhere in the cityscape, which is perceived as a multitude of places within the overall place that is London. These places are lived experiences in all their felt and sensory fullnesses. The stories of play continue to fizz with their moments, their names, their props and routes, be they at the edges or at the zone of in-between. The in-betweens of the city and the destination places are all playable, and all that has happened continues to happen. However, how can we map all this?

WHAT-IF? MAPPING

What transpired in the mapping process was a *what-if?* in itself: playing with maps led this playworker-researcher along a route from observation and discussion during data collection, to conceiving of places and the overall place of the city as a lived-in map, to sketch-mapping play places and starting to amalgamate these in order to find some way of representing all of the above. It must be noted that data are and cannot be comprehensive. Too much has happened, and continues to happen, here. We must, instead, have faith with the idea of places as multi-layered, as the reader will also see with an immersion in, and a sensitive comprehension of, their own neighbourhoods. In the sketches that were made, the preferred play places that were of historical nature were initially represented in red (where these places could be extrapolated from the stories); the contemporary preferred play places were represented in blue. The streets were initially rendered in purple because, conceptually, historical and contemporary play might very well have happened anywhere here (a merging of red and blue). Only playable (more or less) public areas were shown; buildings and private gardens were inverted to white.

True to the observation that *play begets more play*, more playing with the maps resulted in the *what-if?* of inverting the colour scheme. The historical red now became an electric blue, and the contemporary blue became gold; the purple of the possible-play (of then or now) became a moss green.[16]

Here, in the process, a small epiphany made itself apparent. While we know that play is fluid, dynamic, always fizzing, we can also conceptualise a *digging down* into the moss, into the possible places of the city, down through layers of gold and electric blue, down and down in time. It is an archaeological layering, a beautiful object that can be held in the hand. It can be wondered at, just as an historian might wonder at a time-laden artefact, still feeling the energy of those who once made it.

CONCLUSIONS

In this study, local children and older generations alike relayed their favoured places of play in the locale of White City by means of considered playworker observation, drawings, and written and verbal stories. The process of engaging with these means of communication contributed towards raising the tolerance for play, locally, and thereby, in part, increasing those children's opportunities to play in the way that they wanted and needed to. Play observed and discussed resulted in further discussions with local adults, which resulted in greater tolerance towards play, which resulted in more opportunities to play, and so on. The process of collecting, collating, and eventually mapping the data received can also be seen to have contributed towards raising adults' awareness of contemporary children's play.

Play for play's sake is not the prevailing adult perspective on children's primary interactions with the world, but it is the way that children engage with one another, with the natural and built environment, with their streets and cities. Just because they play in this way, without concern for adult agendas, such as learning or development, it does not mean that their play is frivolous and unnecessary. Play leaves its marks on the in-betweens of those city streets, on the destination places, such as parks, on the routes, on the edges, and in the variously playable zones of subjective experience. The research described in this chapter offers a chance to hear about and see these marks on maps. As such, seemingly unimportant or insignificant portions of the city, as perceived by the adult, take on the greater weight of *place*, as recalled by the child (although children may talk about such places in terms of given names or what once happened here). Places, imbued with time and play, remain, infused in the air and embedded in the archaeology of the memory maps of the streets: the places of play are significant to the players long after those players have grown into adulthood and old age. The play that is, and that has been, is all around us: we just have to look more deeply sometimes to see.

What, we ask, might be if more consideration is given towards children in the planning, designing, altering, or rearranging of built environments? What if the understanding of play mapping and places offered by this study contributed knowledge towards the design and management of built environments? If, as Jay Griffiths highlights, the fifteenth to eighteenth-century land

Enclosures in Britain coincided with the increasing disciplining and subjugation of children and childhood – 'the nature of the land and the nature of the child were both to be controlled, fenced in'[17] – then the question that might subsequently be posed is: what appears to be the trend, at the time of writing, in town planners' considerations of play? That is, is play for destination places only or is it for the in-betweens as well?

The understanding of and designing for play in the built environment sector could currently be seen as having its own in-between moment in the United Kingdom. Recent reports such as *Cities Alive: Designing for Urban Childhoods*[18] and *Child Friendly Planning in the UK*[19] prioritise and explore the importance of child-friendly cities and children's play. The Intend to Publish London Plan[20] emphasises the importance of play and recreation space for children and teenagers.

As the presence and focus of these publications indicate, priority is increasingly being placed on children's freedom to play and to move within and enjoy equitable use of public spaces in the United Kingdom. However, the research report *Child Friendly Planning in the U.K.* highlights the gap between, on the one hand, an improved vision for children in cities and, on the other, the lack of opportunities and procedures for children's priorities to be incorporated into the planning systems. The report comments that 'despite commitments to child-friendliness, it can still be difficult to affect planning processes and outcomes'.[21]

The London Plan discusses play provision, talking in terms of increasing opportunities for play and stating that the public realm should provide 'opportunities for social activities, formal and informal play'[22] and 'informal recreation facilities'.[23] Elsewhere in this document, 'incidental play space'[24] is mentioned as being important in larger public realm areas to make them playable for children. This appreciation of *incidental play spaces* could potentially be considered in relation to the in-between places mentioned above: be it that they feature the designed (e.g. wooden stepping-stone logs set into a grassy area) or the incidental (e.g. on pathways, or in and around trees and bushes), inviting children to play. If we place this appreciation alongside the significance of the historical nature of play that this chapter has discussed, and alongside the idea of children's right to play and the extensive use of their everyday places for that play, we could perhaps enhance the current vision for more child-friendly environments, particularly in the public realm.

There could be a vision where the design and management of formal, informal, incidental, and in-between places not only invite children's play in the here and now but also raises adult acceptance of play by speaking to them about historical play memories. Children, being equally valid members of society and their built environments, are deserving of their voices being in receipt of better recognition. The planning document examples outlined above offer a fair start in better providing for children. However, by mapping representations of children's default state of being, their everydayness of play (such as by means described in this chapter), the enhancement of those children's lives and the hearing of their voices can ultimately be further achieved. By such mapping means, children's subjective experiences, their expressions, and their fair inclusion, rights, and adults' understanding of their valid belonging can all be more fully recognised. Mapping play can be a catalyst for a town planning vision of a more equitable built environment.

NOTES

1. Kilvington, Jacky, and Ali Wood. 2010. *Reflective Playwork: For All Who Work With Children*. London: Continuum.
2. Sutton-Smith, Brian. 2003. "Play as a Parody of Emotional Vulnerability." In *Play and Educational Theory and Practice*. Play and Cultural Studies, 5, edited by Don Lytle. (Westpoint, CT: Praeger). Cited in Lester, Stuart, and Wendy Russell. 2008. *Play For a Change. Play, Policy and Practice: A Review of Contemporary Perspectives*. London: Play England.
3. Lester, Stuart, and Wendy Russell. 2008. *Play For a Change. Play, Policy and Practice: A Review of Contemporary Perspectives*. London: Play England (pp. 52).
4. Sutton-Smith, Brian. 1997. *The Ambiguity of Play*. Cambridge, MA: Harvard University Press (pp. 226).
5. Play Wales. 2015. "Play Sufficiency in Wales." Accessed September 16, 2020. https://www.playwales.org.uk/login/uploaded/documents/INFORMATION%20SHEETS/Play%20sufficiency%20in%20Wales.pdf.
6. Unicef. 2020. "Convention on the Rights of the Child Text." Accessed September 16, 2020. https://www.unicef.org/child-rights-convention/convention-text.
7. PPSG. 2005. *The Playwork Principles*. Cardiff: Playwork Principles Steering Group.
8. Play Wales 2015.
9. Amin, Ash. "The Good City." *Urban Studies* 43, no. 5–6 (2006): 1009–1023.
10. Stevens, Quentin. 2007. *The Ludic City: Exploring the Potential of Public Spaces*. Abingdon: Routledge.
11. Lester, Stuart, and Wendy Russell. 2013. "Leopard Skin Wellies, a Top Hat and a Vacuum Cleaner Hose: An Analysis of Wales' Play Sufficiency Assessment Duty (Executive Summary)." Cardiff: Play Wales. Accessed September 18, 2020. https://www.playwales.org.uk/login/uploaded/documents/Play%20sufficiency/PSA%20research%20executive%20summary.pdf.
12. Lester and Russell 2013.
13. Mead, Margaret. 1975. "Children, Culture, and Edith Cobb." In *Children, Nature, and the Urban Environment: Proceedings of a Symposium-Fair: Proceedings of the Symposium-Fair Held 19–23 May, 1975 at the C. H. Marvin Center of the George Washington University, Washington, D.C.*, edited by Calvin W. Stillman,

19–24. Upper Darby, PA: Forest Service, U.S. Department of Agriculture Forest Service, Northeastern Forest Experiment Station. (pp. 23).

14. Korzybski, Alfred. 1994. *Science and Sanity: An Introduction to Non-Aristotelian Systems and General Semantics*. 5th ed. Brooklyn, NY: Institute of Semantics.

15. Griffiths, Jay. 2014. *Kith: The Riddle of the Childscape*. London: Penguin Books (pp. 3).

16. Playworkings. 2018. "The Overlaid Life of Places of Play (Images to Support Article)." Accessed September 16, 2020. https://playworkings.wordpress.com/depository-of-notes/play-cartographies/the-overlaid-life-of-places-of-play-images-to-support-article/.

17. Griffiths, 2014 (pp. 22–23).

18. ARUP. 2017. "Cities Alive: Designing for Urban Childhoods." Accessed September 19, 2020. https://www.arup.com/perspectives/publications/research/section/cities-alive-designing-for-urban-childhoods.

19. Wood, Jenny, Dinah Bornat, and Aude Bicquelet-Lock. 2019. "Child Friendly Planning in the UK: A Review." London: APiC/ZCD Architects/RTPI. Accessed September 19, 2020. https://www.rtpi.org.uk/practice/2019/november/child-friendly-planning-in-the-uk-a-review/.

20. Mayor of London. 2019. "The London Plan: Intend to Publish (Clean Version). Spatial Development Strategy for Greater London." London: Greater London Authority. Accessed September 19, 2020. https://www.london.gov.uk/sites/default/files/intend_to_publish_-_clean.pdf.

21. Wood et al. 2019 (pp. 50).

22. Mayor of London 2019 (pp. 145).

23. Mayor of London 2019 (pp. 251).

24. Mayor of London 2019 (pp. 252).

BIBLIOGRAPHY

Alfred Korzybski. *Science and Sanity: An Introduction to Non-Aristotelian Systems and General Semantics*, 5th ed. Brooklyn, NY: Institute of Semantics, 1994.

ARUP. 'Cities Alive: Designing for Urban Childhoods', 2017. https://www.arup.com/perspectives/publications/research/section/cities-alive-designing-for-urban-childhoods.

Ash Amin. 'The Good City'. *Urban Studies* 43, no. 5–6 (2006): 1009–23. https://doi.org/10.1080/00420980600676717.

Brian Sutton-Smith. 'Play as a Parody of Emotional Vulnerability'. In *Play and Educational Theory and Practice: Play and Cultural Studies*, 5th ed. Santa Barbara: ABC-CLIO, n.d.

———. *The Ambiguity of Play. Cambridge*. MA: Harvard University Press, 1997.

Jacky Kilvington and Ali Wood. *Reflective Playwork: For All Who Work with Children*. London: Continuum, 2010.

Jay Griffiths. *Kith: The Riddle of the Childscape*. London: Penguin Book, 2014.

Jenny Wood, Dinah Bornat and Aude Bicquelet-Lock. 'Child Friendly Planning in the UK: A Review'. London: APiC/ZCD Architects/RTPI, 2019. https://www.rtpi.org.uk/practice/2019/november/child-friendly-planning-in-the-uk-a-review/.

Margaret Mead. 'Children, Culture, and Edith Cobb'. In *Children, Nature, and the Urban Environment: Proceedings of a Symposium-Fair: Proceedings of the Symposium-Fair Held 19-23 May, 1975 at the C. H. Marvin Center of the George Washington University, Washington, D.C*, 19–24. Upper Darby: U.S. Department of Agriculture Forest Servic, 1975.

Mayor of London. 'The London Plan: Intend to Publish (Clean Version). Spatial Development Strategy for Greater London'. London: Mayors Office, 2019. https://www.london.gov.uk/sites/default/files/intend_to_publish_-_clean.pdf.

Play Wales. 'Play Sufficiency in Wales', 2015. https://www.playwales.org.uk/login/uploaded/documents/INFORMATION%20SHEETS/Play%20sufficiency%20in%20Wales.pdf.

Playworkings. 'The Overlaid Life of Places of Play'. Accessed 16 September 2020. https://playworkings.wordpress.com/depository-of-notes/play-cartographies/the-overlaid-life-of-places-of-play-images-to-support-article/.

PPSG - Playwork Principles Steering Group. 'The Playwork Principles.' Cardiff, 2005.

Quentin Stevens. *The Ludic City: Exploring the Potential of Public Spaces.* Abingdon: Routledge, 2007.

Stuart Lester and Wendy Russell. 'Leopard Skin Wellies, a Top Hat and a Vacuum Cleaner Hose: An Analysis of Wales Play Sufficiency Assessment Duty (Executive Summary)'. Play Wales. Accessed 18 September 2020. https://www.playwales.org.uk/login/uploaded/documents/Play%20sufficiency/PSA%20research%20executive%20summary.pdf.

———. *Play For a Change. Play, Policy and Practice: A Review of Contemporary Perspectives*. London: Play England, 2008.

Unicef. 'Convention on the Rights of the Child Text', 2020. https://www.unicef.org/child-rights-convention/convention-text.

Chapter 9

'Like the Palm of my Hand'

Children and Public Space in Central Athens

Christos Varvantakis

INTRODUCTION: CHILDREN'S PARTICIPATION AND THE URBAN PUBLIC SPACE

Public space can be seen as an early and significant, if not uncontested, entry point of children's participation into the public sphere. Indeed, as a number of studies carried out in urban settings suggest, children consider outdoor places to be highly meaningful.[1] Despite the importance placed by children on the outdoors, their access to it in many urban areas is generally very limited, contested, and precarious.[2] Children's views and voices are frequently not heard in discussions over the planning of public space reflecting a wider absence from decision-making processes that affects their lives. Pugh has very poignantly highlighted the costs of ignoring childhood in the theorisation of social life.[3]

Children's participation in public life and decision-making has become a central theme of childhood studies literature in the last quarter of the century, following the UN Convention for the Rights of the Child (1989), in which Article 12 states that children have a right to be listened to and to be consulted on decisions that affect them. In childhood studies, there has been an early and lively response to this declaration, expressed by an interest to de-code, assess, and evaluate the proclamation made by Article 12 and its actual effect on policymaking and children's inclusion.[4]

Despite repeated assertions that children's capacities increase in contexts where they are active participants and despite many institutional attempts to include children's views in policymaking, few steps have been made towards achieving actual children's participation in public life and debate.[5] Much of the literature produced over the past twenty years spells out a multi-faceted critique to policymakers and to the UNCRC about the failure of fulfilment of their proclamations. According to these critiques, at best, attempts for participation and inclusion of children, end up in mere consultation, often with doubtful results.[6]

Scholars have identified a set of issues in attempts to include children in decision-making processes. For instance, according to Thomas, public debate still focuses more on children's 'needs' (however defined) than it does to their wants and wishes.[7] According to Cockburn, one of the main difficulties is that of 'the labels that are attached to children by adults'.[8] Begg criticises children's councils in Norway on the grounds that they are not designed and conducted on children's terms, but, '[i]nstead, children are praised when they behave like small adults and put in their place when they do not'.[9] Percy-Smith, in addressing children's participation in neighbourhood planning, points at a number of difficulties that emerge, such as the failure to reach a diversity of voices, to the 'tension between children having the responsibility for decision making and enjoying their childhood' and to the fact that 'children's voice often doesn't reflect the reality of their place experiences'. Instead, active participation, according to such critics, should move beyond 'adults allowing children to offer their perspectives', and instead should involve young people confronting adult authority and challenging adult assumptions about their competence to make decisions about issues that concern them.[10] According to Woodhead, if we are to develop fully the potential for children and young people to participate in society, we may need to focus more directly on the meaning of participation in everyday life and on how young people (can) practice active citizenship.[11]

Within the context of the Connectors Study, we consider that the conceptual framework for childhood and participation is highly relevant to children's relationship to public space and highlight that children and adults are continually involved in a process of negotiation in cities, from where to play to land use.[12] These negotiations reflect conflict between adults and young people over the use of public space, and it is a common feature of many neighbourhoods.[13] Public spaces are regarded as highly important by children themselves, places to meet and hang out with friends, to simply walk, to play, or engage in various activities.[14]

For adults though, the use of 'the street' by children and young people and their presence in non-specifically designated spaces may be seen as a threat and as a nuisance.[15] In an uncanny enactment of adult's fears, or their desire to provide the best for their children, children in urban contexts are often outlawed from public spaces and effectively corralled within institutions specially designated for them.[16] Such sets of restrictions and contexts of control, militate against children being regarded as autonomous citizens.[17] Nevertheless, our study makes evident that 'children's skills and competence in their use and understanding of public space have been underrated and that children's spatial activities often extend far beyond their parents' awareness'.[18] This indicates the 'potential mismatch' that Elsley notes, between adult ideas about children's relationship to public space and the children's actual experiences.[19]

According to this view, the collision between adults' and children's shared environmental perceptions is inevitable when respect is not given to children's perspectives. It is therefore important that children's and young people's experiences and views on public space are explored within the context of their agency.[20] An exemplary place to address this question is the district of Exarcheia in Athens.

THE EXARCHEIA DISTRICT OF CENTRAL ATHENS

Exarcheia is a district of central Athens with an estimated population of 20,000. As well as being an area of high population density, it is also an area of limited and highly contested public spaces. Furthermore, it has a reputation for being a very lively, rebellious, and semi-autonomous district.[21]

Following the uprising against the Dictatorship in 1973, which took place in local university buildings, the area has attracted leftists and anarchists, as well as several of the underground cultures of Athens. As well as bohemians, artists, and intellectuals, the area is a hangout for junkies – and, often for drug dealers too. Parliamentary and extra-parliamentary left-wing parties have their headquarters here and the district hosts several squatted buildings, leftist and anarchist publishing houses, cafes and bookstores, rock clubs, and other autonomous spaces. It was here in November 1985 that the murder of fifteen-year-old Michalis Kaltezas fired a large wave of protest. In December 2008, the murder of the fifteen-year-old Alexandros Grigoropoulos set the city on fire, giving way to widespread riots.[22]

The area has often been the site of violent clashes between police and protesters, but the district's radical political character and lively cultural life cannot be defined merely by these conflicts, and certainly not by the grotesque and exoticised representations perpetuated by the mainstream media.[23] The district has very powerful local residents' committees. There are several squats, co-operative cafes, a volunteer-run health centre, groups for auto-education, offering seminars and discussion sessions, a self-organised open-air cinema, a citizen-run weekly fair, regular open assemblies in the main square, and a citizen-run park, the self-managed, anti-hierarchical, anti-commercial Navarinou Park, which I shall discuss later.

The neighbourhood lends itself well to an enquiry into how children's perceptions of space which, at least in the eyes of adults, is contested, is affected by an intense activist presence which stresses community participation. Our interest in the Connectors Study is to investigate how children participate in public life, not just within institutional contexts but rather in their everyday interaction with their environments.[24] The focus on the district of Exarcheia provides a valuable opportunity for a discussion of how issues of children's participation and active citizenship are treated not only in official municipal urban planning but also in alternative, citizen-led initiatives. As such, it allows us to consider children's participation within wider processes of social change.[25] The Greek context, where in the course of the past few years strong grassroots movements have emerged as responses to the financial crisis, constitutes a very interesting setting to address such questions.[26]

I will present and discuss data collected in a session with Iason (a pseudonym), who was ten years old at the time of the research. In addition to an open interview and discussion, the session included a child-led walk in the district in which he took me to his favourite places, the making of a map, the use of photography and drawing.[27]

Iason grew up in this neighbourhood and exhibits a striking familiarity with it. He knows each street by heart and rarely thought twice about how to get around. He told me that sometimes he experiments and walks for a while with his eyes closed. I was particularly struck by how often he would use his hands to connect to his surroundings, both literally and in metaphors in his speech. He often touched and felt things with his fingers, including posters, trees and leaves, benches, and other surfaces. He repeatedly used metaphors, like 'I know this neighbourhood like the palm of my hand' or 'I play the streets on my fingers' – expressing an existential bodily connection to the environment he inhabits.[28]

Iason provides a suitable case for an in-depth exploration of the *connections* between children's participation, public space, and public life. I attempt to explore what Elsley understands as the mismatch between children's and adults/societal views on and uses of public space, and to bring forth a child's own understanding and ways of relating to his or her environment, which hints on children's geographies which are often invisible. The study allows us to address questions of actual children's citizenship and agency.[29]

IASON'S EXARCHEIA

Figure 9.1 depicts what may be seen as an official view of Exarcheia. It is a good starting point from which to explore the difference between the views of Iason and that of the municipality of the same area. For example, the main square of Exarcheia is considered to be a green space in the official map (figure 9.1) and other documents and shown as a small park. Indeed, intense efforts have been made in the past by the municipality to redesign the square

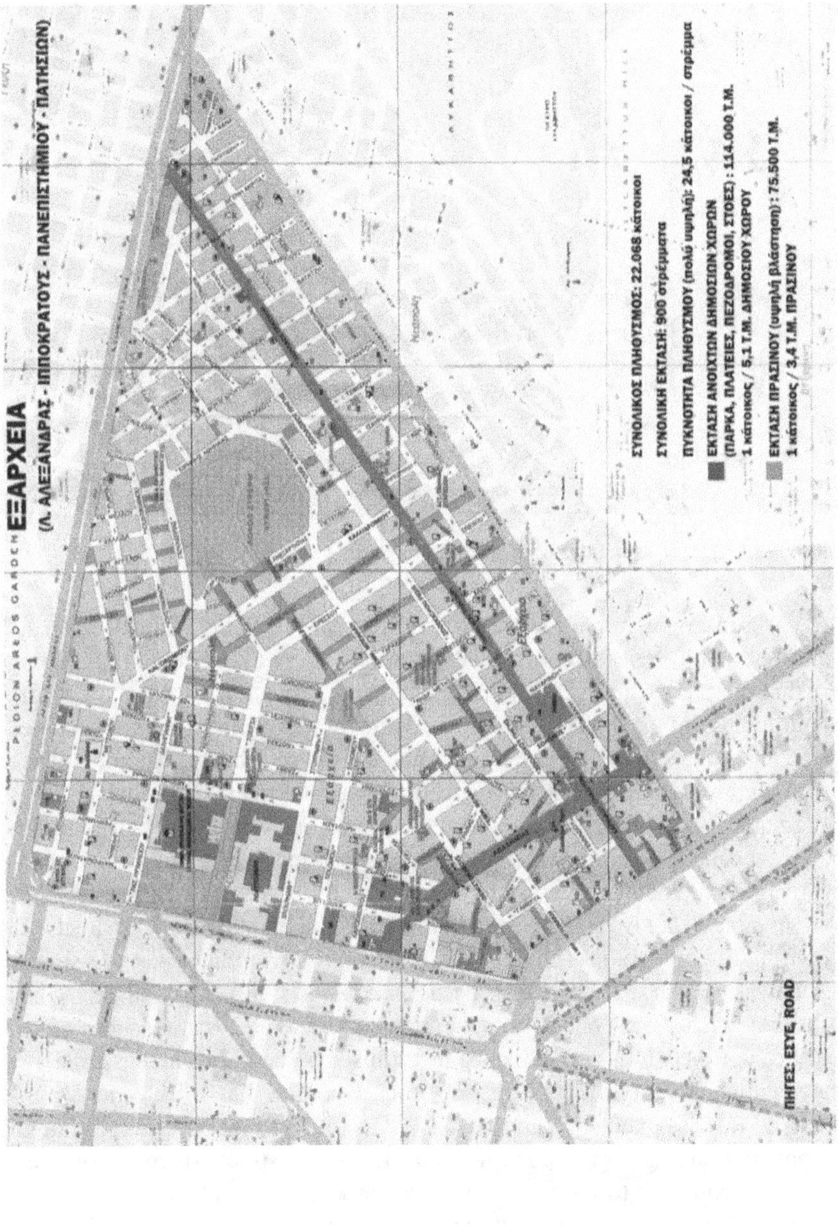

Figure 9.1 Map of Exarcheia (Exarcheia Citizen Initiative). Author's Image.

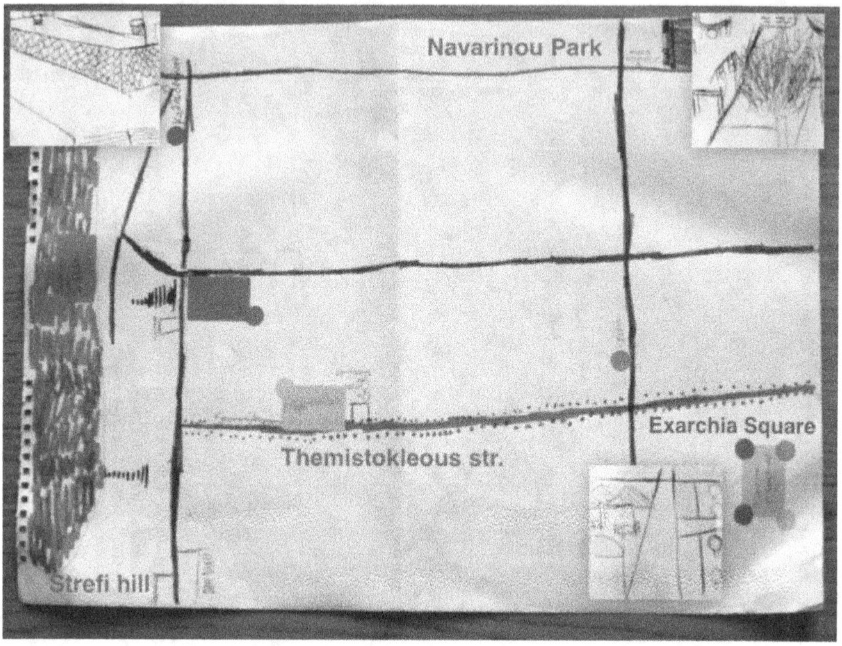

Figure 9.2 Map of Exarcheia Drawn by Iason. Author's Image.

as a park. In his own map, Iason preferred to paint it yellow (figure 9.2) and then to use a blue sheet to draw a detail of it. When I asked about this, telling him that there are some trees in the square, he told me that they are 'not worth mentioning'. He used different coloured stickers to signify the multi-faceted character of the square. He remarked that 'it's a lot of things, but not a park'.

By contrast, the park that does appear on Iason's map, and of which he is particularly fond, is in Navarinou street. As previously mentioned, it is a self-organised park. It is worth remarking here that the spot where the park stands is regarded by the municipality as an illegally occupied space. I will return to this point in more detail later. Furthermore, in the official map (figure 9.1), streets are shown in different colour (originally purple, in this figure light grey) to signify that they are public space, that is, pedestrianised streets. In sharp contrast, Iason's map acknowledges just one street (Themistokleous) as a pedestrian street.

When I asked him about the pedestrianised streets, Iason reassured me that Themistokleous Street is the only one. I pointed out some other streets that are clearly paved for pedestrians. He counter suggested that, although not paved with asphalt these are not *real* pedestrian streets since motorbikes, even cars, continue to pass by as well as park there. In reality, he told me, Themistokleous is the only pedestrianised street. Themistokleous Street is on a hill and has many stairs, which make it practically impossible for cars to cross. Iason's perspective challenges both official designations of pedestrianisation as well as the materials used for pedestrianisation (the paved street). His own view remains truthful to the phenomenological reality he experiences – if cars cross a street, it is not a pedestrian street. But apart from the contradictions and mismatches that emerge overviews on pedestrianised streets, we ought to look also at Iason's appreciation of the one pedestrian street he recognises. He told me several times that he likes this street very much. In his own map, he chose a different colour to signify that it was different, a special road – a pedestrian one. At first, he chose the red marker. But when he started drawing the road, he suddenly stopped and told me that he should use another colour, not red. I asked him why and he answered that 'red is kind of forbidding. Like in the traffic lights, or in the stop signs. And a pedestrian street is anything but [forbidding] . . .'. He finally chose orange, just because it is his favourite colour. Indeed, during the tour of the neighbourhood, which we did with Iason, we hung out a lot on Themistokleous Street. We talked a lot about the street's graffiti and posters, and Iason likes most of the graffiti there. He likes some parts of the street a lot, he just wishes that it was taken better care of. I asked him what he meant, and he brought up the example of the leaves falling from the trees. He thinks that it would be much better if someone would brush away the leaves every once in a while. When I asked him, who should be doing that, he told me that the neighbours themselves should. I asked about the municipal cleaners, and he said that yes, perhaps them too, but the neighbours can do it by themselves.

At some point, Iason showed me an empty house that he and his friends had once broken into. He told me that he and his friends sometimes enter empty houses. He is very good in picking locks. He told me it is his 'special ability'. I asked more about entering empty houses, and he told me that he considers it to be real fun. It is one of his favourite games, because of all the exploration and the mystery. I enquired about permissions to do so. He responded that he knows that he is not supposed to, but doesn't consider it to be a problem actually, because these houses are empty. In his view, they don't disturb anyone.[30]

During our tour, Iason wandered around in an idiosyncratic manner. For instance, as we were walking on the pavement, he would climb the first steps of a building's entrance, or walk over a short wall, playfully. He told me that he always walks like that, that he likes to meander in unusual ways. He would walk around an entire square, just to avoid going 'through the same road again'. I encountered a characteristic example of his alternative ways of walking and being on the street, at Strefi Hill, a green hill near his home. On the edge of the hill, there is an open-air basketball court. He told me that, quite often, when there's no training taking place, he goes there with his friends to play. However, they play *kinigito* (a chase game), or football, rather than basketball. He told me that the court is usually locked when there is no training, but there is an opening on one side of the fence, so they get in through this opening. He took me there, going in and out a couple of times to demonstrate the alternative access to the court. As we walked further, I noticed that actually the door of the court was open, not locked. I didn't say anything.

A SELF-ORGANISED PARK AND THE QUESTION OF ACTIVE CITIZENSHIP

A place that Iason really likes in his neighbourhood is the park on Navarinou street. It was the first place he brought me to, when I asked him about his favourite places. Since the beginning of the twentieth century, in the spot where the park is today, there used to be a clinic, which closed down in the 1970s, when the Technical Chamber of Greece bought the property and later demolished the building.[31] Despite promises made by successive governments to build a park on that spot, the space was rented semi-illegally to an individual who paved it and used it as an open-air car park. In the aftermath of the December 2008 riots, as strong and determined activist and solidarity groups emerged in the city, the open-air car park was occupied in the spring of 2009 by the Exarcheia Residents' Initiative, a group determined to take action and create a park there.[32] The initiative initially invited residents to attend open assemblies where they discussed the actions to be taken. They also defended the space against the riot police who tried to drive them away, on occasions using tear gas. Eventually, resident-led work was undertaken to turn the space into a green park and playground. Today concerts, workshops, theatre performances, film screenings, children's parties, and other activities take place there. Local residents do all the watering and gardening and hold open meetings to plan and discuss issues, activities, and the cultivation of the park.

The park is an on-going, unfinished project, that changes day by day. It was made by the people, as Iason explained to me when we got there. He went on to tell me how the people took over the place and created the park, how they built the benches and planted the trees, how they were working to create the playground, and how they have to work hard to maintain and expand it. He had many stories to tell, stories from when things were in the making, stories about how things have changed, and so on. He talked a lot about gardening, and after he had finished, explaining the fine details of caring for the plants, we played for a while in the playground and then rested in the shadow of some huge trees by the edge of the park. I remarked to Iason that those trees, the big ones, must have existed there before the creation of the park, before the new trees were planted on the residents' initiative. 'Yes,' he agreed, 'but the new ones will also grow'.

I asked him why he liked the park so much. He told me that he liked it because he went there sometimes to play. He also celebrated his birthday there, as did many of his friends and other children in the area. He told me that he likes gardening there himself – but only sometimes, not always. Sometimes he enjoyed participating in other jobs that need to be done in the park. He particularly emphasised the mosaic on the surface of the benches, something he particularly liked. He explained how this was made by breaking and re-using old tiles. He found the process fascinating, and he remarked that someday all the benches would be decorated in that way (there were still many parts of the benches standing bare at the time). I asked him whether he has made some parts of the mosaic himself. He turned a little shy, and told me that 'he tried, but the result was not good'. He repeated how beautiful he thought it was, running his hand over the surface of the tiles.

Several activities for children take place in the park and a local parents' association has embraced the attempt and jointly with the Exarcheia Residents' Initiative organises activities including parties. The extent to which children have participated in the process of the making, designing, and decision-making regarding Navarinou Park is nevertheless unclear. In order to assess and appreciate children's involvement, we need to keep in mind that the whole project is a continuing grassroots attempt to design public space. Iason's participation is limited and informal – it occurs playfully and at his own terms. His relation to gardening evokes a point made by Percy-Smith about the tension between children having the responsibility for decision-making and enjoying their childhood. Additionally, Iason's relation to the making of the mosaic benches resonates with Elsley's comment that participation should occur within the domain of children's agency.[33]

What was different, however, and I think this is the main reason why the park is so appealing to Iason, as well as to the other children I talked to in the neighbourhood, is the fact that they were and are present in the making of the park. Iason's preference and indeed attachment to this park-cum-playground – as opposed to other parks and playgrounds in the area – maybe understood on the grounds of it not having been remotely designed, built behind construction site signs, and later just unveiled to the citizens. Rather, it has been a continuous, work-in-progress run by neighbours. The open process by which the park was made gave Iason and the other children the opportunity to be around in their free time, to connect to the park's making, to come and go, playfully and as desired by them, in accordance with their moods and wishes. Such flows of movement, while falling short of more formal definitions of children's participation and inclusion, are finely tuned with the children's overall relation to their environment.

It was the deinstitutionalised character of the park, the open and improvised ways in which it was created, its anti-hierarchical and organic character, which made the space so appealing for a child to hang around in. It might be what enables a child's playful inclusion and interaction, on his or her own terms, different and less formal compared to those of formal policymakers and institutions. In this sense, and via such ways of participation, the park affords Iason multiple connections to public space and public life. Such encounters may be a fitting point of departure for rethinking and enquiring into active citizenship in childhood instead of merely the institutionally initiated and often artificial invitations to voice opinions.[34]

ACKNOWLEDGEMENTS

I wish to thank Sevasti-Melissa Nolas who carefully read and commented on different versions of this text, as well as walked and discussed Exarcheia with me time and again over the past few years. I would also like to thank Vinnarasan Arudlos, Thalia Dragona, Nelli Askouni, and Pafsanias Karathanasis for their various contributions to the research for this chapter.

Research for this chapter was conducted as part of the ERC funded Connectors Study (ERC-StG-335514, to Sevasti-Melissa Nolas), an international and longitudinal ethnographic research exploring the relationship between childhood and public life. From 2014, we have worked with forty-five children aged six to nine and their families in Athens, Hyderabad, and London. More information about the study can be found at the study's website https://childhoodpublics.org/projects/connectors/

A previous draft of this chapter was presented at the conference: 'Places for Learning Experiences: Think, Make, Change' Aristotle University of Thessaloniki, Greece, and published in the conference proceedings which were edited by Marianthi Liapi and Dimitris Germanos.

NOTES

1. Elsley, S. 'Children's Experience of Public Space', *Children & Society* 18 (2004), pp. 155–164; Horton, J., Christensen, P., Kraftl, P. and Hadfield-Hill, S. '"Walking … just walking": How Children and Young People's Everyday Pedestrian Practices Matter', *Social & Cultural Geography*, 15(1) (2014), pp. 94–115; Morrow, V. *Networks and Neighbourhoods: Children's and Young People's Perspectives* (2001) (Health Development Agency, London); Percy-Smith, B. 'Contested Worlds: Constraints and Opportunities in City and Suburban Environments in an English Midlands Town', in Chawla, Louise, ed. *Growing up in an Urbanizing World* (2001) (London: Earthscan); Ward, C. *The Child in the City* (1978) (London: Pantheon Books).

2. Cahill, S. 'Childhood and Public Life: Reaffirming Biographical Divisions', *Social Problems* 37 (1990), pp. 390–404; Elsley 2004; Percy-Smith, B. 'Contested Worlds: Constraints and Opportunities in City and Suburban Environments in an

English Midlands Town', in Chawla, Louise, ed. *Growing up in an Urbanizing World* (2002) (London: Earthscan); Valentine, G. *Public Space and the Culture of* Childhood (2004) (Aldershot: Ashgate Publishing).

3. Pugh, A. J. 'The Theoretical Costs of Ignoring Childhood: Rethinking Independence, Insecurity, and Inequality', *Theory and Society*, 43 (2013), pp. 71–89.

4. Hart, R. *Children's Experience of Place* (1997) (Irvington Pub Inc, New York); Hart, R. *Children's Participation: From Tokenism to Citizenship* (1992) (UNICEF Earthscan, London); Cockburn, T. 'Partners in Power: A Radically Pluralistic Form of Participative Democracy for Children and Young People', *Children and Society* 21 (2007), pp. 446–457; Nolas, S-M. 'Children's Participation, Childhood Publics and Social Change: A Review', *Children and Society* 29(2) (2015), pp. 157–167; Percy-Smith, 2002; Percy-Smith, B. 2006. 'From Consultation to Social Learning in Community Participation with Young People', *Children, Youth and Environments*, 16(2) (2006), pp. 153–179; Thomas, N. 'Towards a Theory of Children's Participation', *The International Journal of Children's Rights*, 15(2) (2007), pp. 199–218.

5. Barrett, M. and Buchanan-Barrow, E. *Children's Understanding of Society* (2005) (Psychology Press, Hove); Lansdown 2005, Thomas, N. 2007; Kirby, P. and Bryson, S. 2002. *Measuring the Magic? Evaluating and Researching Young People's Participation in Public Decision Making* (2002) (CYPI, London).

6. Percy-Smith, B. and Thomas, N. (eds) *Handbook of Children's Participation* (2010) (London: Routledge).

7. Thomas, N. 2005. 'Interpreting Children's Needs: Contested Assumptions in the Provision of Welfare', in Goddard, J., McNamee, S., James, A. and James, A. (eds) *The Politics of Childhood: International Perspectives, Contemporary Developments* (2007) (Basingstoke: Palgrave Macmillan).

8. Cockburn, T. 'Children's Participation in Social Policy: Inclusion, Chimera or Authenticity?', *Social Policy and Society*, 4(2) (2015), pp. 109–119.

9. Begg, I., 'Participation Rights in Norway', in Crimmens, D. and West, A. (eds) *Having Their Say: Young People and Participation: European Experiences* (2004) (Russell House, Lyme), p. 131.

10. Percy-Smith 2006; Thomas 2007, p. 202.

11. Woodhead, M., Foreword, in Percy-Smith, B. and N. Thomas (eds) (2010) pp. xix–xxii.

12. Elsley 2004; Percy-Smith 2006; Christensen P. and O'Brien M (eds) *Children in the City: Home, Neighbourhood and Community* (2003) (Routledge Falmer: London).

13. This interaction, and the possible conflicts that it entails, are probably better understood within their material dimensions too. Christensen and O'Brien comment that 'living in the city is as much about negotiating relationships with other humans as it is about living in material places and spaces'. Similarly, Askins, K. and Pain, R. 'Contact Zones: Participation, Materiality and the Messiness of Interaction', *Environment and Planning D: Society and Space,* 29(5) (2011), pp. 803–821, call for attention to the materiality of social interaction and the physical nature of encounters

in fostering or foreclosing interaction. See also Cahill 1990; Elsley 2004; Percy-Smith 2002; Valentine 2004, Ward 1979.

14. Elsley 2004; Horton et al. 2014; Morrow, V. 2001.

15. Percy-Smith 2002, 2006; Valentine 2004.

16. O'Brien, M., Jones, D., Sloan, D. and Rustin, M., 'Children's Independent Spatial Mobility in the Urban Public Realm', in *Childhood: Special Issue: Spaces of Childhood*, 7(3) (2000), pp. 255–277; Ennew, J. 'Time for Children or Time for Adults?', in Qvortrup, J., Bardy, M., Sgritta, G., and Wintersberger, H. (eds) *Childhood Matters: Social Theory, Practice and Politics* (1994) (Aldershot Press, Avebury), p. 127.

17. Valentine, G., 'Angels and Devils: Moral Landscapes of Childhood', *Environment and Planning D: Society and Space*, 14 (1996), pp. 581–599.

18. Elsley 2004: 156. See also Hart, R. *Children's Experience of Place* (1979) (Irvington Pub Inc, New York); Matthews, H. *Making Sense of Place: Children's Understanding of Large Scale Environments* (1992) (Hemel Hempstead, Harvester); Ward 1979.

19. Elsley 2004, p, 156.

20. Elsley 2004.

21. I have discussed the neighbourhood of Exarcheia in an entry in the Connectors Study blog, and most of the information provided in this subchapter originate from it, as do the particular information on the Navarinou Park in the following section. Varvantakis, C. 'A Very Particular Kind of Park', *The Connectors Study Blog*, 2014, https://connectorsstudy.wordpress.com/2014/09/11/a-very-particular-kind-of-park/ (Downloaded 26 October 2020).

22. Dalakoglou, D. and Vradis, A. (eds) *Revolt and Crisis in Greece* (2011) (AK Press, Oakland, CA); Karamichas, J. 'The December 2008 Riots in Greece', *Social Movement Studies*, 8(3) (2009), pp. 289–293; Mentinis, M. 'Remember Remember the 6th of December . . . A Rebellion or the Constituting Moment of a Radical Morphoma?', *International Journal of Urban and Regional Research*, 34(1) (2010), pp. 197–202; Sotiris, P. 'Rebels with a Cause: The December 2008 Greek Youth Movement as the Condensation of Deeper Social and Political Contradictions', *International Journal of Urban and Regional Research*, 34(1) (2010), pp. 203–209.

23. It is widely held in Greece that the district is verging on being an independent zone, a no-go area, where the police and the municipality have no real access, and where violence is an everyday phenomenon. To an extent this might be true, and to an extent this may also have happened intentionally by media and conservative political discourse, in order to create an 'inverted state of exception', that would serve the purpose of keeping all the radical elements of the society in one place. For a discussion see: Vradis, A. 'Terminating the Spatial Contract', Libcom.org, October 2012. https://libcom.org/library/terminating-spatial-contract-commentary-greece http://societyandspace.com/material/commentaries/terminating-the-spatial-contract-by-antonis-vradis/ (Downloaded 26 October 2020).

24. Nolas, S.-M., Varvantakis, C. and Aruldoss, V. 'Talking Politics in Everyday Family Life', *Contemporary Social Science*, 12(1) (2017), pp. 68–83; See also Clark,

A. and Percy-Smith, B. 'Beyond Consultation: Participatory Practices in Everyday Spaces', *Children, Youth and Environments* 16(2), (2006), pp. 1–9.

25. Nolas 2015.

26. Varvantakis, C. *Research encounters with activism and solidarity movements in Athens.* 2014, Connectors study blog: https://connectorsstudy.wordpress.com/2014/12/08/research-encounters-with-activism-and-solidarity-in-athens/ (Downloaded 26 October 2020); Varvantakis, C., Dragonas, T., Askouni, N. and Nolas, S-M. 'Grounding Childhood (Trans)national Identities in Everyday Life', *Children & Society,* 33(1) (2018): 68–81, DOI: 10.1111/chso.12299. *See also*: Kaika, M. 'The Economic Crisis Seen from the Everyday', *City: Analysis of Urban Trends, Culture, Theory, Policy, Action,* (August 2013), 37–41; Sotirakopoulos, N. and Sotiropoulos, G. 2013 '"Direct democracy now!": The Greek Indignados and the Present Cycle of Struggles', *Current Sociology,* 61(4) (2013), pp. 443–456.

27. For an extended discussion of the study's multimodal methodological approach, see: Varvantakis, C. and Nolas, S-M. 'Metaphors We Experiment with in Multimodal Ethnography', *International Journal of Social Research Methodologies,* 22(4) (2019): 365–378, DOI: 10.1080/13645579.2019.1574953; Varvantakis, C., Nolas, S-M. & Aruldoss, V. 'Photography, Politics and Childhood: Exploring Children's Multimodal Relations with the Public Sphere', *Visual Studies* (2019), DOI: 10.1080/1472586X.2019.1691049.

28. Gebauer, G. and Wulf, C., *Mimesis: Culture, Art, Society* (1995) (University of California Press: Berkeley); Taussig, M. *Mimesis and Alterity: A Particular History of the Senses* (1991) (New York: Routledge); Varvantakis, C. 'Mourning Deaths, Lamenting Lives: Grief and Transformation in Maniat Laments', *Paragrana,* 20(2) (2011), pp. 140–154.

29. Cockburn, T. *Rethinking Children's Citizenship* (2013) (Palgrave Macmillan, London); Woodhead 2010).

30. Nevertheless, Iason mentioned one incident when they were trying to force open the window of a house that they thought empty and there was someone living inside after all, which led to a lot of trouble.

31. Most of the data presented in this section, about the history and present of Navarinou park, are collected throughout my field research there. I have been contrasting my own findings to the park's blog, run by the Exarcheia Residents Initiative http://parkingparko.espivblogs.net/englishfrench/about-the-park/. For further resources about the Navarinou Park see Ismailidou, E. 'Ναυαρίνου: Το Πάρκο-Πάρκινγκ έκλεισε δύο χρόνια ζωής', *To Vima* (Newspaper) 5 April 2011. http://www.tovima.gr/society/article/?aid=393785 (Downloaded 26/10/2020); Varvantakis 2014.

32. Petropoulou, C. 'From the December Youth Uprising to the Rebirth of Urban Social Movements: A Space-Time Approach', *International Journal of Urban and Regional Research,* 34(1) (2010), pp. 217–224; Rakopoulos, T. 'Resonance of Solidarity: Meanings of a Local Concept in Anti-austerity Greece', *Journal of Modern Greek Studies,* 32(2) (2014), pp. 313–337.

33. Elsey 2004).

34. Nolas, S-M., Aruldoss, V. and Varvantakis, C. 'Learning to Listen: Introducing and Exploring the Idioms of Childhood Publics', *Sociological Review Online,* 24(3)

(2018): 394–413; DOI: 10.1177/1360780418811972. See also: Woodhead 2010, p. xxii; Cockburn 2013; James, A. 'To Be (Come) or Not to Be (Come): Understanding Children's Citizenship', *The Annals of the American Academy of Political and Social Science*, 633(1) (2010), pp. 167–179; Nolas 2015; Woodhead 2010.

BIBLIOGRAPHY

Askins, K. and Pain, R. 'Contact Zones: Participation, Materiality and the Messiness of Interaction', *Environment and Planning D: Society and Space,* 29(5) (2011), pp. 803–821.

Begg, I. 'Participation Rights in Norway', in Crimmens, D. and West, A. (eds) *Having Their Say: Young People and Participation: European Experiences* (2004) (Russell House, Lyme).

Cahill, S. 'Childhood and Public Life: Reaffirming Biographical Divisions', *Social Problems* 37 (1990), pp. 390–404.

Christensen, P. and O'Brien, M. (eds) *Children in the City: Home, Neighbourhood and Community* (2003) (Routledge Falmer: London).

Clark, A. and Percy-Smith, B. 'Beyond Consultation: Participatory Practices in Everyday Spaces', *Children, Youth and Environments* 16(2) (2006), pp. 1–9.

Cockburn, T. *Rethinking Children's Citizenship* (2013) (Palgrave Macmillan, London).

Cockburn, T. 'Partners in Power: A Radically Pluralistic Form of Participative Democracy for Children and Young People', *Children and Society* 21 (2007), pp. 446–457.

Cockburn, T. 'Children's Participation in Social Policy: Inclusion, Chimera or Authenticity?', *Social Policy and Society*, 4(2) (2015), pp. 109–119.

Dalakoglou, D. and Vradis, A. (eds) *Revolt and Crisis in Greece* (2011) (AK Press, Oakland, CA); Karamichas, J. 'The December 2008 Riots in Greece', *Social Movement Studies*, 8(3) (2009), pp. 289–293.

Elsley 2004: 156. See also Hart, R. *Children's Experience of Place* (1979) (Irvington Pub Inc, New York).

Elsley, S. 'Children's Experience of Public Space', *Children & Society* 18 (2004), pp. 155–164.

Ennew, J. 'Time for Children or Time for Adults?', in Qvortrup, J., Bardy, M., Sgritta, G., and Wintersberger, H. (eds) *Childhood Matters: Social Theory, Practice and Politics* (1994) (Aldershot Press, Avebury), p. 127.

Gebauer, G. and Wulf, C. *Mimesis: Culture, Art, Society* (1995) (University of California Press: Berkeley).

Hart, R. *Children's Experience of Place* (1997) (Irvington Pub Inc, New York).

Hart, R. *Children's Participation: From Tokenism to Citizenship* (1992) (UNICEF Earthscan, London).

Horton, J., Christensen, P., Kraftl, P. and Hadfield-Hill, S. "Walking ... just walking": How Children and Young People's Everyday Pedestrian Practices Matter', *Social & Cultural Geography*, 15(1) (2014), pp. 94–115.

Ismailidou, E., "Ναυαρίνου: Το Πάρκο-Πάρκινγκ έκλεισε δύο χρόνια ζωής", *To Vima* (Newspaper) 5 April 2011. http://www.tovima.gr/society/article/?aid=393785 (Downloaded 26/10/2020).

James, A., "To Be (Come) or Not to Be (Come): Understanding Children's Citizenship", *The Annals of the American Academy of Political and Social Science*, 633(1) (2010), pp. 167–179.

Kaika, M. 'The Economic Crisis Seen from the Everyday', *City: Analysis of Urban Trends, Culture, Theory, Policy, Action*, (August 2013), pp. 37–41.

Lansdown 2005, Thomas, N. 2007; Kirby, P. and Bryson, S. 2002. *Measuring the Magic? Evaluating and Researching Young People's Participation in Public Decision Making* (2002) (CYPI, London).

Barrett, M. and E. Buchanan-Barrow *Children's Understanding of Society* (2005) (Psychology Press, Hove).

Matthews, H. *Making Sense of Place: Children's Understanding of Large Scale Environments* (1992) (Hemel Hempstead, Harvester).

Mentinis, M., "Remember Remember the 6th of December . . . A Rebellion or the Constituting Moment of a Radical Morphoma?" *International Journal of Urban and Regional Research*, 34(1) (2010), pp. 197–202.

Morrow, V. *Networks and Neighbourhoods: Children's and Young People's Perspectives* (2001) (Health Development Agency, London).

Nolas, S-M. 'Children's Participation, Childhood Publics and Social Change: A Review', *Children and Society* 29(2) (2015), pp. 157–167.

Nolas, S-M., Aruldoss, V. and Varvantakis, C. 'Learning to Listen: Introducing and Exploring the Idioms of Childhood Publics', *Sociological Review Online*, 24(3) (2018), pp. 394–413; DOI: 10.1177/1360780418811972.

Nolas, S.-M., Varvantakis, C. and Aruldoss, V. 'Talking Politics in Everyday Family Life', *Contemporary Social Science,* 12(1) (2017), pp. 68–83.

O'Brien, M., Jones, D., Sloan, D. and Rustin, M., 'Children's Independent Spatial Mobility in the Urban Public Realm', in *Childhood: Special Issue: Spaces of Childhood*, 7(3) (2000), pp. 255–277.

Percy-Smith, B. 'Contested Worlds: Constraints and Opportunities in City and Suburban Environments in an English Midlands Town', in Chawla, Louise, ed. *Growing up in an Urbanizing World* (2002) (London: Earthscan).

Percy-Smith, B. 'Contested Worlds: Constraints and Opportunities in City and Suburban Environments in an English Midlands Town', in Chawla, Louise, ed. *Growing Up in an Urbanizing World* (2001) (London: Earthscan).

Percy-Smith, B. 2006. 'From Consultation to Social Learning in Community Participation with Young People', *Children, Youth and Environments*, 16(2) (2006), pp. 153–179.

Percy-Smith, B. and Thomas, N. (eds) *Handbook of Children's Participation* (2010) (London: Routledge).

Petropoulou, C., 'From the December Youth Uprising to the Rebirth of Urban Social Movements: A Space-Time Approach', *International Journal of Urban and Regional Research*, 34(1), (2010), pp. 217–224.

Pugh, A. J. 'The Theoretical Costs of Ignoring Childhood: Rethinking Independence, Insecurity, and Inequality', *Theory and Society*, 43 (2013), pp. 71–89.

Rakopoulos, T. 'Resonance of Solidarity: Meanings of a Local Concept in Anti-austerity Greece', *Journal of Modern Greek Studies*, 32(2) (2014), pp. 313–337.

Sotirakopoulos, N. and Sotiropoulos, G. 2013 '"Direct democracy now!": The Greek Indignados and the Present Cycle of Struggles', *Current Sociology*, 61(4) (2013), pp. 443–456.

Sotiris, P. "Rebels with a Cause: The December 2008 Greek Youth Movement as the Condensation of Deeper Social and Political Contradictions", *International Journal of Urban and Regional Research*, 34(1) (2010), pp. 203–209.

Taussig, M. *Mimesis and Alterity: A Particular History of the Senses* (1991) (New York: Routledge).

Thomas, N. 'Towards a Theory of Children's Participation', *The International Journal of Children's Rights*, 15(2) (2007), pp. 199–218.

Thomas, N. 2005. 'Interpreting Children's Needs: Contested Assumptions in the Provision of Welfare', in Goddard, J., McNamee, S., James, A. and James, A. (eds) *The Politics of Childhood: International Perspectives, Contemporary Developments* (2007) (Basingstoke: Palgrave Macmillan).

Valentine, G. 'Angels and Devils: Moral Landscapes of Childhood', *Environment and Planning D: Society and Space*, 14 (1996), pp. 581–599.

Valentine, G. *Public Space and the Culture of Childhood* (2004) (Aldershot: Ashgate Publishing).

Varvantakis, C. *Research encounters with activism and solidarity movements in Athens*. 2014. Connectors study blog: https://connectorsstudy.wordpress.com/2014/12/08/research-encounters-with-activism-and-solidarity-in-athens/ (Downloaded 26 October 2020).

Varvantakis, C. and Nolas, S-M. 'Metaphors We Experiment with in Multimodal Ethnography', *International Journal of Social Research Methodologies* 22(4) (2019), pp. 365–378, DOI: 10.1080/13645579.2019.1574953.

Varvantakis, C., 'Mourning Deaths, Lamenting Lives: Grief and Transformation in Maniat Laments', *Paragrana*, 20(2) (2011), pp. 140–154.

Varvantakis, C. 2014. 'A Very Particular Kind of Park', *The Connectors Study Blog* https://connectorsstudy.wordpress.com/2014/09/11/a-very-particular-kind-of-park/ (Downloaded 26 October 2020).

Varvantakis, C., Dragonas, T., Askouni, N. and Nolas, S-M. 'Grounding Childhood (Trans)national Identities in Everyday Life', *Children & Society,* 33(1) (2018), pp. 68–81, DOI: 10.1111/chso.12299.

Varvantakis, C., Nolas, S-M. & Aruldoss, V. 'Photography, Politics and Childhood: Exploring Children's Multimodal Relations with the Public Sphere', *Visual Studies* (2019), DOI: 10.1080/1472586X.2019.1691049.

Vradis, A. 'Terminating the Spatial Contract', Libcom.org, October 2012. https://libcom.org/library/terminating-spatial-contract-commentary-greece http://societyandspace.com/material/commentaries/terminating-the-spatial-contract-by-antonis-vradis/ (Downloaded 26 October 2020).

Ward, C. *The Child in the City* (1978) (London: Pantheon Books).

Woodhead, M., Foreword, in Percy-Smith, B. and N. Thomas (eds) *Handbook of Children's Participation* (2010) (London: Routledge).

Part 3

WATCH THIS SPACE

Chapter 10

Just Mapping for Civic Action

Inclusive Neighbourhood Planning in the Elephant and Walworth

Barbara Brayshay and Nicolas Fonty

> Maps of urban (public) assets uncover untapped resources performing the existence and availability of that resources, and, therefore, a new field of possibility that enables new forms of collective action. In that sense, the map creates a new reality rather than describes an existing one.
>
> Adrien Labaeye, 2018

In this chapter, we discuss the findings from a community-based mapping project in which we explore the role of participatory mapping as a tool for facilitating local community engagement and civic action in an urban planning and redevelopment context.

The chapter is divided into four sections, the first introduces JustMap and the project setting in the context of communities experiencing regeneration in London. This is followed by an overview of the theoretical and methodological frame underpinning our approach to participatory collaborative mapping. We then illustrate this with a case study, a project commissioned by the Elephant and Walworth Community Forum, in Southwark, south London to assist with their development of a Neighbourhood Plan for the area. We conclude with a summary of the challenges and rewards of participatory mapping as a tool for activists and groups campaigning for a say in the (re) development of their neighbourhoods.

JUST MAPPING AND REGENERATION IN LONDON

JustMap is an on-going counter-mapping cartography project dedicated to the collaborative mapping of community resources, campaigns, and projects in

London. Our methodology is based on public workshops organised at community events or festivals with the aim of collecting the grassroots voices of those who would traditionally be ignored in their campaigning for a fairer city. We use the map as an 'agent provocateur' opening up new spaces of representation, enabling citizens to document and discover their resources (both tangible and less tangible), to create conversations, share knowledge, and build cooperation and alliances, highlighting community assets and networks and connecting actors campaigning for change. Mapping is our activism.

Much of our work has been facilitated by collaboration with Just Space,[1] an informal alliance of community groups, campaigns, and concerned independent organisations that have come together to act as a voice for Londoners affected by the impacts of housing regeneration in their neighbourhoods. Just Space provides advice and support for communities, improving grassroots public participation in the planning process and ensuring that their interests are heard in a system increasingly dominated by the interests of private developers, aims that resonate well with JustMap's commitment to coproduction and citizen empowerment.

Estate communities across London have undergone decades of significant upheaval and change that began with the 'right to buy' policy of Margaret Thatcher introduced in the Housing Act 1980 and have been on-going in various guises since.[2] In more recent times, the impact of austerity and a

Figure 10.1 JustMap Community Consultation Leaflet. Author's Image.

commitment to deconstruct social housing and the perceived social ills associated with it has led to a model of estate regeneration and renewal that has resulted in the demolition of many aging estates and the dispersal of their associated traditional London communities.[3] Local authorities have increasingly moved the renovation of estates out of the public realm into the private sector, enabling developers to demolish existing social housing and replace it with mixed tenure housing. The outcome has been a significant loss of social housing in the city and a corresponding increase in private ownership and properties let at 'affordable rents', defined as 80 per cent of the local market rent.[4] Recent research by Lees and Hubbard demonstrates the scale of change.[5] They identified 161 estates in London of more than 100 households where there have been schemes to demolish and redevelop housing since 1997. Their findings estimated that 190 regeneration schemes on the 161 estates resulted in the demolition of at least 55,000 homes, with the average scheme involving the decanting of 274 households prior to demolition and rebuild.

In a weeklong residency in 2017, at the Institute for Contemporary Arts in London, JustMap worked with Architects for Social Housing (ASH), to create a large-scale map installation showing the extent of the estate demolition and rebuilding/renewal programme across London boroughs.[6] For the purposes of the mapping, the definition of 'regeneration' was restricted to privatisation in the form of a stock transfer to a housing association, refurbishment (usually with the prior decanting of residents) and/or either partial or full demolition; when all three forms of regeneration result in the loss of homes for social rent and the corresponding relocation of all or part of the existing estate community. The resulting map showed that the policy of estate regeneration involving demolition and the sale of publicly owned land to the private sector was being actively pursued in all London boroughs irrespective of the political party allegiance of the Local Authority. All Labour controlled councils, a majority in the city, were found to be adopting this model of renewal.

Our study area in Southwark is no exception to this process, it has seen the demolition of the Heygate and Aylesbury Estates and the accompanying displacement of the majority of social housing tenants, who were unable to return after the redevelopment of their estate.[7] With virtually no social housing on offer in the new developments and the expensive 'affordable' rental properties beyond their reach, the majority of social housing tenants have been forced to relocate. Leaseholders have fared little better, similarly unable to purchase the expensive new built properties with the low-value compulsory purchase prices they received for their old homes, resulting in an exodus out of London to more affordable locations. The corresponding impact of the breakup of family support networks and the loss of traditional communities have been documented elsewhere.[8]

An investigation of the London housing market by Transparency International (2017) found that a majority of 'off-plan' sales of the new estate homes were to overseas investors. Demolition of the Heygate estate's 1,212 council homes began in 2013, the replacement development, Elephant Park (along with Trafalgar Place) will have just 100 social rented homes out of a total of 2,924 new homes. When the estate was fully decanted in preparation for redevelopment at the beginning of 2008, existing social housing tenants were promised the 'right to return' but the lack of social housing provision in the new development made this impossible for the majority. Jerry Flynn, a local housing activist and community organiser, describes it as planned estate removal rather than regeneration.[9]

Mapping of the displacement out of the Heygate Estate in, and mapping of the flows of tenants and leaseholders out of, the Aylesbury Estate[10] clearly demonstrate the extent of the dispersal of the original communities.[11] The phased redevelopment of the Aylesbury estate is on-going and due for completion in 2025/6. Redevelopment of the Elephant and Castle has moved on from the estates, with plans for the demolition and gentrification of the shopping centre and the relocation of existing market traders – plans that are being vigorously contested by Latin Elephant,[12] a consortium of local traders campaigning for the rights of BAME communities and demanding relocation spaces for traditional market traders displaced by the demolition.

Lees and White summarise the complex political, economic, and cultural processes that have contributed, over time to the current crisis confronting estate communities facing regeneration:

> London's council estates and their residents are under threat like never before. Council tenants are being forced out of their homes due to estate renewal, welfare reforms, poverty, and the precarity of low-income work . . . This dispossession is not a singular process, rather it relates to several coevolving and indeed accumulative processes including but not limited to: council estate renewal/gentrification.[13]

Faced with the prospect of their homes being demolished and communities decimated by redevelopment has led residents to organise and campaign against the proposals. All too frequently resident's express frustration that statutory consultation undertaken by the local authority are simply tick-box exercises, in which their views are ignored and development plans go ahead irrespective of residents' opposition. Feelings of disempowerment and helplessness in the face of the planning 'juggernaut' has prompted residents to come together to campaign for their right to remain and demand representation in the planning process to prevent the demolition of their homes and communities.

Formal policy tools for public consultation in urban planning date back to the 1960s when a Planning Advisory Group report on The Future of Development Plans in 1965 led to the setting up of the Skeffington Committee to look at the participation of the public in local development plans. This led to the publication of the Skeffington Report in 1969. The report stated that plans should be subject to full public scrutiny and debate. Since then, the possibilities for public participation in regeneration schemes has evolved with increasing demands from the public to be involved in the design of redevelopment rather than just having the opportunity to respond retrospectively to proposals. Sendra and Fitzpatrick provide an overview of the options available to citizens and planners, ranging from national- to local-level scales, illustrated with a range of case studies demonstrating how communities have utilised a range of statutory and non-statutory instruments to influence the planning process and fight back against the destruction of their homes.[14] Key to the case study we present here is the policy of local-level Neighbourhood Planning (NP).

NEIGHBOURHOOD PLANNING

NP was introduced through the Localism Act (2011) to give communities a right to have a say in shaping development in their area. The main instrument available to them is a Neighbourhood Development Plan, described as:

> a community-led planning framework for guiding the future development, regeneration and conservation of an area, empowering communities to have a direct say in how their neighbourhood is shaped in the future.[15]

Citizens wishing to have a say in the future development of their neighbourhood and/or carry out a community-led regeneration scheme can use NP as a tool to have their voices heard in development proposals. However, it is a complex procedure, fraught with difficulties at each step of the way. Sendra and Fitzpatrick detail the process and discuss both the costs and benefits for communities embarking on a NP project.[16] The first stage requires a group of at least twenty-one people to apply to the Local Authority for the designation of a neighbourhood area and the convening of a Neighbourhood Forum. Once designated, they can then move on to develop the Neighbourhood Plan. There is a requirement that neighbourhood planners undertake community consultation and engagement and demonstrate an evidence base for their plans and ensure that they conform to national and local policy requirements.

In theory, NP gives communities direct powers to develop a shared vision for their locality and shape the development and growth of their area. This might include choosing where they want new homes, shops, and offices to be built or having their say on what new buildings should look like and what infrastructure should be provided. Powers include granting planning permission for the developments they want and confirming that the aspirations of the neighbourhood are aligned with the strategic needs and priorities of the wider local area.

The UK Government's guidance states that:

> Neighbourhood planning provides the opportunity for communities to set out a positive vision for how they want their community to develop over the next 10, 15, 20 years in ways that meet identified local need and make sense for local people. They can put in place planning policies that will help deliver that vision or grant planning permission for the development they want to see.[17]

However, there is criticism over how far it genuinely promotes democratisation of local planning. Critics argue that it is poorly funded and time-consuming and that 'rather than ushering in a new era of local engagement, perhaps we are simply granting power to those with deep pockets and a lot of spare time on their hands'[18] and that in practice, people get caught up in a complex, time-consuming process, leaving them with little time for campaigning and direct action which may be more productive in realising their aspirations.

Importantly, the powers given to communities through NP are limited, in that they are not able to affect local authorities' strategic priorities for the scale of development in the area. The community has to approve the same level of development already agreed by the local authority. Less development is not an option even if this is the majority view of the residents. Neighbourhood plans cannot contradict the local authority local plan, and this means that in reality the power of the Neighbourhood Forum is very limited when councils intend to redevelop an area. Creating a neighbourhood plan is not only time-consuming but requires high levels of expertise and knowledge of the planning system, as well as considerable commitment from residents. This can be especially challenging for communities who may lack both the time and resources to undertake such a complex process. The positives are that at the very least it provides an opportunity for community representation in the planning system and for citizens to 'sit at the table' when decisions are being made about their neighbourhood. It is also a platform for collective action where people can come together and re-imagine their future.

The requirement for community consultation and engagement in the NP process is where the role of participatory mapping comes into play.

A THEORETICAL FRAME: PARTICIPATORY MAPPING FOR COMMUNITY CONSULTATION

Sometimes termed collaborative or community mapping, we have chosen to use the term 'participatory mapping' very deliberately here as it foregrounds our research methodology which is deeply embedded in the principles of Participatory Action Research (PAR). PAR is a research approach in which researchers and participants work together to understand a problematic situation and, most importantly, change it for the better. Essentially a 'bottom up' rather than a 'top down' approach PAR aims not just to gather knowledge for the sake of it but rather that knowledge produced through PAR practice inherently contains an agenda for social change. Its aims include mobilising and empowering local communities to have a say in identifying and solving their problems, advocating for the inclusion of grassroots stakeholder experiences in the development of policy interventions as well as attempting to correct power imbalances in knowledge and information flows.[19] Our approach to activist-oriented research and reflexivity are grounded in a sense of shared effort and owners, one that aims to build participant's agency and support collective action, with a view to shifting the power dynamics of inequitable situations.

The varying levels of citizen participation in PAR projects have been defined in typologies of engagement. The most cited example is possibly Sherry Arnstein's ladder of participation,[20] of particular relevance here as it is a model based on citizen engagement in planning and renewal programmes. The rungs of the 'ladder' correspond in ascending order to increasing levels of participation but also increasing levels of empowerment and control in decision-making and autonomy with – 'manipulation' on the bottom rung and 'citizen control' at the top.

In Arnstein's formulation, the quality and depth of citizen participation in planning is rooted in access to power. Although she never defines power, Arnstein maintains the control of power has significant implications to the socio-economic advancement of 'have-nots' and thus embodies the potential to transform 'nobodies' into 'somebodies'.[21]

In the context of our participatory mapping project, the concept of a typology of participation provides a useful checklist – for building inclusivity, purpose, and empowerment into the project design. PAR is an approach that has been adapted to a variety of situations, as it is context-specific and targets the needs of different social and cultural groups, frequently located in a specific geographic area. It is this spatiality that lends itself particularly well to the use of mapping as a community consultation tool.

Advances in the availability of Participatory Geographic Information Systems (PGIS) technology has led to the increasing use of maps as tools for both data gathering and data visualisation. Many studies and examples have demonstrated the role of participatory mapping as a creative way to engage communities and increase citizen participation in local campaigns and action.[22]

With PAR as the theoretical foundation, our methodology has evolved through our practice. It references both the participatory geographic mapping of Haklay and Francis and the CommunitySensor model of De Moor.[23] Fundamental to both approaches is the concept of iterative cycles of map-making with a community to collectively define its issues, priorities, and actions.

Haklay and Francis deploy a six-stage process in their community-mapping methodology in projects that frequently include a citizen science element.[24] The six stages of the methodology involve gathering pre-existing information, discussion and priority setting, general perception mapping – which involves capturing qualitative local knowledge such as local history, memories, or feelings about places – and data collection to produce an array of qualitative and quantitative information to inform community plans of action. The resulting data is visualised as on-line geographic maps and made available as an information resource for the community.

Aldo de Moor's CommunitySensor methodology proposes a similarly cyclical approach that consists of two interconnected cycles, both driven by the mapping practice and named after the main purpose of each cycle – a Community Network Development Cycle together with a Community Sense-Making Cycle (see figures 10.2 and 10.3).[25]

Using the on-line network mapping tool Kumu, rather than geographic mapping, to visualise the data places greater emphasis on the complexities of the many and varied stakeholder interactions of the community. De Moor describes community network mapping as a core *communal sensemaking activity*, a participatory process of capturing, visualising, and analysing community network relationships and interactions for community sense-making, building, and evaluation purposes.[26]

The Community Network Development Cycle can be summarised as follows:

1. mapping the community network through visualising the most relevant pieces of the community into map elements, connections, and views.
2. using these mapping artefacts to make sense of the collaborative common ground of community network in terms of issues, priorities, and next actions.
3. designing and implementing community network interventions needed to carry out these actions, and so building the community and making its collaboration grow in the context of its wider network.

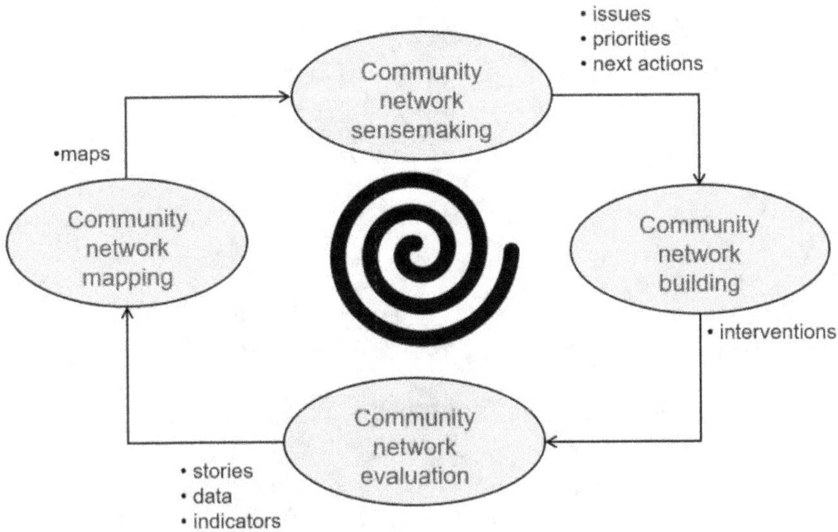

Figure 10.2 The Community Network Development Cycle (De Moor, 2017). Author's Image.

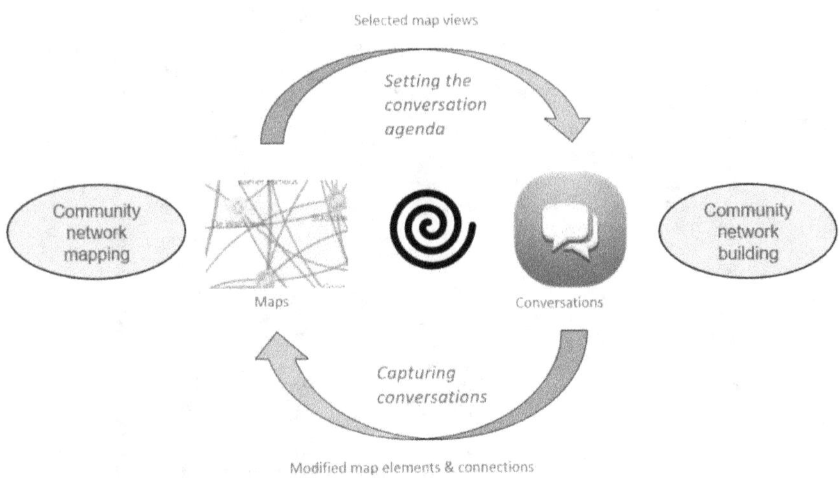

Figure 10.3 The Community Network Sense-Making Cycle (De Moor, 2017). Author's Image.

4. evaluating the effects of these interventions in terms of collecting stories, data, and indicators to provide the inputs for the next round of mapping. This process is to be repeated continuously, resulting in ever richer and more situated maps, a deeper joint sense of awareness and ownership of the collaboration ecosystem the community network consists of, and

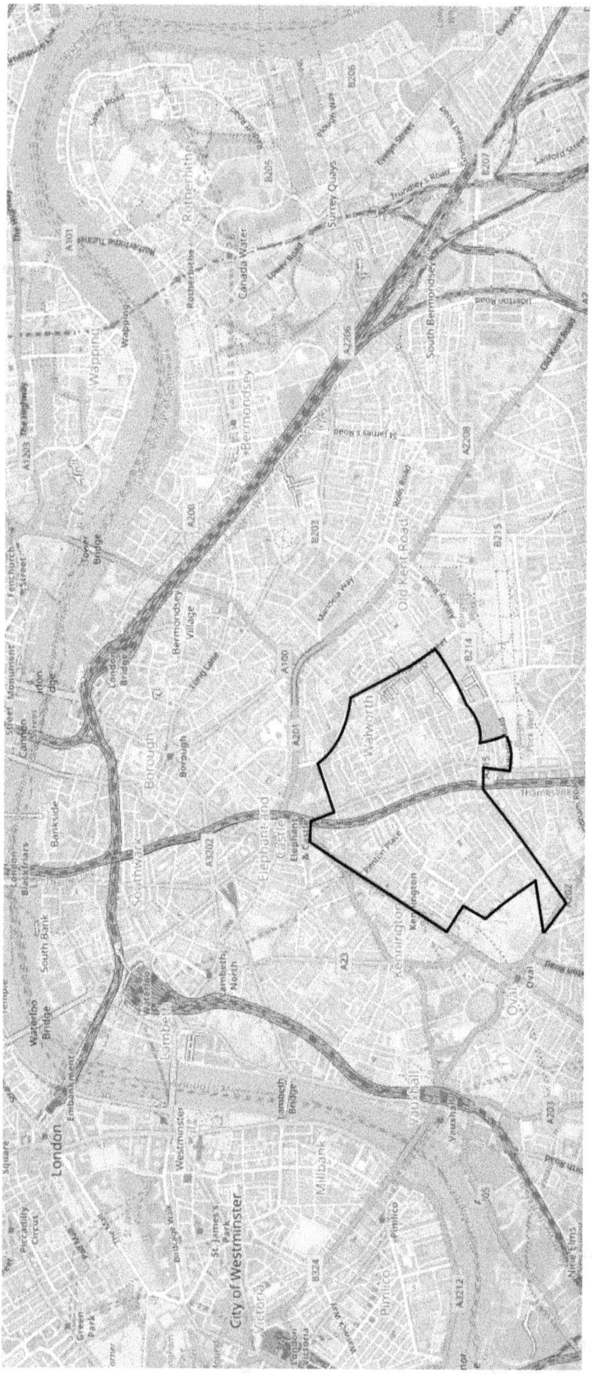

Figure 10.4 The Elephant and Walworth Neighbourhood Partnership Area. Author's Image.

> more effective community-building interventions, the results of which are measured in terms of stories, data, and indicators that help inspire the next round of development.

This iterative method of mapping collectively discussing the map and mapping again has been conceptualised by De Moor as the practice of *sense-making*, a process in which a community maps its resources, networks, and objectives to give meaning to their collective experiences and to gain an understanding of who they are and what they aspire to. It is therefore essential as part of the map-making to include aspirations as well as identifying assets and resources, so that the mapping becomes a collective re-imagining not only to make sense of the neighbourhood but to build a collective vision for the future.

The Community Sensemaking Cycle is a sub-cycle of the Development Cycle and is made up of four elements used to create an initial 'seed map' to spark dialogue with the community:

1. Setting the conversation agenda
2. Capturing conversations
3. Community network mapping
4. Community sense-making – the process by which people give meaning to their collective experiences

JustMap's methodology brings together both network and geographic mapping tools in community-mapping projects such as the Elephant and Walworth case study discussed here. The seed mapping for our projects utilises geographic mapping both as an engagement tool in our public consultation events and workshops and then to visualise the contributions made by participants to the mapping conversations. It's easy enough to find public utilities such as schools, hospitals, and post-offices using Google maps, what we seek to do in our participatory mapping practice is to go below the surface of conventional maps to discover the places that are important to the community, to record places under threat and those that have been lost. The resulting map is one that is personalised by, and for, the community and importantly uses the language of the community (De Moor, 2018a)

THE ELEPHANT AND WALWORTH NEIGHBOURHOOD CASE STUDY

It is against a backdrop of redevelopment, confronted by the huge development pressure in the area that a group of residents of Elephant and Walworth came together to form the Elephant and Walworth Neighbourhood Forum. Walworth is located to the South of Elephant and Castle, with Kennington to the West, Old Kent Road to the East, and Burgess Park and Camberwell to the south. It has many of the characteristics of south London's multi-cultural communities facing the pressures of gentrification and change.

The forum's first application for designation was submitted in January 2013. Following extensive consultation, a revised application was submitted in 2014. The formal consultation period on the application for designation ended in July 2016 and in September 2016, Southwark Council finally approved the designation of the forum and the Walworth Neighbourhood Area for the purposes of NP.

The Neighbourhood Forum aims to promote inclusive NP and to draw up a NP that will 'ensure that the benefits of the redevelopment can reach existing residents, existing businesses and existing groups'.[27]

The key aims of the initial work are to ensure that the plan is genuinely community-led and that as many as possible of the network of Walworth's diverse communities of both place and identity are involved in the consultations, so that it represents the needs and aspirations of Walworth's citizens.

The challenges identified by the forum in achieving these goals are:

- How to identify and record the resources and assets of the community?
- How to identify local needs?
- How to capture the aspirations of the many diverse local communities going forward?

In response to these challenges, JustMap was commissioned by the forum to organise a series of mapping workshops at local festivals and events during the late summer of 2017, culminating in a public meeting held in October 2017.

SETTING THE CONVERSATION AGENDA

Objectives were to use mapping as a tool to raise public awareness of the forum, increase community participation, and inform the NP by gathering local knowledge and feedback from residents. The community festival and event settings provided an opportunity to engage with a wide range of participants, especially those not formally involved with the forum and those residents who would not normally attend community meetings or workshops.[28]

The meetings were held at:

- The Walworth Society Festival, Westmoreland Road (31 August 2017)
- Informal workshop with members from Pembroke House (1 September 2017)

- The Beehive pub (8th September 2017)
- The Pasley Park Festival (16 September 2017)
- The Nursery Row Park Autumn Fair (23 September 2017)
- The Elephant and Walworth NP Forum Annual General Meeting (14 October 2017)

Mapping stalls were set up at each event inviting all those who live, work, or utilise the area to pin their places, stories, and proposals to a large-scale paper map with colour-coded flags which were added to a data reordering sheet.

Following the outreach workshops, the data was transposed from recording sheets to a spreadsheet and uploaded on-line to produce a geographic map using Carto mapping tools.[29] Printed editions of the map were taken to each festival workshop so that people could add more data and discuss the map markers gathered during the previous events. The map grew richer with each iteration as more places and stories were added, in an evolving process, engaging residents' interests and increasing local awareness and interest in the NP.

The same data was then used to create an affinity map clustered by the six topics of the draft Neighbourhood Plan.

Figure 10.5 Elephant and Walworth Neighbourhood Forum mapping stall at the Westmoreland Road festival. Author's Image.

COMMUNITY SENSE-MAKING

On 14 October 2017, we presented the fifth iteration of the geographic map at the Annual General Meeting of the Forum, to share findings with the community and as a tool to provoke further discussion for setting priorities and a shared vision for the NP. The meeting was facilitated by Tony Burton of Neighbourhood Planners London,[30] bringing together residents, local businesses, charitable organisations, and community members. The meeting was focused around a three-step plan for protecting and nurturing community assets. First, each action area co-ordinator was asked to make a case for their priority theme of the NP:

Heritage and Community Assets: Protect and enhance the use of heritage buildings and other community assets which are of enduring importance for the local community.

Housing: Protect the affordable housing stock in an area undergoing deep development pressure and advocate for social housing quota in new developments.

Green Spaces and Links: Link the area's myriad small parks and amenities together to create a network of traffic-light biodiversity-friendly walking and cycling routes.

Food Growing: Support local people reclaiming land for growing food across the Walworth and Elephant area.

Local Economy: Ensure that existing local and independent businesses are able to thrive in the area. Encourage the creation of affordable business space.

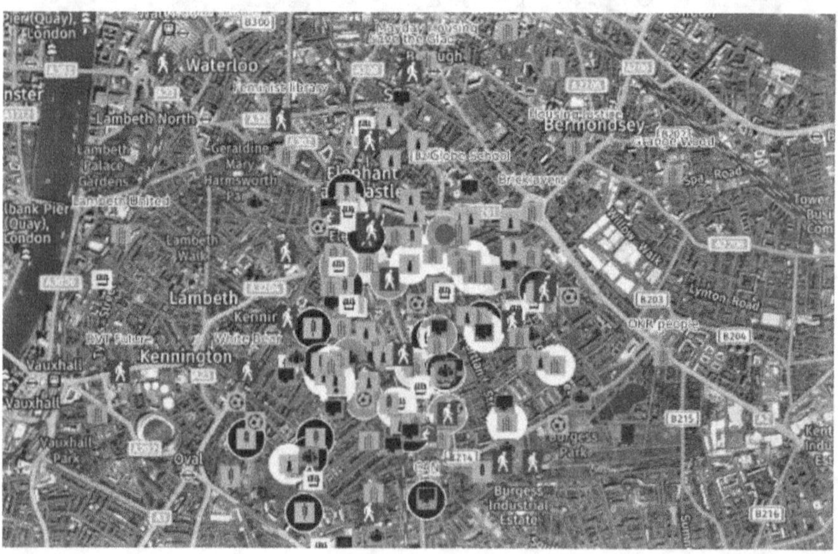

Figure 10.6 Elephant & Walworth Community Map. Author's Image; Link to Elephant and Walworth On-line Community Maps on the JustMap website http://justplace-london.blogspot.com/p/elephant.html.

In the second activity of the day, people separated into working groups with the aim of translating their shared visions for the NP into actions. Each working group was asked to generate three short-term campaigns and three long-term goals for their action area.

> *The passion of participants and facilitators alike was overwhelming – each working group struggled to round up discussions within the half hour!*
>
> (E&W Forum member)

Groups then returned to the plenary, where each action area co-ordinator presented their six priorities. Participants were then invited to vote for three campaigns and three long-term goals to take forward. Votes were counted and the results of the priority voting were:

2017/18 Campaign Priorities

- Town Hall Campaign (Heritage & Community Assets)
- Green Map and Contact Directory (Green Spaces & Links)
- SME Business Map and Contact Directory (Local Economy)

The Forum's long-term priorities:

- 'Economic activities as social heritage' (Heritage & Community Assets)
- Tackling dormant assets (Heritage & Community Assets)
- Stop demolition of council housing (Housing)

Setting these immediate priorities is just one part of a much larger task of building the NP. The results of the mapping and voting were then visualised as concept maps with Kumu, showing the six topics as clusters depicting the different long-term and short-term projects that were proposed and their relative popularity. The short-term proposals are highlighted in red and the long-term ones in pink. The size of the bubbles represents the number of votes, campaigns, and inspiring stories which were shared during the conference linked to each theme.

Priority setting in this way is just a small part of the much larger task of building the Neighbourhood Plan, but it provides a democratising process for the next steps.

CONCLUSION: PARTICIPATORY MAPPING FOR COMMUNITY SENSE-MAKING

Reflecting on the use of participatory mapping as a tool for consulting communities in urban planning contexts we found that the PAR framing of the project methodology proved to be a very effective framework for our ideas about working collaboratively *with* the community and to make certain that as far as possible both the process and the outcomes were community-led and community-owned.

The forum had already agreed on the six main priorities of the draft Neighbourhood Plan that formed the framework for thematic mapping and data analysis; we made no intervention in that. However, if that framework had not already been in place, we would have worked with them to create a thematic structure from the seed mapping data. Workshops were held at community events and social venues, such as pubs and local festivals, to ensure that the project had the widest possible reach, aiming to engage residents across all age groups, genders, and ethnicity within the relatively limited time frame of the six workshop sessions. The outcomes in terms of priority setting and next actions were generated and agreed upon by residents participating in the workshop at the Forums Annual Conference. Throughout the project, the role of JustMap was to contribute our knowledge of map-making and technical expertise in terms of organising the mapping events and managing and disseminating the maps.

The outcomes from the project demonstrate that participatory mapping can make a valuable contribution to inclusive and cooperative planning. Bringing together geographic-mapping and network mapping methodologies to create both place-based and conceptual data visualisations provoke conversations and most importantly the *process* of mapping contributes to a multitude of ways to community-building – whatever its aims.

Returning to Arnstein's ladder of participation introduced in our methodology section, we would suggest that participatory mapping at the very least offers neighbourhood planners a creative and imaginative way to meet the requirement to run lawful consultation. But much more than that the Elephant and Walworth case study has demonstrated that it also may go some way to helping communities reach the top three 'citizen control' steps of the ladder.

The mapping we carried out in Elephant and Walworth revealed very clearly that social capital in the area is being lost. This is particularly true of community spaces which are under pressure from redevelopment. However, it also shows the emergence of a collective and positive vision for the future generated in part by the map-making.

The methodology proves that to be effective as a tool for community empowerment, participatory mapping must be collaborative, with a physical presence in grassroots public spaces and events. This is essential if the NP is to deliver a fairer Elephant and Walworth, where residents will have access to the services they need and where people of different ages, genders, and ethnicities work together to make decisions about the places in which they live.

The participatory mapping game helped the Elephant and Walworth Forum AGM to define proposals and priorities for the future and succeeded in directly involving local people in decision-making. The affinity (cluster) maps were a particularly effective way of visualising the data. An additional benefit was that it also enabled residents to find information about local campaigns and projects that they didn't know about and to get involved in taking action themselves, for example, joining a community fruit-growing project or a campaign to save the library from redevelopment.

> The council plans to close the library and reopen it as a private gallery made me so angry about what was happening in my community, but I didn't know what to do about it until I did the map. I learned about the campaign, and other groups in my area and found a way to get my voice heard. Now I can be part of something and not a person alone.
>
> Walworth resident (M/62)

Returning to our opening quote from Adrien Labaeye, we found that performing the existence and availability of resources opens new fields of possibility that in turn enable new forms of collective action. Experiences such as that of the Walworth resident above demonstrated that for many people participating in conversations around the map, finding new spaces of representation and becoming engaged with the issues was much more important than making beautiful cartographs.

By following De Moor's CommunitySensor approach, we outlined key issues of interest to the community (figure 10.1) that set out the agenda for the conversations that followed, collected local stories, and began the process of community sense-making.

Challenges as always revolved around achieving diversity of citizen participation and inclusion within the limitations of the timeframe for the project, as well as the mapping workshops in public spaces a further round of mapping with specific targeted groups would go some way to making sure more voices are heard.

It is also worth mentioning the democratisation of map-making. We have found that the limitations of time and resources available for many community projects make it difficult to share and train participants in mapping software skills to enable them to take over the management and further development of the maps for themselves. Freely available software, such as Google My Maps, makes mapping more accessible to communities, however some training would be needed in creating network maps with Kumu. In this project, we were only able to begin the process of network mapping by creating the thematic affinity maps around the clusters of the six themes of the NP (figures 10.7 and 10.8). The next steps had time and resources allowed would be to begin to map connections and collaborations within and between stakeholders and the NP themes.

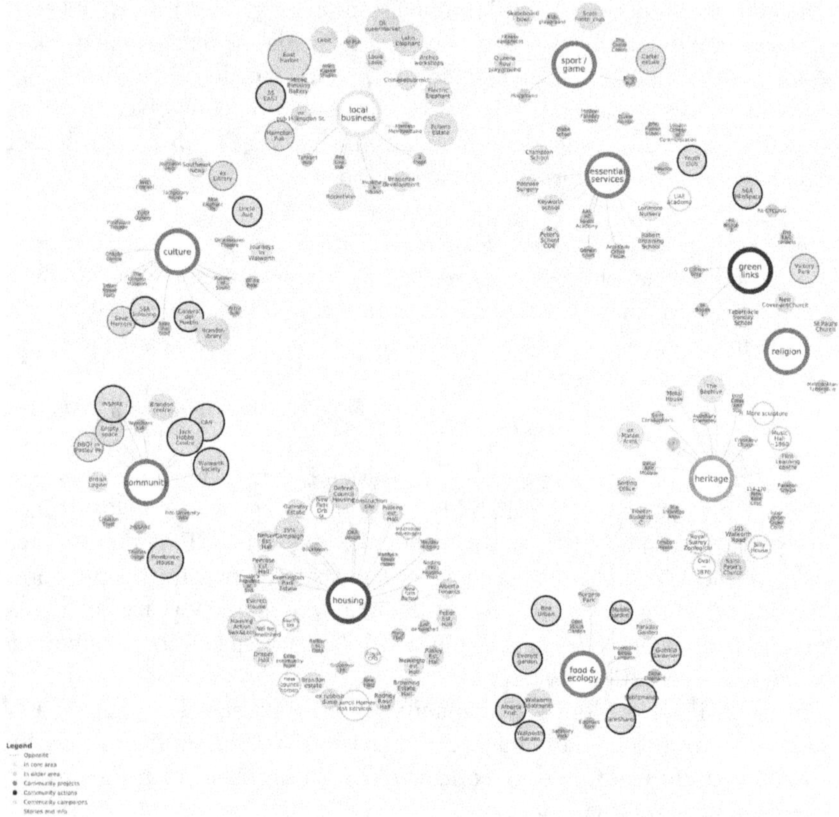

Figure 10.7 The Data Clustered by the Topics and Priorities of the Draft Neighbourhood Plan. Author's Image; Link to Elephant and Walworth On-line Community Maps on the JustMap website http://justplace-london.blogspot.com/p/elephant.html.

However, to conclude, we really would not want any community to be deterred from using participatory mapping as a consultation tool because of a lack of digital technology and/or skills. Geographic paper maps and some pins are all that is needed to run a workshop and a sheet of paper and some pens are all that is needed to create affinity or network maps. As De Moor suggests:

> Get your mapping hands dirty. No need to use fancy tools, just start capturing and reflecting upon what you see, using only a sheet of paper, if need be. You can always convert those paper representations into electronic form later.
>
> De Moor, (2018b)

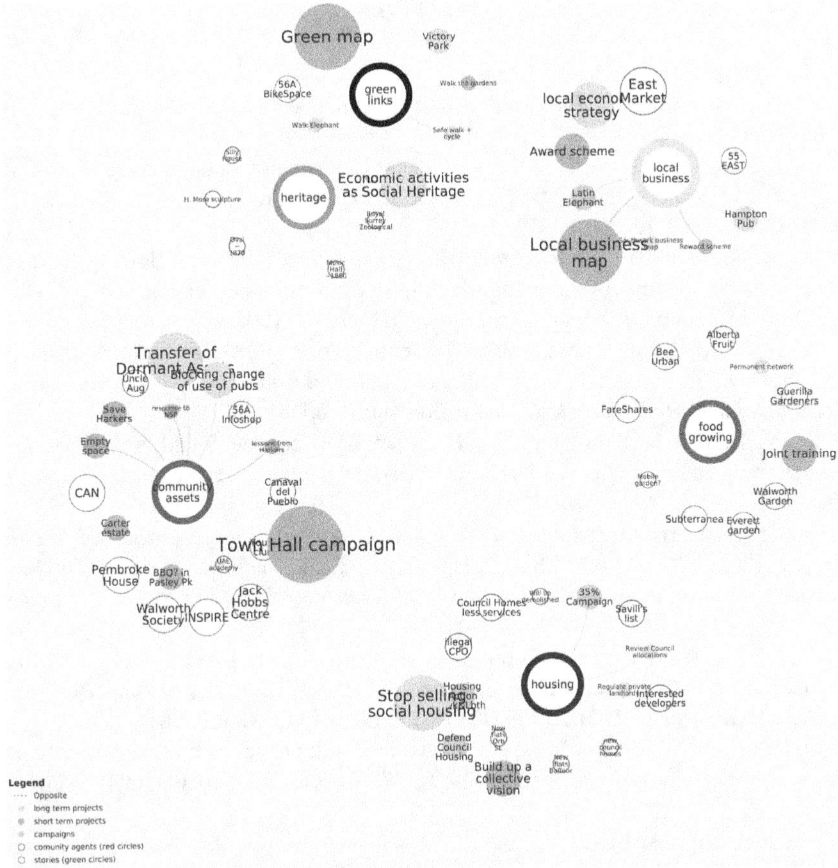

Figure 10.8 Affinity Mapping from the AGM Showing Places, Groups and Stories from the Workshops Classified According to the Topics of the Draft Neighbourhood Plan. Author's Image; Link to Elephant and Walworth On-line Community Maps on the JustMap website http://justplace-london.blogspot.com/p/elephant.html.

The resulting maps from the Elephant and Walworth project are a collection of views from local residents which expressed many different points of view, including issues of tension and debate which were subsequently openly discussed and addressed to resolve conflict. We hope that the collection of conversations revealed through the participatory community-mapping process is an authentic representation of the community's voice and perception and has made a contribution to their efforts to re-imagine the future of their community.

NOTES

1. Just Space https://justspace.org.uk/.
2. Beckett, A. 2015. *Promised You a Miracle: Why 1980–82 Made Modern Britain*. Penguin.
3. Lees, L. and White, H. 2019. The social cleansing of London council estates: contemporary experiences of 'accumulative dispossession', *Housing Studies*. DOI: 10.1080/02673037.2019.1680814.
4. Minton, A. 2017. *Big Capital: Who is London for?* Penguin Books; Watt, P. 2018. This pain of moving, moving, moving: Evictions, displacement and logics of expulsion in London, *L'Année Sociologique*, 68, pp. 67–100.
5. Hubbard, P., and Lees, L. 2020. Gentrification, displacement, and council estate renewal in London, in Baeten, G. et al. (eds.), *Beyond Housing Displacement: Conceptual and Methodological Issues*, London: Routledge.
6. A detailed account of the ASH Estate Map can be found here https://architectsforsocialhousing.co.uk/2017/09/10/mapping-londons-estate-regeneration-programme/.
7. Campkin, B. 2013. *Remaking London: Decline and Regeneration in Urban Culture*. London/New York: I.B. Tauris; Lees, L. 2014. The urban injustices of New Labour's 'new urban renewal': the case of the Aylesbury Estate in London, *Antipode*, 46(4), 921–947.
8. See for example, Elliot-Cooper, A., Hubbard, P., and Lees, L. 2019. Moving beyond Marcuse: Gentrification, displacement and the violence of un-homing, *Progress in Human Geography*. DOI:10.1177/0309132519830511; Watt, P. 2018.
9. Flynn, J. 2016. Complete control, developers, financial viability and regeneration at the Elephant and Castle. *City*, 20(2), 278–286, DOI:10.1080/13604813.2016.1143685.
10. Hubbard, P. and Lees, L. 2020.
11. The project findings can be found on https://estatewatch.london/.
12. See https://latinelephant.org/ for details of the campaign.
13. Lees, L. and White, H. 2019. The social cleansing of London council estates: contemporary experiences of 'accumulative dispossession', *Housing Studies* 35(10), 1701.
14. Sendra, P. and Fitzpatrick, D. 2020. Community-led regeneration: A Toolkit for Residents and Planners UCL Press (free to download at https://www.uclpress.co.uk/products/125697).
15. Locality. 2016. 'Neighbourhood Plans Roadmap Guide'. An updated version of this guide can be found at: https://neighbourhoodplanning.org/wp-content/uploads/NP_Roadmap_online_full.pdf (Accessed 8/12/2021).
16. Sendra, P. and Fitzpatrick, D. 2020.
17. UK Government Neighbourhood Planning Guidance (Accessed 20/01/2021).
18. Neighbourhood plans – democracy in action or just a sham? https://www.theguardian.com/local-government-network/2011/nov/28/neighbourhood-plans-democracy-action-sham (Accessed 20/01/2021).

19. Wakeford, T. and Sanchez Rodriquez, J. 2018. Participatory action research: towards a more fruitful knowledge, in Facer, K. and Dunleavy, K. (eds.), *Connected Communities Foundation Series*. Bristol: University of Bristol/AHRC Connected Communities Programme. Watt, P., and Minton, A. 2016. London's housing crisis and its activisms, *City* 20(2): 204–221. ISSN 1360-4813; Pant, M. 2014. Participatory action research, in Coghlan, David and Brydon-Miller, Mary (eds.), *The SAGE Encyclopaedia of Action Research* (Vols. 1–2). London: SAGE Publications Ltd. DOI: 10.4135/9781446294406.

20. Arnstein, S. R. 1969. A ladder of citizen participation, *Journal of the American Planning Association* 35(4).

21. Cardullo, P. and Kitchin, R. 2019. Being a 'citizen' in the smart city: up and down the scaffold of smart citizen participation in Dublin, Ireland, *GeoJournal* 84, 1–13. https://doi.org/10.1007/s10708-018-9845-8.

22. See for example, Chambers, R. 2006. Participatory mapping and geographic information systems: Whose map? Who is empowered and who disempowered? Who gains and who loses?, *The Electronic Journal of Information Systems in Developing Countries*, 25; Sieber, R. 2006. Public participation geographic information systems: A literature review and framework, *Annals of the Association of American Geographers* 96(3), 491–507.

23. Haklay, M., and Francis, L. 2018. Participatory GIS and community-based citizen science for environmental justice action, in Chakraborty, J., Walker, G. and Holifield, R. (eds.), *The Routledge Handbook of Environmental Justice*. Abingdon: Routledge, pp. 297–308; De Moor, A. 2017. Community Sensor: towards a participatory community network mapping methodology, *The Journal of Community Informatics* 13(2), 35–58.

24. Haklay, M. and Francis, L. 2018. pp. 4.

25. De Moor, A. 2017.

26. De Moor, A. 2017. pp. 37.

27. Elephant & Walworth Neighbourhood Forum Web Site http://ewnforum.org/.

28. Elephant & Walworth Neighbourhood Forum Web Site http://ewnforum .org/.

29. CARTO is a cloud computing platform that provides GIS and web mapping tools for display in a web browser; URL https://carto.com.

30. Neighbourhood Planners London a support organisation for communities undertaking NP https://www.neighbourhoodplanners.london.

BIBLIOGRAPHY

Arnstein, S. R. 1969. A ladder of citizen participation. *Journal of the American Planning Association*, 35(4).

Beckett, A. 2015. *Promised You a Miracle: Why 1980-82 Made Modern Britain*. Penguin.

Campkin, B. 2013. *Remaking London: Decline and Regeneration in Urban Culture*. London/New York: I.B. Tauris.

Cardullo, P., Kitchin, R. 2019. Being a 'citizen' in the smart city: up and down the scaffold of smart citizen participation in Dublin, Ireland. *GeoJournal* 84: 1–13. https://doi.org/10.1007/s10708-018-9845-8.

Chambers, R. 2006. Participatory mapping and geographic information systems: whose map? Who is empowered and who disempowered? Who gains and who loses? *The Electronic Journal of Information Systems in Developing Countries* 25.

De Moor, A. 2017. Community Sensor: towards a participatory community network mapping methodology. *The Journal of Community Informatics* 13(2), 35–58.

De Moor, A. 2018a. A community network ontology for participatory collaboration mapping: towards collective impact. *Information* 9(7): 151. https://doi.org/10.3390/info9070151.

De Moor, A. 2018b. Community Network Mapping – Just Get Going. Community Sensor. https://www.communitysense.nl/2018/02/12/community-network-mapping-just-get-going/ (accessed 16/03/2021).

Derounian, J. 2011. Neighbourhood Plans – democracy in action or just a sham? Guardian Newspaper https://www.theguardian.com/local-government-network/2011/nov/28/neighbourhood-plans-democracy-action-sham (accessed 20/01/2021).

Elliot-Cooper, A., Hubbard, P., and Lees, L. 2019. Moving beyond Marcuse: Gentrification, displacement and the violence of un-homing. *Progress in Human Geography*. doi:10.1177/0309132519830511.

Flynn, J. 2016. Complete control, developers, financial viability and regeneration at the Elephant and Castle. *City*, 20(2): 278–286, DOI:10.1080/13604813.2016.1143685.

https://doi.org/10.1080/13604813.2016.1143685.

Haklay, M. 2013. Citizen Science and Volunteered Geographic Information – overview and typology of participation. In Sui, D.Z., Elwood, S. and M.F. Goodchild (eds.), *Crowdsourcing Geographic Knowledge: Volunteered Geographic Information (VGI) in Theory and Practice*. Berlin: Springer, pp. 105–122, DOI: 10.1007/978-94-007-4587-2_7.

Haklay, M., and Francis, L. 2018. Participatory GIS and community-based citizen science for environmental justice action. In Chakraborty, J., Walker, G. and Holifield, R. (eds.), *The Routledge Handbook of Environmental Justice*. Abingdon: Routledge, pp. 297–308.

Hubbard, P. and Lees, L. 2020. Gentrification, displacement, and council estate renewal in London. In Baeten, G. et al. (eds.), *Beyond Housing Displacement: Conceptual and Methodological Issues*. London: Routledge.

Labaeye, L. 2017. Collaboratively mapping alternative economies, co-producing transformative knowledge. *Netcom*. http://journals.openedition.org/netcom/2647.

Lees, L. and White, H. 2019. The social cleansing of London council estates: contemporary experiences of 'accumulative dispossession'. *Housing Studies*, 35(10): 1701–1722, DOI: 10.1080/02673037.2019.168081.

Lees, L. 2014. The urban injustices of New Labour's 'new urban renewal': the case of the Aylesbury Estate in London. *Antipode* 46(4): 921–947.

London Tenants Federation, Just Space and Lees, L. 2020. Estate Watch London & beyond. *Radical Housing Journal*, 2(2).

Minton, A. 2017. *Big Capital: Who is London for?* Penguin Books.

Pant, M. 2014. Participatory action research. In David Coghlan & Mary Brydon-Miller (eds.), *The SAGE Encyclopaedia of Action Research* (Vols. 1–2). London: SAGE Publications Ltd. DOI: 10.4135/9781446294406.

Sendra, P. and Fitzpatrick, D. 2020. Community-led regeneration: A Toolkit for Residents and Planners UCL Press (free to download at https://www.uclpress.co.uk/products/125697).

Sieber, R. 2006. Public participation geographic information systems: A literature review and framework. *Annals of the Association of American Geographers*, 96(3): 491–507.

Transparency International UK. 2017. Faulty Towers: Understanding the impact of overseas corruption on the London property market. https://www.trustforlondon.org.uk/publications/faulty-towers-understanding-impact-overseas-corruption-london-property-market/ (accessed 10/01/2010).

Wakeford, T. and Sanchez Rodriquez, J. 2018. Participatory action research: towards a more fruitful knowledge. In Facer, K. and Dunleavy, K. (eds.), *Connected Communities Foundation Series*. Bristol: University of Bristol/AHRC Connected Communities Programme.

Watt, P. 2018. This pain of moving, moving, moving: Evictions, displacement and logics of expulsion in London. *L'Année Sociologique* 68: 67–100.

Watt, P., and Minton, A. 2016. London's housing crisis and its activisms. *City* 20(2): 204–221. ISSN 1360-4813.

Chapter 11

Empathy Walks

Leticia Sabino, Sofia Croso Mazzuco, Julie Plichon, Debanil Pramanik, and Sonja Baralic

Project Team: Leticia Sabino, Sofia Croso Mazzuco, Julie Plichon, Debanil Pramanik, Sonja Baralic, Maria Ernanny, Kavita Tailor

INTRODUCTION AND OBJECTIVE

'Empathy Walks' is a project that propels alternative reflections and ways of 'reading' cities by means of collective and interactive walking experiences that reveal different narratives and coexisting social layers. It was inspired by the work and discourse of academics and practitioners engaging in people-centered city planning and urban design. These include: geographer David Harvey, relating to his concept of 'the right to the city';[1] journalist Jane Jacobs' grassroot neighbourhood vision of city planning; architect and urban designer Jan Gehl who advocates for 'cities for people' and the value of public space;[2] sociologist Saskia Sassen's emphasis on the multicultural contribution of different citizens to the diversity of cities;[3] Suzanne Hall, who works with the ethnography of the everyday;[4] and Monica Degen, whose work examines how sensory experiences mediate and are shaped by urban space and urban life.[5]

Empathy Walks emerged from a multidisciplinary team of architects, geographers, and social scientists graduating with an MSc in Urban Design and City Planning from the Bartlett School of Planning, London. The team's aim has been to create a methodological platform and tool that supports greater social inclusion and justice through the creation of qualitative and collaborative cartographies, opening up spaces – both physical ones, that is, common spaces of encounter, and subjective ones, namely discussions and reflections – and the opportunity of encounter and dialogue between citizens being underrepresented (or found in vulnerable situations) and the decision-makers and urban experts who altogether shape the transformation of cities. Empathy Walks promote walks led by people from underrepresented communities, who might find themselves in a situation of vulnerability (including migrants, maintainers, people who find themselves unemployed, or facing displacement) aggravated by exclusive urban design.

From these walks, shaped by new perceptions and awareness of the different social layers of the city, social cartographies are narrated and then mapped – initially by the walk leaders, and then by the walk participants. Walking together not only helps build empathy, it also gives people a closer understanding of realities and social layers that coexist in the same city. Empathy Walks reveal narratives and open a platform for diverse voices to speak and be heard, paths to be walked, and social maps to be overlaid.

Empathy is a relatively new term in English and other Latin languages, the first time it appeared in English was in 1909, created by Edward Tichner to translate the term 'einfuhlung' in German, which means 'feeling into'. From the 1930s, it was described as taking the role of other persons to understand a situation.[6] It is both a cognitive and affective process – which means being aware of the presence of the Other – perceive them, care for the other without judging and with respect. It can build momentary bridges between strangers with no other expectation than understanding, nurture careful awareness about otherness, and trigger identification.

Through the *walks* citizens can thus become more aware of each other's perspectives and needs for flourishing, which in turn develops a more complex perspective on cities, urban diversity, and justice in planning and development. The methodology tailored to these *walks*, and the collective production of grassroot data, aims to provoke and highlight the shortcomings of a planning process nurtured by cold data depicted onto maps that show borders and divisions often not reflective of lived experience. Empathy Walks practices a process that builds on the cartographic tools available for adding depth and emotion to the analysis of spaces. The aim is to challenge the supposed objectivity of adopted datasets and mainstream mapping, highlighting that there are real people and communities with both individual and shared histories, stories, and perceptions which are valid in any attempts to understand a place or move to proposing intervention.

The Covid-19 pandemic has had a multiplicity of effects on how people relate to each other. In the midst of social distancing, an era of neighbourhood solidarity and technology-powered sociability has flourished. Mutual Aid groups have emerged to support vulnerable shielding members of the community, people have been using video chat platforms to keep in contact, while physical gatherings have been strictly controlled or prohibited. Within this challenging context, it is important to re-contextualise how cartographies can connect people and share stories that are able to stimulate connection and mutual support.

This chapter will first introduce the problems being tackled and the intended outcomes of Empathy Walks, outlining the methodology developed and presenting case studies of the approach and finally develop a critical look at the outcomes and ways in which the methodology can be further evolved.

THE PROBLEM

Maps, as traditionally used in city planning, are built by layers that depict the different structures of the urban fabric: streets, open spaces, land uses, and so on. Urban maps are tools that narrate division between different urban elements and socio-economic differences. As abstract representations, traditional maps leave an open door for important and meaningful information about citizens and communities to be lost – with the pitfall of failing to capture the richness and complexity of urban diversity and temporal change. The choices made in constructing traditional maps often fail to register the experiences and value generated by different groups who use the same city but in different ways. This lack of depth in mapping made Empathy Walks consider how alternative mapping can provide a platform for narratives which stimulate reflection on social issues.

In addition to the data represented on maps, there are other layers of significance which conventional maps cannot capture primarily, because they are made from information collected via a top-down process, and namely not by the people who directly experience the places being mapped. For example, feminist sociologist Frank Tonkiss highlights the marginalisation of women's experience of the city in mainstream maps (which also overlook the needs of vulnerable groups such as children, migrants, elderly, and ethnic minorities). Tonkiss evidenced how gendered fear affects personal geographies of the city, especially around freedom of movement:

> Women's ambivalent relation to the city in terms of freedom and danger goes beyond the symbolic coding of space. It boils down to critical decisions about spatial practice. In this context, the geography of gender in the city involves not only margins of freedom but maps of danger.[7]

Lefebvre furthermore defined that the production of spaces is composed by a triad of layers: lived, conceived, and perceived.[8] Mainstream maps only cover and represent the conceived spaces (that is, how they look like physically), thus failing to translate how they are perceived and lived by different people, or groups, who would interact with these spaces in different and personal ways. On the same idea, Tonkiss emphasises that

> if the official order of the city is written down as so many rules, codes, maps and plans, the individual's version is a spatial story told as if out loud in the streets of the city, leaving no trace other than a movement in the air.[9]

As people's existence and experience in the city can indeed be reduced to their movements, leaving no trace, we are left wondering: how can we trace these movements to ensure inclusive planning? If city planning and public policies are influenced by maps and data, how could these represent individuals' spatial practices and reflect diverse needs? How can we trace and make visible the 'invisible' movement of people to support inclusive planning through the way that cities are mapped and spatially structured?

More worryingly still, big data, as large automatically generated datasets, is ever more widely used to develop knowledge of different communities or places, as if it was unbiased and complete evidence which can draw conclusions of people's movement or behaviour. The limitations of this approach are those of exclusion and blindness to the movements and values generated by the activities that do not relate to people's spending power or, for example, the many mobilities beyond commuting to work.

On the other hand, the democratisation of the ownership of data is an opportunity for how communities can take control of their own story, voicing the narratives that have the power to shape local processes in a specific and wanted direction. Considering this opportunity, Empathy Walks was developed to offer a tool to negotiate urban decisions by means of making visible the different valuable narratives and humanising urban data.

Storytelling is a powerful tool to access narratives and guide urban transformations. How the story of a place is narrated can influence the way decisions are made and the pace of change. Hence, it is crucial that local stories are narrated by local communities and bring awareness to the different actors that influence urban transformations. For example, a community might be dissatisfied with the narrative created by a decision maker to define their community. Empathy walks can contribute to the emergence of voices that challenge that official narrative to reshape it collectively, as a tool to democratise decision-making. This process aims for a high level of participation, which could relate to Sherry Arnstein's ladder of participation level of 'citizen control'.[10]

METHODOLOGY

The Empathy Walks methodology combines urban design and spatial analysis tools with social tools such as 'Non-Violent Communication' (NVC) to stimulate connection and empathy and engage citizens in innovative and replicable ways of generating inclusive narratives of cities. This approach is valuable not only in the production of alternative mapping but also for the empathic interactions and the possibility of dialogue that opens up during the walk – prompting potential collaborations and building social capital (see figures 11.1 – 11.7).

NVC is an empathic communication approach developed by Marshall Rosenberg in the 1960s to build dialogue based on the capability of acknowledging and recognising people's own needs and those of others. This approach is based on the understanding that violence (and exclusion) is generated by the lack of ability to identify strategies for meeting these needs. It can be applied at different levels, including our inner selves, one-to-one interactions, and in groups, following four key steps: intention, reflection, observation, and need. The intention step means wanting to connect to the subject or feeling, thus appreciating and being open to understand. Reflecting involves communicating back to the speaker what you hear them saying. The

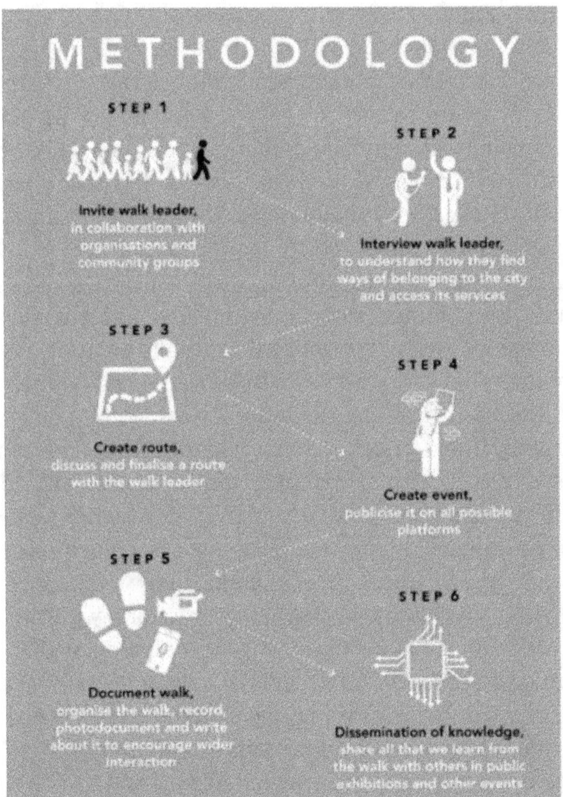

Figure 11.1 Empathy Walks (Title Image). Author's Image.

observation step comes by holding back our judgement or blaming, and the understanding step is about acknowledging the needs behind a given speech that are present and asking to be met.

The methodology of the walks (described in more depth below) is based on six main steps:

- Step 1. Identify a walk leader and make the invitation to create an Empathy Walk.
- Step 2. Interact with the walk leader and listen to their stories to identify their main needs and concerns in relationship to the city.
- Step 3. Design the walk route, in a way that captures the essence of the walk leader's needs.
- Step 4. Set a date/time, spread the word, invite participants who somehow relate to the needs being posed by the walk leader.
- Step 5. Document the walk and create the empathic maps.
- Step 6. Share the knowledge gained to generate a discussion on the topic proposed by the walk.

Every walk has a different walk leader who guides participants through their own ways of living in the city. Walk Leaders are invited from communities experiencing unfavourable conditions: migrants who seek means of belonging to the city, displaced workers who see their workplace disappear with rapid urban change, city maintainers whose contribution to the urban dynamic goes unacknowledged, and so on. The walk leader is invited to share their daily walking paths and experiences with walk participants in order to, through empathy, build mutual respect and understanding while acknowledging the needs of people who experience the city differently.

To structure the walks, the walk leader co-creates and co-maps their everyday routes with support from the Empathy Walks team – who work through the method of mental and affective mapping, tools that rely on memory and feelings towards place. Mental mapping allows senses and feelings to be predominant in the approach to space analysis, revealing perceptions, temporalities, and meaningful elements. This stage is guided with a structured questionnaire covering different ways of connecting with places and meaning to understand the walk leader's personal interactions with the city's social and spatial layers. Before the walk event, the Empathy Walks team experiences the co-mapped route with the walk leader in order to better grasp the understanding of what issues surround the experience of the walk leader, and therefore help the team curate the walk and extend invites.

To enrich the experience and generate knowledge and data for future actions and community-led planning, the *walks* instigate mapping tools to be used by all participants. Precisely, during the walks participants are given an OS map

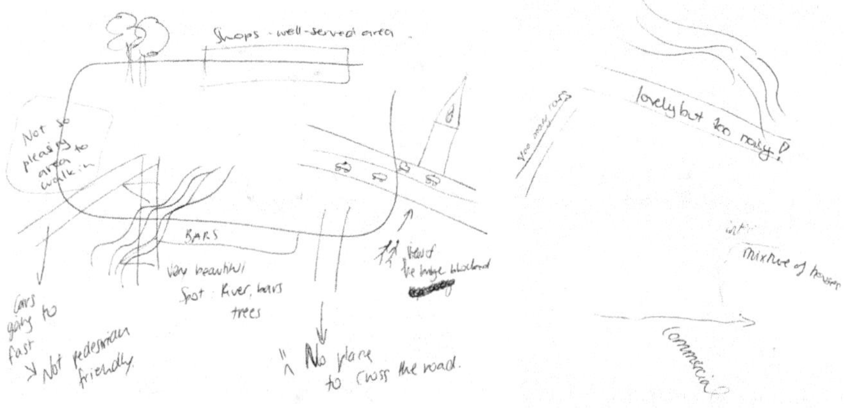

Figure 11.12a and b Drawings of Two Mental Maps Developed during a Walkabout around Hammersmith, London, in 2016. Author's Image.

(a city map showing the outline of streets and buildings) to annotate their own perceptions and feelings in relation to spaces and stories narrated in the guided route and thus creating a sensory map (a tool much inspired by '*Jane Jacobs' walks*' http://www.janejacobswalk.org/). For the sensory maps, participants are provided with simple icons meant to represent some feelings or perceptions that may arise during the walk, such as noise, danger, comfort, surprise, beauty, and so on. The icons act as a starting point to encourage participants to reflect on how experiencing other people's paths and narratives makes them feel.

Besides the walking methodology, Empathy Walks is engaged in developing tools, materials and strategies supporting grassroot and collective cartographies. The sensory maps generated during the walks by each participant (and previously by the walk leader) intends to translate qualitative information (namely, feelings and emotions) linked to city spaces – a valuable layer which can add depth to traditional maps (such as OS maps) that focus mainly on picturing the spatial elements and boundaries of cities. See examples of the sensory maps below:

Following the walk, the maps produced by the different participants are read, analysed, and layered over one another to generate maps from empirical and personal experiences. The walks are valuable for they allow participants to access local knowledge of specific urban spaces and form their own opinion about how urban issues and social inclusion could be alternatively tackled. This exercise and the use of icons can help discover connections and patterns between people's sensations, experiences, and personal layers (places of belonging, places of conflict, social places, and workplaces, etc.) of the city. It is a direct experience of 'otherness' that forms not only

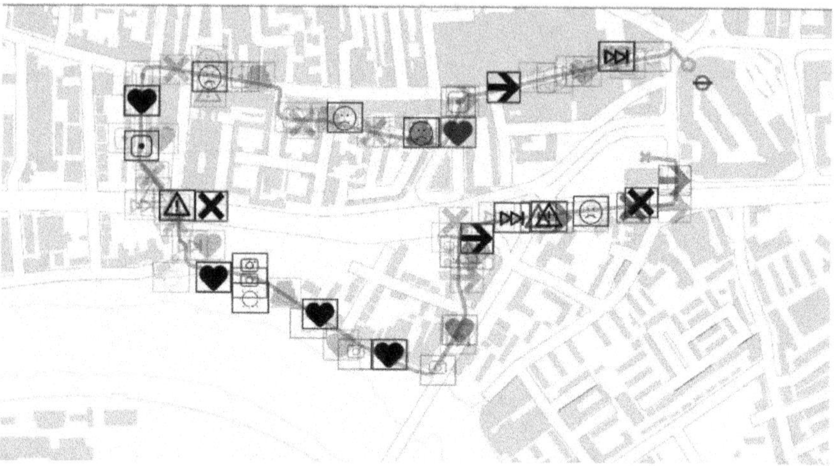

Figure 11.3 Annotation of the Sensory Maps Overlaid to Identify Feeling Patterns (in the spirit of Jane Jacob walks). Author's Image.

alternative knowledge to be fed into new maps but also a more direct process of decision-making and co-creation of the urban.

Finally, as the walks aim to foment new discussions on city planning through alternative cartographies, the knowledge generated from the walks is also directed at inspiring city makers to engage a wider pool of citizens in the research and knowledge of cities.

It is also important to highlight that the proposed methodology is designed to be open-source, allowing anyone to apply it within different communities and contexts according to the need of generating empathy.

> 'You don't need to be a voice for the voiceless. Just pass the mic'.
>
> Su'ad Abdul Khabeer, 2017[11]

Empathy Walks is not designed to act as a spokesperson for different communities: instead, it is designed to be used as a tool and a civic platform where different and diverse voices can be heard directly by other people, including citizens and city planners, who are constantly shaping the city. Hence, walk leaders are storytellers during the walks. Empathy Walks is intended to connect people and broadcast urban narratives and cartographies in order to influence the way people interact and change the city, allowing them to see that a good city is one which considers the needs of people and has something to offer to everyone. As Jane Jacobs wisely said: 'Cities have the capability of providing something for everybody, only because, and only when, they are created by everybody'.[12]

EXAMPLES OF WALKS

Two case studies are presented and analysed below to show Empathy Walks in practice.

CASE STUDY 1: EMPATHY WALK 'THERE IS NOTHING HERE'

On Saturday 1st of July 2017, Empathy Walks and walk leader Christian organised a walk entitled 'there is nothing here'. Christian works in an industrial estate along Camley Street, in Cedar Way, north of King's Cross, where he discussed issues of urban regeneration.

The title of the walk, 'there is nothing here' critically focused on the way that the industrial and employment land in King's Cross is under significant pressure and threat of displacement from the wider regeneration of the area. Up-market apartments and student accommodation came to complete the new cultural quarter around Granary Square, adjacent to King's Cross train station. Christian began the walk with a story about when he visited the marketing suite for the up-market apartments around the new 'Gasholders Park' site. He introduced himself and his workplace just located a few yards north-west and, surprisingly, the agent's reaction was – 'oh there? I thought there was nothing there'. From this starting point, the walk is an invitation to explore the construction of the neighbourhood and its new identity and the voids of history and its surroundings. By designating Cedar Way Industrial Estate as 'Brownfield Land' (in other words ripe for development), the planning system has built the belief that there was nothing going on beyond the masterplanned scheme for King's Cross.

Following Christian's daily route to work, the walk participants were invited to explore the social, political, and sensorial elements of the industrial estate and discover the richness of activities happening there. The narration created understanding about hidden and subtle realities in the area. Beyond the new and flashy version of King's Cross – depicted by seductive images and marketing material – a whole world seemed to crumble and disappear: it is the *old* world, seen as empty, obsolete, awaiting redevelopment. The walk emphasised the diversity of productive activity, skills, and services essential to the running of the city (from the warehouse containing all the salt for the borough's roads, the fishmongers for the city's restaurants, to the many car repair businesses). The value and strategic importance of these activities to the city is one that is vastly overshadowed by the forces of land values and masterplanning processes across the capital.

The forty-five-minutes walk gathered a group of ten people and finished at a local pub – strategically creating an opportunity for informal discussion and sharing thoughts about the area and the walk experience.

Commenting on the walk, one participant stated:

> I was surprised to see the wide variety of exciting businesses on my doorstep. I felt a bit ignorant and enlightened. I think it would be a real shame to lose the fishmongers, butchers, car shops, etc. Favourite spots were the Elan Muesli building with the fruit trees, etc., and learning about the healing ideas for housing above the meat and fish businesses, using the unwanted heat to warm the houses.

As for the walk leader, he reported that the small format of the walk had allowed him to practice and develop public speaking skills. These skills are then transferable to other formats of events and can contribute to building skills within the community.

This walk case study shows an area that has already undergone change and disregarding many local narratives generating exclusion and the loss of jobs and can build awareness on the value of the hidden stories and hopefully influence urban interventions of this kind in other areas in a way that considers its past, memory, and community.

CASE STUDY 2: EMPATHY WALK OF SOUTH KILBURN, NOVEMBER 2018

On Sunday 4th November 2018, an Empathy Walk took place in South Kilburn, London, led by Leslie and Dee. Around thirty people attended the walk. Like many of London's neighbourhoods, a post-war housing estate in South Kilburn is undergoing rapid change.

Leslie and Dee are community activists from the Granville Community Kitchen, a key community asset in South Kilburn. We walked with them to get a taste of the character of the neighbourhood, getting an insight into its existing diverse communities, valuable assets, and identity, as well as understanding the pipeline of regeneration.

The walk started at Kilburn Park Station. Leslie presented the group with an article seen in the Evening Standard newspaper, which reads:

> Lined with red-brick mansion blocks and white stucco houses in an elite corner of west London, the wide, gracious avenues of Maida Vale sweep northward from Regent's Canal to meet the cosier Victorian streets of Queen's Park. And

> in between the two, like the rotten core of an apple, is South Kilburn, dominated by its eponymous post-war council estate.

The walk uncovered existing precious community assets in the neighbourhood, from a contested park to community shops at the Peel Precinct.

What became very clear was the conflicted nature of the area, a story of regeneration where commercial space on the estate is overtaking the community's and social spaces. We got to experience some of that conflict directly at the end of the walk at the Granville community centre. The centre is run by the South Kilburn Trust, an organisation that the walk leaders criticised for what they perceive to be commodifying community assets in the area. We had rented the space to have tea and cakes, and at the end what was meant to be a communal reflection, over the walk together, turned out to be a rather odd situation – as the management of the centre had instructed staff not to provide us with tables and chairs, forcing the group to stand up or sit on the floor. Local staff known personally to Leslie and Dee found themselves conflicted. That conflict was eventually resolved when local staff decided to overlook management instructions and provide us with chairs. Even though the chair story seems rather anecdotal, it does tell a story of an area which is being fragmented when the process of regeneration creates competition and tension amongst the local community through the division and scarcity of resources. Donations for the walk went towards the Granville Community Kitchen and the rental fee for the Granville community centre.

One unexpected outcome of this experience is that one walk participant, UCL Bartlett School of Planning's Associate Professor Pablo Sendra, met with Leslie after the walk to develop a Community Plan for two of the building of the estate together with the residents. The Community Plan includes a detailed financial viability study and social impact assessment.

The walk in South Kilburn is illustrative of the collaborative power generated by the walks. The walk leader is able to tell their side of the story, which opens up a space of dialogue, one which is horizontal and conducive to community building.

TYPES OF MAPS

By combining participant maps it is possible to discover different layers of sensorial and spatial perception. This is done by overlaying the participants' annotated maps with the walk leaders' personal maps, to show how personal experience narrated has impacted on the participants' perceptions of the area.

Figure 11.4a and b Context Map – the 'New King Cross' versus the Old, Designated as Brownfield. *Source*: Empathy Walks archive. Author's Image.

PERCEIVED LIMITATIONS AND HOW TO FURTHER DEVELOP THE EMPATHY WALKS METHODOLOGY

Besides stimulating the creation of new cartographies through empathy amongst citizens, Empathy Walks is able to

- bring people closer in an interactive way that also acts as a social gathering and an opportunity to engage with people from different backgrounds;
- apply mapping as a way to invite people to reflect on their experience of the spatial and social qualities of their environment;
- engage people who are not (directly) connected in the city planning process to reflect on urban design matters, by providing accessible tools to read and analyse the built environment through hearing different stories and social narratives. Empathy Walks can help people understand urban design through their everyday stories;
- the walks and the mapping work are enablers that turn abstract planning concepts (such as 'regeneration' or 'brownfield land') into the tangible realities of people narrating their stories.

After walk events, the Empathy Walk team is able to point out some limitations of the methodology in tackling the issues it intends to solve and discuss how methods can be further developed. Besides improving the experience of the walk itself, the team reflects on how additional tools can be built to make this empathic tool more useful for the communities themselves and the planning and development process in general. This can be done by producing a helpful output, recording, and distributing findings.

In our reflection based on the walks and on the project development, the following are some of emerging questions and perceptions:

What Works well?

Small-sized walking groups allow a good level of interaction between all participants, making the narrative of the walk leader easier to follow, and more striking and efficient. Walks lasting up to one hour and ending in a social place (such as a cafe) keeps up the group dynamic with interesting conversations. In this 'informal' atmosphere, precious information can circulate and stimulate the practice of inclusive city planning.

What Could Be better?

Annotating maps (namely, those that are to be completed by walk participants) can be made easier, by having simplified icons (which represent

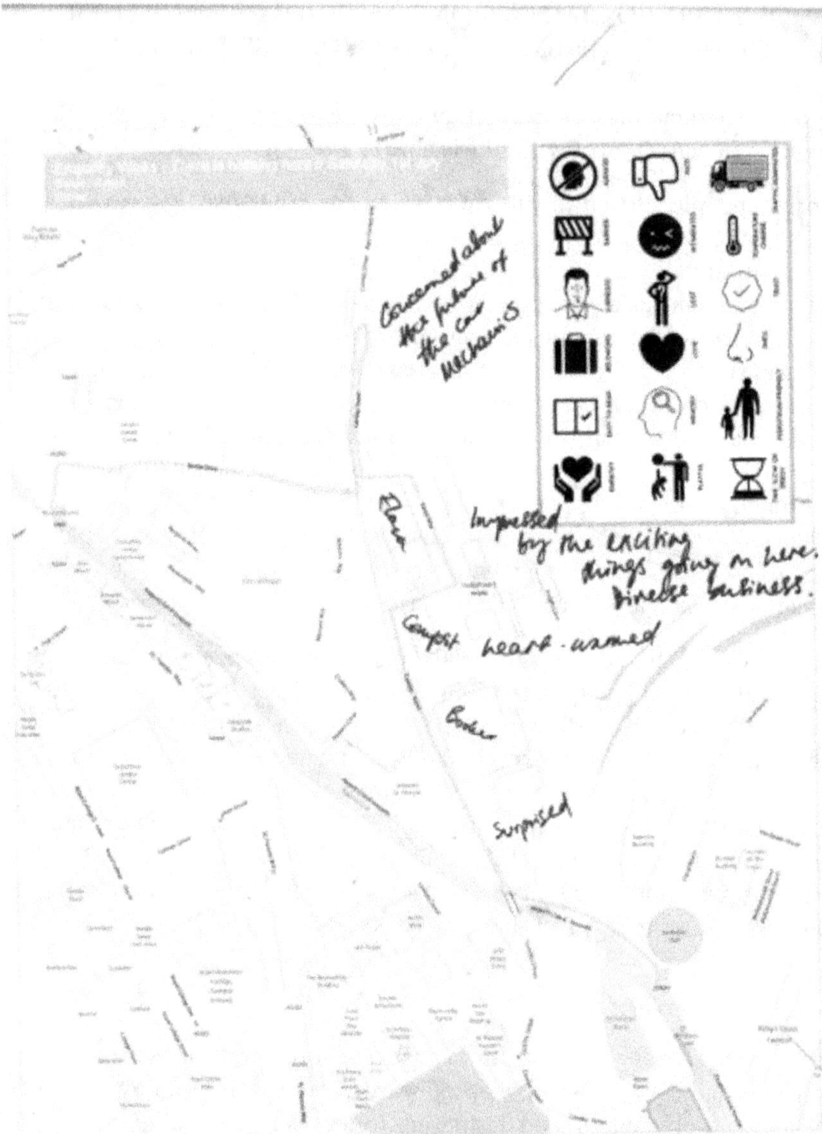

Figure 11.5 Empathy Walks Map by a Participant, Annotated during the Walk through King Cross and Cedar Way/Camley Street Industrial Estate. Author's Image.

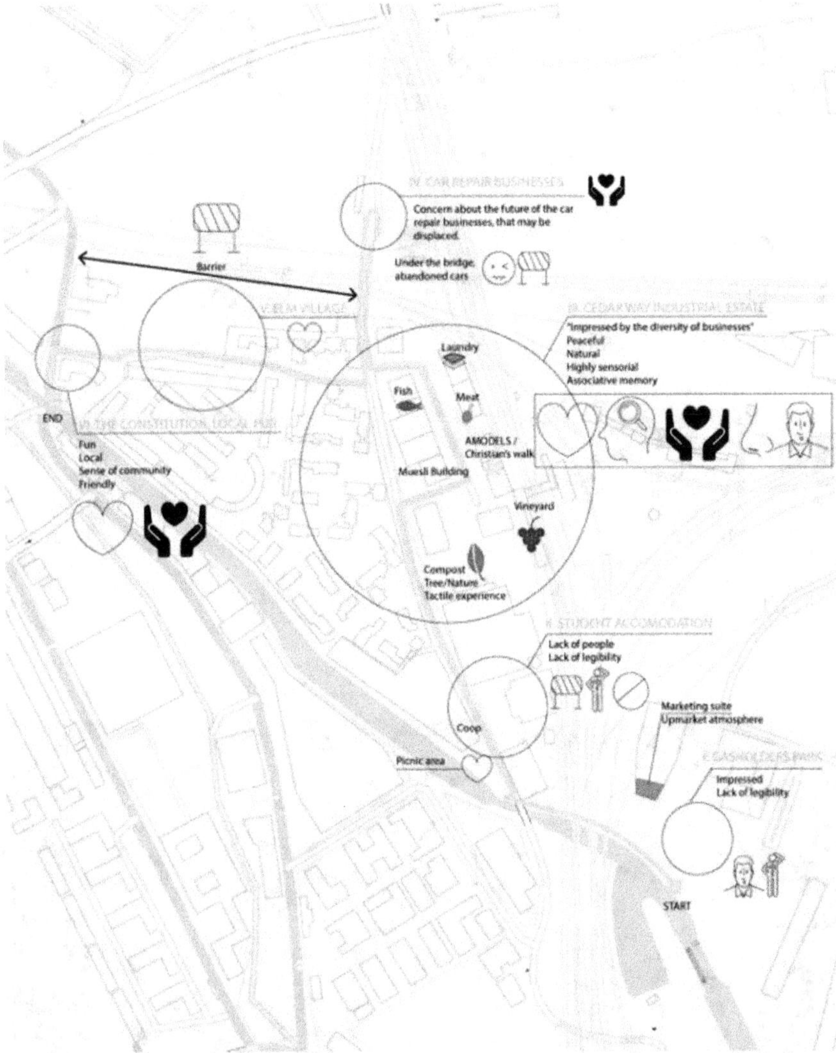

Figure 11.6a and b Empathy Walks Layered Participants' Map, Gathering the Different Reactions to the Walk and Annotations. Large Icons Gather Reactions that have been Shared between Most Participants. Author's Image.

feelings/perceptions) that allow the annotation process to become more fluid. Digital platforms could enable participants to select and drop icons in real time, as well as visualise other participants' mapping – which would create a 'real-time' empathy sharing map.

Regarding the identification of walk leaders, in order to build trust and relationship within new communities, it has been key to work closely with

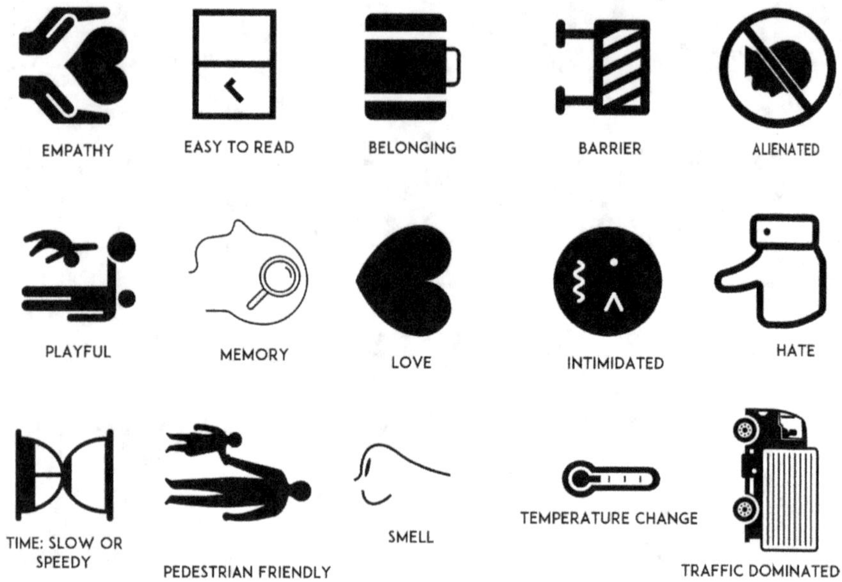

Figure 11.6a and b (Continued)

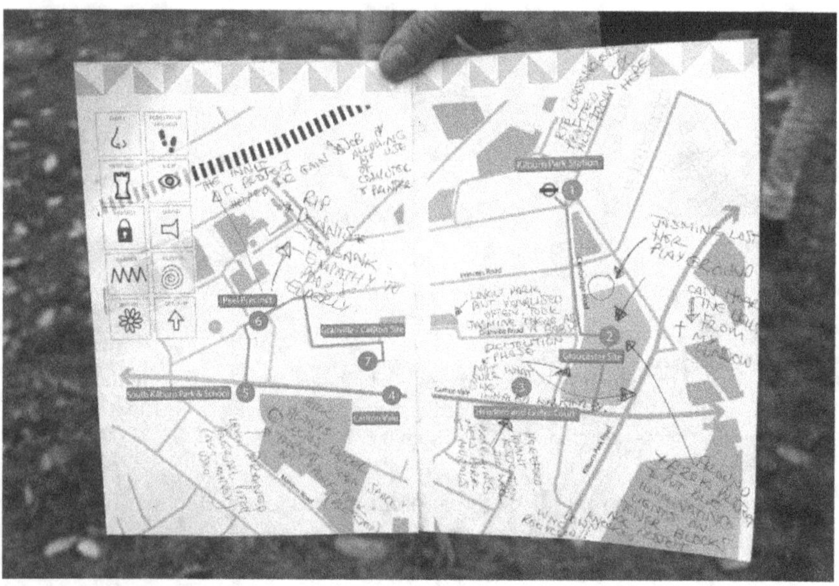

Figure 11.7 Annotated Map by a South Kilburn Walk Participant. Author's Image.

already constituted community groups – establishing networks and empowering people in sharing their stories and perspectives. Walk leaders frequently articulate the benefit they have experienced in leading a walk which include the development of public speaking skills and making new connections and networks.

In order for Empathy Walks to have better results, such as stimulating inclusion in the city planning process, the methodology could also focus on key actors, such as developers that are in the process of developing their intervention, or even focused on informing communities and city planners during local planning process, considering appropriate timing.

Another possibility is to advocate for legislation in which this kind of qualitative and participative process is required for every urban transformation proposal. As an example, Spain Inclusive Law has a requirement of gender safety walks audits for every major urban project.

CONCLUSION

Maps are the crucial tools and products generated by the walks – they are both outputs and a method to engage people and connect them with each other, thus generating empathy. Serendipity is also a result of the walks, since it is never possible to predict what interaction will result. The resulting maps can be used for studying an area by revealing different spatial and soft layers (sensorial or abstract) that are mapped both collectively and from personal experiences enriched by empathy. Overlaying those fine-grained maps with more 'formal' city planning maps (for instance, OS or planning policy maps) is intended to turn cold data into complex, people-centered data. Let us map our cities together, because collective and grassroot mapping is also a way to shape inclusive cities.

The question of inclusion has come into sharper focus with the Covid-19 pandemic, which has made more visible socio-economic disparities, and how social and spatial inequalities affect people's ability to access essential services. On one hand, the new notion of 'key workers' has given a name to the modern 'working class' – people working in the essential social and health services, delivery drivers, people whose work involves the everyday maintenance of the city, and who could not therefore stop or stay at home. One's health and employment status, living situation and access to computers and the internet for example have defined their level of comfort, ability to access essential services, or participate in social life during the pandemic. While many have managed to stay connected and share their experience with others, the experience of many during this period will have been hidden by their level of social isolation or digital exclusion.

In this challenging context, it seems now more than ever necessary to provide a platform for diverse stories to be told, and ask: can a map be a levelling tool for city planning? Can maps represent the contested nature of places? We believe that the mapping process can democratise city planning and plant the seed for new ways for citizens to regard each other's presence.

More can be found at: https://empathywalks.wordpress.com/

NOTES

1. Harvey, D. 2012. *From the Right to the City to the Urban Revolution*. Verso.
2. Jacobs, J 1961. *The Death and Life of Great American Cities*. Blackwells.
3. Sassen, S. 1991. *The Global City*. Princeton University Press.
4. Hall, S. 2012. *City, Street and Citizen: The Measure of the Ordinary*. Routledge.
5. Degen, M. 2008. *Sensing Cities: Regenerating Public Life in Barcelona and Manchester*. Routledge.
6. Mead, George H. 1934. *Mind, Self, and Society*. The University of Chicago Press; Piaget, J. 1932. *The Moral Judgment of the Child*. By Jean Piaget. (International Library of Psychology and Philosophy. London: Kegan Paul, Trench, Trübner & Co.
7. Tonkiss, Frank. 2005. *Space, the City and Social Theory*. Polity Press. pp. 102.
8. Lefebvre, Henri. 1991. *The Production of Space*. Wiley.
9. Tonkiss, Frank. 2005. pp. 115.
10. Arnstein, S. R. 1969. A Ladder of Citizen Participation. *Journal of the American Institute of Planners*, 35(4): 216–224.
11. https://twitter.com/drsuad/status/830838928403988480?lang=en.
12. Jacobs, J. 1961. pp. 238.

BIBLIOGRAPHY

Arnstein, S. R. 1969. A Ladder of Citizen Participation. *Journal of the American Institute of Planners*, 35(4): 216–224.

Degen, M. 2008. *Sensing Cities: Regenerating Public Life in Barcelona and Manchester*. Routledge.

Hall, S. 2012. *City, Street and Citizen: The Measure of the Ordinary*. Routledge.

Harvey, D. 2012. *From the Right to the City to the Urban Revolution*. Verso.

Jacobs, Jane. 1961. *The Death and Life of Great American Cities*. Blackwells.

Lefebvre, Henri. 1991. *The Production of Space*. Wiley.

Mead, George H. 1934. *Mind, Self, and Society*. The University of Chicago Press.

Piaget, J. 1932. *The Moral Judgment of the Child.* By Jean Piaget. (International Library of Psychology and Philosophy. London: Kegan Paul, Trench, Trübner & Co.

Sassen, S. 1991. *The Global City*. Princeton University Press.

Tonkiss, Frank. 2005. *Space, the City and Social Theory*. Polity Press.

Chapter 12

WILD CITY | FIADH-BHAILE | ORASUL SALBATIC

Mapping the Wild in the Greeny Howe of Glasgow

Alec Finlay, Deirdre Heddon, and Misha Myers

WILD CITY (2018), a collaboration between artist Alec Finlay and artist-researchers Deirdre Heddon and Misha Myers, explored what 'wildness' means in an urban context.[1] Combining the interests and practices of the collaborators, it aimed to encourage place-awareness through participatory walking and reading, place-names, translations, and community maps.

WILD CITY drew on Finlay's previous work in the field of ecopoetics and place-awareness in the Scottish wilderness, primarily the Cairngorms and Glenmoriston. Finlay had explored the hill crazes of stalking and climbing, surveying rewilding and innovative pinewood regeneration projects, and working with place-names as records of past, present, and future ecology. He had also been using place-awareness to deal with constraints on his walking due to long-term illness. Finlay sought to challenge and extend this substantial body of work, which includes five books, *The Road North* (2014), *A Company of Mountains* (2013), *Some Colour Trends* (2014), *A Far Off Land* (2017) and *Gathering* (2018), by relocating the concept of rewilding to an urban context, supported by *The Walking Library*.

The Walking Library, launched in 2012, is a long-term creative-research collaboration between Heddon and Myers, which explores the relationships between walking, literature, and environment. Inspired by repeated references throughout the late-eighteenth and nineteenth centuries to books carried on long (pleasure) walks, it asks variations on the question 'What book would you take on a walk?' in order to curate collections of books and then walk with them. The first *Walking Library* accompanied a month-long peripatetic arts festival, Sideways, which, by walking the many but disappearing slow pathways of the Flanders region of Belgium, aimed to inspire local publics to walk more and drive less. Receiving more than 200 suggestions to the question, 'What book would you take for a walk?', a stock of nearly 100 books was purchased, with books and extracts shared en route across more than 300 km.

The inaugural *Walking Library* has been followed by other editions, including *The Walking Library for Sweeney's Bothy* (2013), *The Walking Library for Women Walking* (2016), and *The Walking Library for a Wild City* (2018). To curate each edition's one-of-a-kind assemblage of books, the project follows the method established for Sideways. Each new edition starts with a question refined and tailored to pursue an inquiry within the context for which it is produced. This question is directed to audiences of each project, who then offer their suggestions or donations for books to be carried and walked, along with a rationale for their choice. A catalogue is compiled and published on the project website and added to the collection gathered for each project.[2] The collection is displayed at facilitated events, and members of the public are invited to browse and then select a book or two to walk with, in a small group of other readers-walkers. Prompted by the landscapes through which the library walks, walkers pause together to share aloud extracts from the books carried, facilitating dynamic exchanges between environments and text. At the end of each walk, walking companions are invited to reflect on their experience by creating a hand-drawn map and/or written reflection, recalling what stood out for them along the walk (Figures 12.1 and 12.2). Participants receive a *Walking Library* rucksack patch, sometimes redesigned with new colours to suit a particular edition.

Reflecting on its many editions, Heddon and Myers have discovered that *The Walking Library* creates a space that is simultaneously collective and contemplative in bringing awareness to the environment.[3] Through collective shared readings from the books carried, the books themselves shift from a background for the walk to come in and out of focus as a lens that shifts perception from elsewhere and when to here and now. The readings engender an atmosphere and ambience, bringing things from our surroundings to awareness. Likewise, the landscape guides attention, the configurations of people with places, directions of the walk, and the words read and spoken in conversations that happen along the way. The interaction of collective, book and landscape to guide and direct the walk and perception undoes the notion of any centre – of a guide, performer, or anthropocentrism.

The Walking Library for a Wild City was curated and walked in the spring and summer of 2018, taking eight different routes which covered all points on the compass. The wild city in this instance was Glasgow, and the questions posed to prompt suggestions for books for this edition were:

What book reveals wildness in the city?
What book would you rewild by walking?

The books that people suggested ranged wildly, from those which helped people see what is sometimes overlooked – for example, the variety of wild things growing in vacant lots – to dystopian apocalyptic fiction – where nature makes a comeback – alongside helpful pocket guides to urban foraging. As we walked and read and looked and sensed together, we remapped the so-called Dear Green City of Glasgow and rediscovered the significance of old names – Sauchiehall Street, *willowhaugh*; the Kelvin, *the reedy river*. Walking down Greendyke Street, through Glasgow Green, up Greenhead Street, onto Arcadia Street, continuing to Green Street, we renamed these streets Dandelion Daunder and Piss-the-Bed Verge, Windowed Wilding Garden and Copper Beech Row, Kite Green and Buddleia Bog.

Walking through the infamous Barras market, we encountered remnants of mussel shells on the pavements, the remains of a seagull's dinner (unlikely), or a wild city reveler tempted by the nearby Loch Fyne Shellfish Bar, which is quite some distance from Loch Fyne. There used to be cockle and mussel stalls in the Barras, but they have mostly all gone now. A childhood memory is shared by one of our walkers of walking through this part of the city years

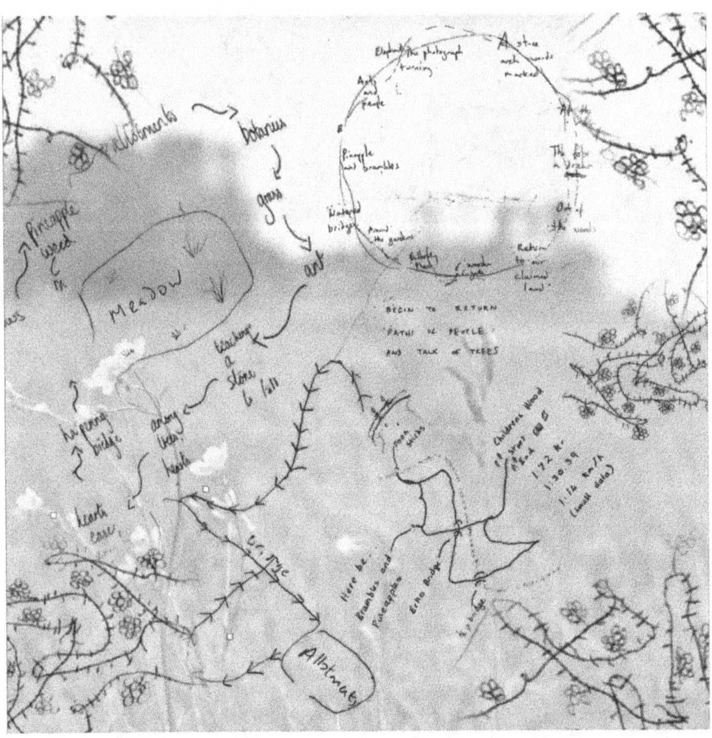

Figure 12.1 Collage of Walking Library Participants' Maps of the 'Reedy River Walk' a Walk to the Kelvin River from the Children Wood, 11 August 2018. Author's Image.

ago and the crunching of shells underfoot like walking on the beach. Sophie reads an extract from *The Seabird's Cry* by Adam Nicolson:

> *It's a disturbing realization: the gulls on the Hastings seafront, largely fed on fish and chips and on the rubbish from people's bins, have been transformed by the food we have been giving them. They are no longer wild creatures, but a reflection of who we are and how we live. They are the gulls of the Anthropocene. The impression you get, the instinctive gut feel of these birds on the seafront, that they are somehow degraded and impoverished, is borne out by the science. We have created a race of gulls that reflects the worst of us.* [4]

The maps generated by participants after the walks offer collective poems of place; collections of memories, records of the senses, readings of environments, sharing of knowledge of place, plants, and animals (Figures 12.1 and 12.2). Some are words plus image, others purely word or visual. Repetitions across the maps reveal what stood out most to the group as a collective; moments,

Figure 12.2 Collage of Walking Library Participants' Maps of the Low Walk Slow Walk' from Govanhill Baths to the Hidden Gardens, 30 June 2018. Author's Image.

things said and shared that then stuck in their collective memory. Like inside jokes or the stories a family tells and retells to one another that do not have the same significance for those outside, they evoke that bond and have the weight of something significant. Creating word clouds from across the collection of maps, buddleia and seagulls are revealed as the most often repeated.

The physical and social environment of the walker-reader in *The Walking Library for a Wild City* is the post-industrial landscape and the drawings depict environments that are far from the bucolic imaginary. The maps perform an ambulatory poetics surfacing an ambience that is wildly mixed (Figure 12.2). Tower blocks, motorways, ruinous buildings, and the detritus of human lives lived in the city are entwined with and often the sites for the city's non-human wildness, the so-called weeds pushing through the cracks in walls and

Figure 12.3 'wild city/urban crofts, after Evie Murray': Alec Finlay, ink on paper, 2018. Author's Image.

pavements, clinging on to damp overflows and rooting into piles of rubbish. Nature is not some distant, fetishized object over there but is everywhere, hanging on in there, flourishing even, despite the environmental degradation caused by the once heavily industrial city. The wildness of the human city dwellers is prevalent too in the copious piles of empty tinnies and discarded bottles. Human and more-than-human collaborate in making the cityscape. The maps are lively, invoking, in Timothy Morton's words, 'the underside of ecomimesis, the pulsing, shifting qualities of ambient poetics.' These are not 'pretty or sublime pictures of nature.'[5] And in many of the maps, there we stand, we humans, grouped together as a species within the environment.

Finlay's re-imagined maps of the cityscape offer another rendering of *WILD CITY*'s ambient and ambulant poetics, Glasgow remapped through evocative reminders of persistent ecologies that can still be felt with just a slight shift in perspective and orientation or through the different rhythms stepped by a dauner (See Figures 12.3 and 12.4). The River Clyde, bisecting the city, remains a *forceful flower*:

Figure 12.4 'wild city manifesto': Alec Finlay, riso poster; photography Mhairi Law, 2018. Author's Image.

many versions of the wild survive in place-names
place-names lead us into the wild, but names also tame the wild
the Scottish situationist doesn't drift, she dauners
bring back the seasons!
note everything wild around you;
now note the relationships between these wild elements.[6]

NOTES

1. *Wild City* was commissioned by Glasgow Life for the 2018 European Championships Festival.
2. See https://walkinglibraryproject.wordpress.com/ for all editions and catalogues.
3. Heddon and Myers have published extensively on the project: 'Pedestrian pedagogy: The Walking Library for Women Walking', *Journal of Public Pedagogies*, 2019(4), 108–117; 'Walking Library for Women Walking', in *Walking, Landscape and Environment*, eds. Pippa Marland et al. (London: Routledge, 2020); 'The Walking Library for a Wild City', *Performance Research*, 23(7), 2019, 48–49; 'The Walking Library Collections: The Convivial Logic of a Library Made for Walking', *Interartive,* 2018; 'The Walking Library: Mobilising Books, Places, Readers and Reading', *Performance Research*, 22(1), 2017, 32–48; 'Stories from the Walking Library', *Cultural Geographies*, 21(4), 2014, 639–655.
4. Adam Nicolson, *The Seabird's Cry* (London: William Collins, 2017), 145.
5. Timothy Morton, *The Ecological Thought* (Cambridge, MA: Harvard University Press, 2010), 159.
6. Finlay has published a book drawing on the collective explorations of *Wild City*, *WILD CITY/FIADH-BHAILE/ORASUL SALBATIC: Mapping the wild in the greeny howe* (Edinburgh: Morning Star Publications, 2018). This contains further maps by Finlay, along with accounts of each of the eight walks undertaken by *The Walking Library.*

BIBLIOGRAPHY

Finlay, Alec, *Wild City*, *WILD CITY/FIADH-BHAILE/ORASUL SALBATIC: Mapping the wild in the greeny howe* (Edinburgh: Morning Star Publications, 2018).

———. *Gathering* (Zurich: Hauser & Wirth, 2018).

———. *A Far Off Land* (Edinburgh: Morning Star Publications, 2017).

———. *The Road North* (Edinburgh: Morning Star Publications, 2014).

———. *Some Colour Trends* (Huntly: Deveron Arts, 2014).

———. *A Company of Mountains* (Isle of Skye: ATLAS, 2013).

Heddon, Deirdre and Misha Myers, 'Pedestrian Pedagogy: The Walking Library for Women Walking', *Journal of Public Pedagogies*, 2019(4), 2019, 108–117.

———. 'Walking Library for Women Walking', in *Walking, Landscape and Environment*, eds. Pippa Marland et al. (London: Routledge, 2020).

———. 'The Walking Library for a Wild City', *Performance Research*, 23(7), 2019, 48–49.

———. 'The Walking Library: Mobilising Books, Places, Readers and Reading', *Performance Research*, 22(1), 2017, 32–48.

———. 'Stories from the Walking Library', *Cultural Geographies*, 21(4), 2014, 639–655.

Myers, Misha and Deirdre Heddon, 'The Walking Library Collections: The Convivial Logic of a Library Made for Walking', *Interartive,* 2018.

Morton, Timothy, *The Ecological Thought* (Cambridge, MA: Harvard University Press, 2010).

Nicolson, Adam, *The Seabird's Cry* (London: William Collins, 2017).

Chapter 13

Performing Cartographies

Getting Inside and Beyond the Map

Misha Myers and Lucy Frears

INTRODUCTION

> All set, then let's begin . . . by imagining this. By imagining yourself jumping, jumping and falling, free falling in the 15,000 feet of empty airspace above you now adjusting pitch and roll and yawl with outstretched arms and legs, for sixty seconds, face down an aerial view of this place, birdseye . . .[1]

This is how the Hibaldstow track of Mike Pearson's mp3 audio walk 'Carrlands' begins. In the first sixty seconds of the walk, he describes the landscape seen from this birdseye view, the patchwork fields, hedges, roads and man-made waterways, the carr, then immerses the walker in a close-up, intricate, and intimate experience within it as they hit the ground and walk the 'bottom lands'.

The aerial view of the landscape's three dimensions represented in the two-dimensional perspective of the topographic map is disrupted or augmented by work such as Pearson's Carrlands. *Performance cartographies* offer an experiential movement through the textured embodied and storied depth of place, whereas traditional forms of map-making neglect the presence of people and their embodied experiences of place. Using performance cartographies, one is *inside* the map both physically in the landscape and sometimes simultaneously, virtually, if technologies are used, for example, through hearing and seeing a multi-media and multi-layered 'narrative archaeology' of places that can be located and experienced within the map.[2]

This chapter focuses on embodied geographies of landscape, *deep mapping* and embodied forms of mapping, wayfinding, performance, and art, including locative media art or 'remote performance', performances involving live or recorded performers mediated by or present through technology.[3] Two case studies of cartographic performance works are presented, each created by one of the co-authors and each from a different period of time in the development of locative media art technologies.

The immersive lived experiences that new forms of digital storytelling offer suggest a cartography that extends beyond the surface of the map and both metaphorical and geographical constructs of space and place. The lively and lived layers underneath the plotted lines and symbols of the map surface are revealed through an embodied encounter with place.

DEEP MAPPING FUTURES

William Least Heat-Moon coined the term *deep maps* to describe his book *PrairyErth*, an assemblage of mixed sources about Chase County, Kansas.[4] Performer Mike Pearson and archaeologist Michael Shanks developed d*eep mapping* of place further as a method of site-specific performance, which pursued textual and live performative assemblages that offered 'juxtapositons and interpretations of the historical and the contemporary, the political and the poetic, the factual and the fictional, the discursive and the sensual . . . as an alternative kind of site-report'.[5]

Through layering of remnants gathered from different eras and social standpoints in this mode of performance, the story of places becomes polyvocal and 'a stratified amalgam of relationships amongst parts appears', one which evokes rather than describes place.[6] The story is not a hegemonic nor a mimetic representation of place. Rather, it is as Clifford McLucas described, 'a conversation and not a statement'.[7]

In the late 1990s, feminist art historian and theorist Lucy Lippard called for a way to experience the layering of history in place, that deep mapping and subsequently locative media, seem to have answered: 'We need more fluid ways of perceiving the layers that are everywhere, and new ways of calling attention to the passages between old and new, of weaving the old place into the new place'.[8]

McLucas, Pearson, and Shanks' former colleague from the Welsh theatre company *Brith Gof*, developed his version of *deep mapping* and this interweaving of temporal stratigraphy by using text and image traces to document a site-specific theatre work and film to capture the creative process of his large layered graphic *deep map*, called *Stalking the San Andreas Fault*.[9] McLucas describes three basic elements of his deep maps 'a graphic work (large, horizontal or vertical), a time-based media component (film, video, performance), and a database or archival system that remains open and unfinished'.[10] In addition, he suggests they should be big, slow, include the representation of a mixture of views represented, both insider and outsider, amateur and expert, that require negotiation and contestation over who and what is represented and how; they should provoke debate about the documentation and portrayal of people and places. Increasingly, geographer artists, such as Iain Biggs, Anthony Lyons, Jane Bailey, and Owain Jones, have followed McLucas' exploration to make their own *deep maps* using text, visual arts, and media assemblage rather than performance. The eighth of McLucas' ten principles on deep maps points towards future digital possibilities that could only be imagined in 2001.[11]

CARTOGRAPHIC PERSONALISATION WITH LOCATIVE MEDIA

Locative media is often described as location-aware media, such as sound and image, triggered automatically by GPS (Global Positioning System) as in Lucy Frears' *Hayle Churks* app,[12] the second case study discussed further in this chapter. However, locative media may facilitate locational experiences with place-based content using audio players, such as in Misha Myers' *Take me to a place*, the first case study included below, as well as radio microphones,[13] live phone calls via mobile phone,[14] SMS, robocalls, and pervasive media platforms.[15]

Locative media is also described by Tuters and Varnelis as both annotative, 'virtually tagging the world', and phenomenological, 'tracing the action of the individual in the world'.[16] For example, the annotative locative media work *Transborder Immigrant Tool* created in 2008 by Electronic Disturbance Theatre 2.0/ b.a.n.g lab[17] offers audio poetry along with location of water caches for immigrants attempting to cross the United States/Mexico border on their smartphones using GPS, thereby providing not only inspiration but also information for survival. The phenomenological locative media work *Milk*, created by Esther Polak and Ieva Auzina in 2003,[18] tracked the movements of nine people involved in the production, transportation, and consumption of cheese through the use of GPS devices and then presented the information gathered in a multi-media installation.

Large works made with locative media using a smartphone and GPS satellites enriches *deep mapping* (remote) performance through the smartphone's ability to store, locate, display, and play large amounts of multi-media content across a large area. Textured collages of stories and artefacts (such as photographs) enfold the community and landscape of the past into the present, to 'create a sense of self and belonging'.[19] Locative media is able to use all media's interactive capacities through audio, still and moving image with a live map and in areas with good Wi-Fi or 4G connectivity. Participants are able to upload their own content to an unfinished open database. The uploaded participant's story could sit next to that of an expert's which would naturally raise discussion on who participates/is represented as participants self-select to add content.

In his video diary about making his deep map McLucas talks about 'trying to put people back in' the map.[20] While McLucas included images of his colleagues in his maps, locative media offers at least three other possibilities of such personalisation: authoring, tracking, and triggering. In addition to being able to upload images, personal comments and recordings can be added to enable (1) authoring of content and further interactivity. A live on-screen map can (2) track the position of the walker (usually represented as a human figure) as they move across the landscape. The voices and stories of people who inhabit or work within an area can become an aural dimension of the map and the landscape. They become embodied by the listener-walker-participant as they are encountered and (3) can be triggered as participants' bodies move through invisible GPS zones or, as in earlier forms of locative media, they press play.

Locative media works combine and transform the three elements of deep mapping (graphic, time-based, open-ended data) to bring cartographic performance back into deep mapping through interactivity, mobility, and remote or recorded presence. Locative media calls upon the participant to be both user and producer of the map, experienced as both annotative and phenomenological. The participant's own movement becomes a part of the work. They add place-based or placed content to it as they move or drift through landscape such as in Proboscis and Shawn Micallef's *Urban Tapestries* software platform which allows people to author, exchange, and experience 'virtual annotations of the city',[21] or Blast Theory's *Rider Spoke*, during which audience members' secrets are locked into locations that they contribute to and discover along a bicycle ride.[22]

While some locative media involves other modes of mobility, as with *Rider Spoke*, it is usually experienced at walking pace. This slowing down of the body–mind allows the information received both through the media and direct experience within place to be absorbed, noticed, and attended to. The multiple forms of media found in locative media involve different sensorial engagements with locations, which makes non-visual modes of mapping and engaging with landscape equally important as visual ones. For example, in hearing polyvocal perspectives of place (to which sometimes the participant can add their own voice) played through an app or MP3 player, the body acts as a 'resonance chamber'[23] as it responds to both digital content and the characteristics of the physical environment experienced visually including 'kinaesthesia (the sense of movement), proprioception (felt muscular position) and the vestibular system (sense of balance)'.[24]

While 'locomotion and cognition are inseparable, and an account of the mind must be as much concerned with the work of the feet as with that of the head and hands', all senses are involved together in this form of interactive performance.[25] It may also deepen awareness of the emotional responses to places, as pursued in the dérives of the Situationist International in Paris in the 1950s, events organised as peripatetic drifts through the city following emotional responses to locations encountered along the way to discover new routes through and modes of experiencing the city.[26] This kinaesthetic and psychophysical response to place facilitated through the combination of locative media with deep mapping presents a fourth dimension to the map.

THE FOURTH DIMENSION OF THE MAP

In a locative media context, *deep mapping* is experienced as four-dimensional (with the fourth dimension being time): it is a digging down into the map through the strata of place, histories, stories, and meanings, such as during an archaeological dig (carefully using sharp observation and various tools). In this way, it differs from conventional two-dimensional or three-dimensional mapping. Use of 360° binaural sound recordings are proposed as a way to expand and enhance the senses and deepen the experience of space. Card '7' of 22 manifesto cards presented at the Ambient Literature Symposium suggests the significance of audio: '[s]ound allows you to identify distances between you and objects, constructing the world. Reflections from surfaces characterise the size and nature of the space'.[27]

In MP3 sound walks and locative media experiences, audio opens the body to the hidden or unseen – sounds from underground or voices from the past for example and mixes digital/virtual and physical/material worlds. The recorded sounds may synchronise with those in the immediate location to reveal patterns, such as an airplane played in headphones as the walker passes under a flight path and looks up to see a plane passing overhead, as in Janet Cardiff's audio walk *The Missing Voice* (*Case Study B*; 1999), which layers a fictional reality within the everyday landscape of the city.[28] This concurrence blurs the gap between digital and physical spaces in everyday life, past and present, here and there, that is already an everyday experience for many smartphone users who access 'Internet on the move'. Cardiff, also known for her immersive video walks, describes how audiences desire a closer and more intimate engagement with art, with the 'experience of the photographic image, bigger screens, more immersive sound, until as an audience we want to go further and be inside them with our bodies [. . .] narrowing this gap'.[29]

With the 'spatial turn' across disciplines of contemporary art, media, and the social sciences, specificity of location and context has been recognised as significant in the construction of knowledge and art.[30] The renewal of interest in geographic and physical location across humanities scholarship has led to the development of 'spatial humanities' as a discipline in itself.[31] Embodied landscape geographies[32] and the '*Media Turn* in geography'[33] have also led to a shift from a focus on surfaces, physical objects, and graphic forms towards dynamic interactions, processes, and interrelationships in space. Ideas on spatiality have challenged cognitive ways of perceiving meaning in the landscape, understanding the integration of knowledge and place-making and explaining mapping practices.[34]

Spatial practices and site-specific methods of performance and art that may be understood as performance cartographies, both live and mediated, have made a significant contribution to this turn and emerging discipline, particularly through non-representational or embodied methods of evoking place. While GIS is spatially deterministic and requires landscapes and societal patterns and processes to be tied to the 'spatial geometrical primitives of point, line, polygon and pixel', it affords the ability to integrate information from a common location, to link and view simultaneously quantitative, qualitative, and visual data in conversation with one another in spaces where they occur.[35] The use of qualitative data, such as memory and narrative has enabled complexity, ambiguity, multiplicity, and contingency in the evocation of landscape and place especially while being witnessed and embodied within locations such as in site-specific performance or audio/multi-media walks discussed so far. These narratives refresh understandings of place within time encouraging reflection on the relationship between time and place.

The aerial view of the map is replaced with the embodied experience of place from the ground as an entangling intersection of multiple trajectories of movement, not as location. Geographer Doreen Massey suggests 'a "place" is formed out of the particular set of social relations which interact at a particular location'.[36] Places are collections of 'stories-so-far', suggesting that they are unfinished and on-going, which she argues 'is a prerequisite for history to be open'.[37]

REVELATION OF PLACE AS UNFINISHED

The experience of place as unfinished is enabled by locative media's affordances of openness. Its capacities for open-ended databases, interactivity and continuous authoring, and contribution of place-based content allows multiple contrasting and contested stories from different eras to be layered onto one location and be experienced in situ. Instead of being organised according to the perceptions of one cultural world view, voices once hidden or ignored are heard alongside more dominant narratives of places, thus offering alternatives to the repressive or coercive tendencies of map-making. In this way, locative media affords temporal and cultural dissonance.

The two case studies that follow are maps created by each author, one an example of an early stage of locative media pre-GPS and the other a GPS-enabled locative app with which media content was experienced and triggered automatically and more seamlessly with the body's interconnection and movement within a particular place. Both works are potentially lost as the technologies that originally enabled them have become obsolete or replaced, or as the places that they map become erased or transformed beyond recognition.

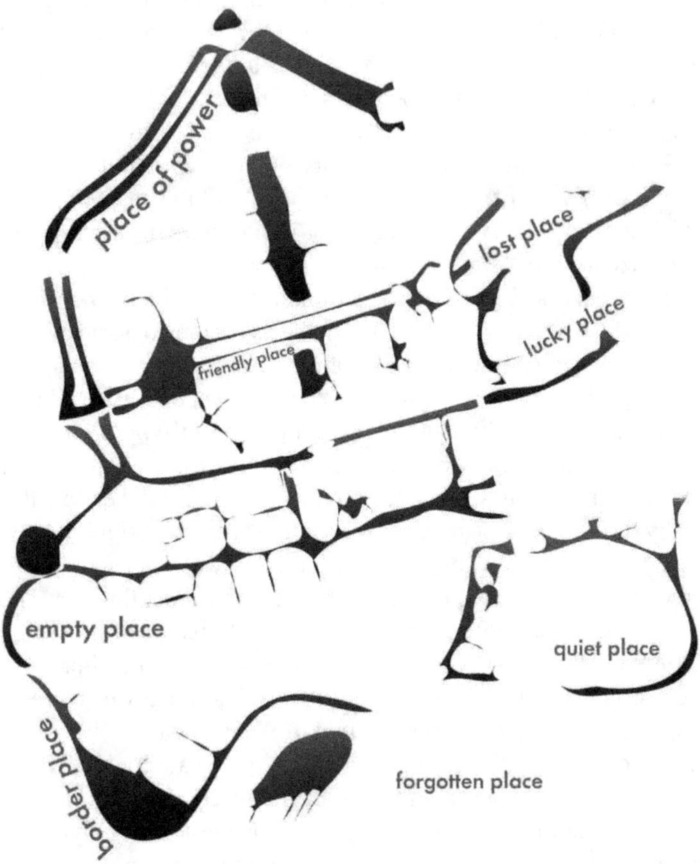

Figure 13.1 Maps from Take Me to a Place, an Audio Work Directed by Misha Myers, Artwork Design by Dan Harris, 2004. Author's Image.

CASE STUDY #1: *TAKE ME TO A PLACE* (2004)

Take me to a place is a map of overlooked, invisible, displaced or disappeared places, and the map itself is unseen (Figure 13.1). It is heard through voices and through song. *Take me to a place* was a song map created and composed with a group of trans-national inhabitants of Plymouth and theatre and social work students working together in collaboration and directed by Misha Myers.[38]

Take me to a place was originally created as twenty audio tracks including nine interludes and eleven original songs to be experienced in different ways: (1) as a guided audio walk, which the listener-walker-participant could take for a walk with a portable audio player/listening device and headphones and activate the tracks at locations as directed within the map; (2) as a collection of songs to be listened to anywhere; and (3) as a live performance performed by the creators and performers of the song map at the Barbican Theatre, Plymouth (Figure 13.2). It was recorded in the professional recording studios of Music Zone, a community music charity based in Plymouth.

Maps are always a snapshot, or earshot in the case of this map, of a moment in time within the dynamic flow of change that is happening in place. As much as maps may make apparent what is there, this map conveys what is not there or is no longer there. This can be a source of frustration when a landmark necessary for orientation has been removed or a road or path has become obstructed. However, this experience of temporal dissonance, of disparity between the map and lived experience of place, reveals a palimpsest of tenses, temporalities, and paces.

If you take a walk in Plymouth today guided by *Take me to a place* as an audio walk, something it was originally designed for, you would experience this temporal dissonance. You would find that the Plymouth and the route through it described by the map has changed and in some places quite significantly due to redevelopment of the city's commercial centre that was in progress at the time of the map's making and which it guides the walker-listener through. This redevelopment, launched in 1993 as the Plymouth 2000 plan and then relaunched again in 1999 as Plymouth 2020, is the city's second major re-imagining after it was reconstructed following the destruction of it's medieval centre with the Plymouth Blitz of 1941, A German air raid in World War II.

The songs of *Take me to a place* were not geolocated with GPS and the audio map does not always align with a precise coordinate on a map, but with a general location, such as Central Park or the Hoe in Plymouth, both expansive parks in the city. However, one song does transport the listener to an iconic landmark in the British landscape, a red phone box, a visual reference point that too may be disappeared with technological transformation.

However, this alteration does not just erase landmarks in the landscapes that locative media works map. Where other earlier forms of non-digital maps have survived with preservation, these digitally enabled maps may disappear as their operating systems and mechanisms become obsolete. This is concerning, given the point made about their significance in revealing different histories of place. This is one impetus to revisit this work here as a case study nearly two decades after it was created and to not just describe it, but to present it in a way that it could be re-experienced. The simplicity of the technology used in its making allows for this. It exists as audio files. At the time it was made, these files were disseminated via an audio CD to be experienced as an audio walk, and it was also performed live as a map to be listened to and followed in the imagination.

Take me to a place includes multiple cultural perceptions of place expressed through diverse voices and musical styles, demonstrating the cultural dissonance afforded by locative media's open-ended capacities. The members of the group that created *Take me to a place* were originally from Algeria, Afghanistan, Estonia, Germany, Jordan, Kurdistan, Russia, the United States, and the United Kingdom. The group held not only diverse cultural worldviews but also personal experiences of place and place-belonging, some with transnational, migrant or refugee experience and others as locals with life-long connections to the city. Along with those different cultural perspectives, they also brought different musical styles and languages to the project. The process of creating the map invited the group to pair up or create songs about particular places in small groups to share these perspectives and musical styles, to bring them together to find expression for their different experiences of places.

The group created a typography of places in the city – *friendly places, lucky places, lost places, border places, places of power, quiet places, avoided places,* and *places of strength.* Together they searched for these places in the city and shared their experiences, emotional associations, and perceptions of the locations they found. Together they developed lyrics about the locations that interested them most. Sometimes the group would support the development of the song around one person's story or perspective of a place, in other instances, the song included conflicting or diverse views or reflected the mutual understanding that came out of a conversation about a place. This conversation is also heard in the diversity of musical tonality and vocal styles. In the songs, the listener can hear the negotiation of different cultural perspectives as the singers and musicians attempt to make sense of, come to a mutual understanding of their different musical styles. Sometimes that is heard musically as dissonance or as polyphony with one independent melody interweaving with another.

What follows is the tracklist of the song map with links to the sound files stored on Soundcloud. These audio reference points can still guide a listener on a walk in Plymouth or in the imagination from wherever you are, at least for as long as the platform hosting them survives. Each song track is followed by 'Interludes' which either give a short introduction to a song or provide a poetic description of landmarks to look for to find the next location of the next song. The scripts of these Interludes are included here in italics. Some more pragmatic directions are included, which were originally printed on an insert included with a CD, the first method of dissemination of the map. However, as mentioned previously, both technology and places change, and these directions are no longer as helpful as they once were as a guide to actual routes through the city. The earpoints and the pathways and landmarks they describe may have since been removed or obstructed. They, along with the textual traces of them here, reveal Plymouth's 'story-so-far' and guide the walker through a ghost landscape, through the cracks and fissures made by temporal dissonance.

Take Me to a Place

We start on the Hoe, beneath Smeaton's Tower. From here there are so many views of Plymouth Sound to choose from. Why not try them all while listening? Walk and listen or sit down for a while looking out to sea.

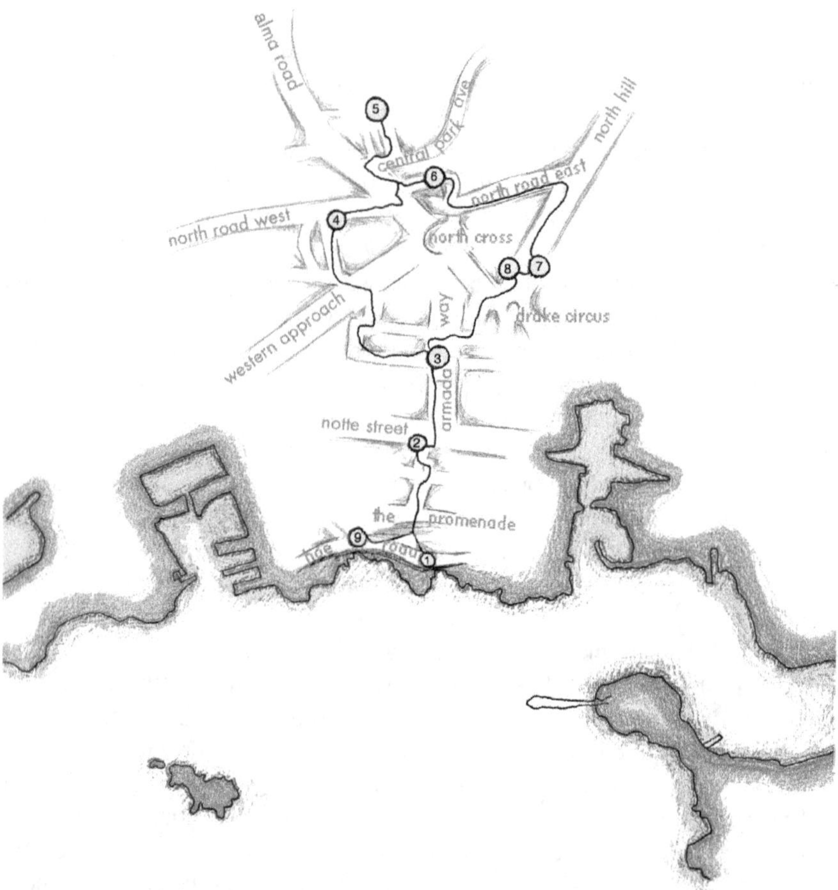

Figure 13.2 Map from Take Me to a Place, an Audio Work Directed by Misha Myers, Artwork Design by Dan Harris, 2004. Author's Image

TRACK 1. TAKE ME TO A PLACE

An improvised performance by Hamond and Rachel

The text for this song came from different members of the ensemble and was used as inspiration for the walks we shared together throughout our devising process.

https://soundcloud.com/homingplace/vocalatitude-take-me-to-a

TRACK 2. INTERLUDE 1

Beneath a floral introduction to the city floats our private island, uniting inhabitants and visitors to Plymouth alike. Smeaton's Tower beams down, watching over us, while the piano calls from the New Waterfront Café. From here we can watch all types of people passing, and understand this country and culture we are living in. Oh I do like to be beside the seaside. Don't dwell in the past.

'I take you to a Friendly place'

https://soundcloud.com/homingplace/vocalatitude-walking-interlude

TRACK 3. 'C'EST LA MER' 1

Written by Gulio, Joe, Katharine and Mohammed and performed by the ensemble

A Friendly place.

https://soundcloud.com/homingplace/vocalatitude-cest-la-mer-1

TRACK 4. INTERLUDE 2

In this border place between sea and land, far and away, home from home, what do you feel?

'I feel at home'

https://soundcloud.com/homingplace/vocalatitude-walking

TRACK 5. 'BEAUTIFUL EYES'

Written and performed by Clementine, Doris, Joe, Joy, Mohammed, and Monika

This song is an English translation of Mohammed's song from Afghanistan, in which he sings the original.
https://soundcloud.com/homingplace/vocalatitude-beautiful-eyes

TRACK 6. INTERLUDE 3

From the Hoe, cross the Promenade to the left, and follow Osbourne Place. Turn right to Citadel Road. On the left is Armada Way, then turn left again into Windsor place, which leads you onto Notte Street.

Here are happy times, but behind Drake's steady gaze lies turbulent history. Cross the Citadel Road boundary and follow the Armada's colonising Way to Windsor Place and today's ruling class in Gill Akaster House.

'I take you to a Lost place'
https://soundcloud.com/homingplace/vocalatitude-walking-1

TRACK 7. 'SO FAR FROM YOU'

Written by Ramazan with Lauren and Katie and performed by Ramazan, Lauren, Katharine, Sophie and Gulio

Ramazan's Lost place.
https://soundcloud.com/homingplace/vocalatitude-so-far-from-you

TRACK 8. INTERLUDE 4

Turn right along Notte Street, then left into Princess Street Ope. Pass the fountain and the law courts, then cross under the subway to Armada Way. Take the weight off your feet at the sundial.

'Take me to a Lucky place'

Throw a penny, make a wish
https://soundcloud.com/homingplace/vocalatitude-walking-2

TRACK 9. 'ECHO'

Written by Igor, Joy, Rachel, Sophie and Yelena and performed by the ensemble

Yelena's Lucky place.
https://soundcloud.com/homingplace/vocalatitude-echo

TRACK 10. INTERLUDE 5

Walk left down New George Street, then turn up Market Avenue. Take Market Way on the left, and cross the dual carriageway. Take the path past the school and up Penrose Street to North Road West. Turn right, and at the end of the road turn left to cross over the footbridge. Turn to your right up Saltash Road, and right at the roundabout into Central Park Avenue. Take the second right to the park on Holdsworth Avenue.

Under the subway Barry sings for his supper, while traffic at the old junction carries on by on foot to the new markets of Topshop, Disney, Dingles, and Barratt's. The sun still rises and sets, as the big hand slowly counts the hours.

'It reminds me of . . . '

Yet this is a city of contrasts. You can buy everything here – even people. At times you have to watch your back. Approaching from the West, a far cry from London, this Oxford Street leads to North Road West.
https://soundcloud.com/homingplace/vocalatitude-walking-3

TRACK 11. 'CENTRAL PARK'

Written and performed by Igor, Joy, Rachel, Sam, Sophie, and Yelena

This song arose from the discussion over contrasting views of this particular part of the city. It is also a positive affirmation of strength and individuality in the face of displacement. Yelena sings the verse in Russian.
https://soundcloud.com/homingplace/vocalatitude-central-park

'I wake up today and go for a walk.
Sun shining into my eyes but I'm dancing.
I go on the street like a little bit crazy
Because I have a very good mood.
People looking in the window and can't understand me
But it's all the same for me and I will go further.

I will go around Central Park
And sing about beautiful days.
Friends singing my song and dancing with me
In spite of that rainy day.
People pass by and can't understand us
But it's all the same for us and we will go further.

TRACK 12. INTERLUDE 6

Retrace your steps back to the train station beyond the footbridge.

'When I first arrived I felt . . . '

https://soundcloud.com/homingplace/vocalatitude-walking-4

TRACK 13. 'PENZANCE TO PLYMOUTH'

Written and performed by Igor, Joy, Katharine, Sam, Sophie, and Yelena

This song is based on the story of Igor's arrival in the city ..."always running".

https://soundcloud.com/homingplace/vocalatitude-penzance-to

TRACK 14. INTERLUDE 7

Past the car park and up the hill, then turn left along North Road East. At the end, turn right, and cross the road to the Central Museum and Library.

After crossing this border into Plymouth, Wake up and find what Worth it Holds. The streets behind the station can show you the whole city. Penny, come quick! Another clock points right to a district rich with Parks, and we revisit Oxford on our way to a public retreat where the Thorn has no sting.

'I take you to a Friendly place'

https://soundcloud.com/homingplace/vocalatitude-walking-5

TRACK 15. 'CENTRAL LIBRARY'

An improvised performance by Gulio, Hamond, Ramazan, and Sam

Ramazan sings about Igor's Homely place, using text from different members of the ensemble based on their experiences of the library.

https://soundcloud.com/homingplace/vocalatitude-the-libary

TRACK 16. INTERLUDE 8

Leave the library, turn right, and cross the road at the pedestrian crossing to the phone box opposite, facing the city centre.

'I take you to a Homely place'

(Why do you go to the library?) 'Because. . . '

Down the Man-made Road, off your soap-box at Hyde Park corner, and through the second city centre. There is nothing Plain about this – see the Haart of Plymouth. Sherwell United Church really is, on the first Friday of the month, but don't hang about outside. We'll show you another way home.

When one door shuts, another one opens.
https://soundcloud.com/homingplace/vocalatitude-walking-6

TRACK 17. 'THE PHONE BOX'

Written and performed by Clementine, Doris, Mohammed, and Monika with backing from Joe, Joy, Lauren, Ramazan, Sam, and Sophie

This song is written about Doris and Monika's first experiences of the city, and the many occasions spent on the telephone trying to settle into a new life.
https://soundcloud.com/homingplace/vocalatitude-phone-box

TRACK 18. INTERLUDE 9

Come out of the phone box, and take the subway under Drake's Circus to the precinct. Turn right into Cornwall Street, then left down Armada Way and under the subway. Turn right out of Princess Street Ope, then cross Notte Street to turn left into Windsor Place and Armada Way. Back on the Hoe, turn right and follow the seafront. On your right are some steps up to the War Memorial, opposite the ice-cream stall. Take a seat and reflect.

Oh what a Circus. If Drake had lost his game of boules, what could we have lost? Or gained? ¿Donde est costa? Come full circle to the safety of our liminal coast, neither here nor there, but somewhere in between. History is never as simple as it sounds. Etch our names in silver, 2004.
https://soundcloud.com/homingplace/vocalatitude-walking-7

TRACK 19. 'JOHN BROWN'

Written and performed by Clementine, Doris, Joe, Lauren, and Monika with backing by the ensemble

This song was written about the War Memorial on the Hoe, and the suggestions of other lives that come from reading the names on the monument, the ones that are missing.
https://soundcloud.com/homingplace/vocalatitude-john-browns-place

TRACK 20. 'C'EST LA MER' 2

Written by Gulio, Joe, Katharine, and Mohammed, performed by the ensemble
https://soundcloud.com/homingplace/vocalatitude-cest-la-mer-2

CASE STUDY #2: HAYLE CHURKS APP (2013)

The *Hayle Churks* app, created by Lucy Frears, was set in a marginal semi-rural coastal location in Cornwall, the United Kingdom, on the edge of a post-industrial town suffering from visible neglect and, similar to *Take me to a place,* during a period of rapid redevelopment. Where *Take me to a place* brought to the foreground voices that may be neglected and paved over with the redevelopment in Plymouth, the regeneration in Hayle became a major focus or impetus for the development of the app. Initial insensitive redevelopment threatened the town's World Heritage status as part of the Cornwall and West Devon Mining Landscape World Heritage Site. The app offered a new way of experiencing, understanding, appreciating, and potentially caring for land that had an imperceptible heritage. Using media, historically perceived as haunted,[39] and sound that lends itself to 'amplifying the [. . .] uncanny qualities of places' located disembodied voices told their memories of the place pre- pre-ruin, demolition, and gentrification.[40]

Free to all users, the app was created using Appfurnace by Calvium as part of Frears' interdisciplinary practice-based PhD studentship initiated by Dr. Misha Myers and Professor Desilvey, supported by the *European Social Fund* (ESF).[41] Partly funded by Heritage Lottery Fund, the app was published on iTunes (December 2013) and won a national Collections Trust award (June 2014). Designed to be experienced on location, the audio welcome started automatically as soon as the app opened (see figure 13.3). Once the 'Walk in Hayle' button was pressed (figure 13.4) GPS triggered audio and images as the walker moved transforming GPS – the global positioning system – into a 'Geo-Poetic System'. The intention was that embodiment, immersion, the participant's body–mind and the magic of locative media would interact and connect the listener-walker-participant to landscape.

The smartphone screen was small so the eye was led from displayed images and a live map (and ear through audio to prompt investigation and

Figure 13.3 Using Hayle Community Archive Images for the Welcome Screen. Author's Image.

Figure 13.4 Selection Screen. Author's Image.

imagination) out into the material landscape. As a telephone, the smartphone is the conduit in a private exchange with a disembodied voice; one can feel closer to a person on the telephone than those physically present.[42] By contrast during this locative media experience, the phone connected the user not to a person at a distance as with a telephone call but with place, the location where they are standing. The person speaking was not there on the line, indeed they might not still be alive, but their absent and present voice was heard.

Audio content included oral histories from the archive of the Hayle Oral History Project coordinated by Frears 2008–2010. Memories, containing more than personal stories of place, were an essential element of the app.

They capture dying accents, colloquial expressions, vocabulary, the pace, pronunciation, and tone of speech – the essence of place. App users walked through voices and the sound of place they contain within them re-placing them back into the community and landscape to create 'living history',[43] an 'active archive'.[44] The app also included memories and expert interviews made specifically for the app, a seven-part fictional narrative called *Minnie's Story* that made public the racism between American servicemen based in the town during the war, plus a five-part guerrilla takeover by the '*Counter Tourist*' written and performed by *Crab Man*/ Phil Smith. The '*Counter Tourist*', recorded to sound as though talking through the phone, spoke directly to the participant and encouraged alternative approaches, both performative and playful, to explore the heritage site through interaction and imagination and open up new paths to knowing.

The app also contained three interactive questions, (for example, what could this land be used for); a musical loop etc (containing Hayle field recordings); three original songs; images or a series of images that showed with each new story while an on-screen map geolocated the participant at all times (figure 13.3).[45] A soundscape loop was built with layered sounds gathered around Hayle, many recorded with binaural mics. Binaural sound was used to open up a hyper-real three-dimensional aural sphere around the participant's body.[46] 360° binaural sound heard through headphones can be deeply immersive, but the aim was also to pitch the senses out into the landscape, to aid embodiment and the intertwining and interaction of human and landscape. Ambient environmental sounds (such as seabirds calling and waves) bled in past the headphones to mix with the binaural recorded digital sound to create what George Bures Miller refers to as 'a third reality'.[47] Reid and Hull describe the 'synaesthetic confusion caused when you are not sure if a sound is real or virtual' as a locative media 'magic moment'.[48] Deep immersion is temporarily interrupted when the participant looks around and listens to the physical world to confirm whether the waves or seabirds are real or on the recording.

The way memories are recounted is often nonlinear, jumping between events and themes rather than following a 'linear chronological sequence'.[49] The app was also nonlinear, the order in which each participant heard the content depended on her personal route around the site. Even though the drift (undirected movement) transformed into a suggested marked route in later iterations (see figure 13.5) there was no guaranteed sequence of clips and no narrator to provide context, coherence, or a linear history. The *Churks* app wasn't a historical tour of Hayle then, it was an experiential haunting of place. Remnants of memories fade in and out. Personal stories connect with listeners to share 'a private moment in a public space',[50] changing 'who gets to tell the story'.[51] Experiments within the app were designed to shift the

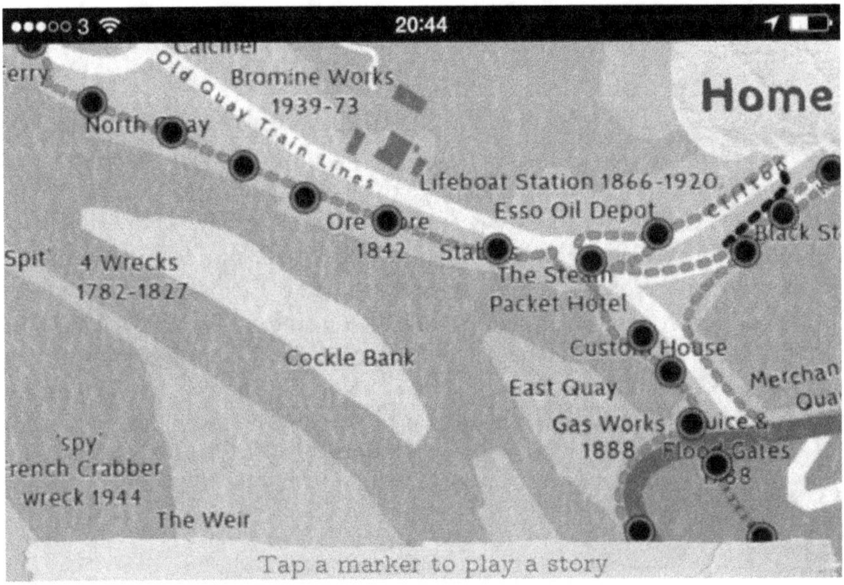

Figure 13.5 Hayle Churks Smartphone App Created by Lucy Frears Using Appfurnace by Calvium with Graphic Design by Chris Price for the Map Screen. Author's Image.

participant through layers of immersion and embodiment to connect users with both the digital storyworld and physical landscape – to connect participants to the layered stories of an industrial landscape.

CONCLUSION

If landscape is taken as a *taskscape* as described by Tim Ingold, a place of activity and movement of people – connection to people could be interpreted as also a connection to place.[52] McLucas spoke of people being removed from maps,[53] but the strategies and principles of deep mapping combined with locative media discussed here along with the two locative media case studies *Take me to a* place and *Hayle Churks* put people back into the map.

Digital cartography and spatial storytelling have become more integrated into everyday life with use of social media platforms enabled with personalised map-making tools. Comments, images, or stories pinned to specific places and times can be uploaded and shared with a mobile phone, apps and social media, promoting an experience of being located all the time. While performance cartographies may capitalise and draw upon these propensities and behaviours of everyday locatedness, they do so by activating and instatiating experiences of place that are not arbitrary.

As examples of performance cartographies both *Take me to a place* and *Hayle Churks* do this in two ways, first by locating voices and the spoken memories in landscape, then by attracting people to engage with and pay attention to hidden, neglected, vulnerable, or threatened histories and experiences in two places undergoing rapid change. The new knowledge and understanding of places that cartographic performance offers may promote attachment and belonging to place, sense of care for its past and future. However, they themselves are vulnerable and threatened by the pace of technological development and neglect of preservation. So the question arises about who will care for them and what value do they have. They may bring people into the map, but also people in place and the destinies of both are intertwined.

NOTES

1. Mike Pearson and John Hardy (2006) *Carrlands* [locative media: three MP3 audio walks; River Ancholme valley, UK]. Available at: http://www.carrlands.org.uk/ [Accessed 20 January 2015].
2. Jeremy Hight (2006) Views from Above: Locative Narrative and the Landscape. *Leonardo Electronic Almanac*, 14(7–8): 1–10.
3. Mike Pearson (2010) *Site-Specific Performance*. New York: Palgrave Macmillan, p. 82. See also, Mike Pearson (2011) Deserted Places, Remote Voices: Performing Landscape. In Steve Daniels et al., eds. *Envisioning Landscapes, Making Worlds: Geography and the Humanities*. New York: Routledge, p. 282.
4. William Least Heat-Moon (1999) *PrairyErth: (a deep map)*. Boston: Mariner Books. [Originally published: Boston: Houghton Mifflin, 1991.] See also Mike Pearson and Mike Shanks (1997) Performing a Visit: Archaeologies of the Contemporary Past. *Performance Research*, 2(2): 51.
5. Pearson and Shanks (1997: 41–53). See also Mike Pearson and Mike Shanks (2001) *Theatre/Archaeology*. London: Routledge, pp. 158, 162. For other examples of Pearson and Shanks *deep mapping* see also *Deep Maps or Esgair Fraith* (1993-1998), ‘Performing a Visit’ (1997), Pearson’s performance ‘Bubbling Tom’ (2000) and Pearson’s *In Comes I* (2006).
6. Joseph Corner (1999) The Agency of Mapping: Speculation, Critique and Invention. In Dennis Cosgrove, ed. *Mappings*. Critical views. London: Reaktion, p. 235.
7. Clifford McLucas (2001) *Deep Mapping*. Available at: http://cliffordmclucas.info/deep-mapping.html [Accessed 3 February 2013].
8. Lucy Lippard (1997) *The Lure of the Local: Senses of Place in a Multicentered Society*. New York: New Press, p. 85.

9. Brith Gof was founded in 1981 by Mike Pearson and Lis Hughes Jones. With Pearson and Clifford McLucas the theatre company became well known for large scale site-specific work (National Library of Wales 2013). Clifford McLucas (2001) *Deep Mapping.* Available at: http://cliffordmclucas.info/deep-mapping.html [Accessed 3 February 2013]; (2000-2001) *Stalking the San Andreas fault* [mixed media deep map]. Available at: http://metamedia.stanford.edu/~mshanks/threelandscapes/Map-on-a-wall-video.html [Accessed 3 February 2013].

10. Clifford McLucas (2001) '10 things..' Available at: http://cliffordmclucas.info/deep-mapping.html [Accessed 3 February 2013].

11. Unfortunately, as McLucas died in 2002, he missed the opportunity to pursue this potential of deep mapping through locative media.

12. Lucy Frears (2013) *Hayle Churks app* [locative media smartphone app; Hayle, UK]. Available at: https://itunes.apple.com/gb/app/hayle-churks/id777596532?mt=8 [Accessed 20 December 2013].

13. See Duncan Speakman (2006) *Sounds from above the Ground.*

14. See Rimini Protokoll (2005) *Call Cutta* and Misha Myers (2004) *Yodel Rodeo,* Homelands Exhibition, Spacex Gallery, Exeter, UK.

15. See Misha Myers (2016) *Nobody's Ocean,* Monash Academy of Performing Arts and (2020) The Waves Carve Their Own Desires: The Affects and Agencies of Intramedial Performance. *Research in Drama Education: The Journal of Applied Theatre and Performance* 25(3), 364–377. Also see Robert Walton and Jason Milling *(2016) Vanitas,* Available at: http://www.vanitas.online/.

16. Marc Tuters and Kazys Varnelis (2006) Beyond Locative Media: Giving Shape to the Internet of Things. *Leonardo* 39(4): 359.

17. Electronic Disturbance Theater et al. (2008) *The Transborder Immigrant Tool* [locative media: a smartphone plays poetry & directs immigrants crossing the desert between Mexico & USA to water]. Available at: http://www.lareplay.net/artists/electronic-disturbance/ [Accessed 16 March 2012].

18. Esther Polak and Ieva Auzina (2003) *MILK* [locative media: GPS maps Latvian milk to its destination – cheese on a Dutch table]. Available at: http://www.milkproject.net/en/contact.html [Accessed 2 February 2012].

19. Barbara Bender (2001) Introduction. In Barbara Bender and Margot Winer, eds. *Contested Landscapes: Movement, Exile and Place*. Oxford: Berg Publishers, p. 5.

20. See notes from McLucas' video included on: http://cliffordmclucas.info/deep-mapping.html.

21. Proboscis (2002–2004) *Urban Tapestries* [locative media: participatory story project]. Available at: http://proboscis.org.uk/projects/2000-2005/urban-tapestries/ [Accessed 17 December 2012].

22. Blast Theory (2007) *Rider Spoke* [locative media: leaving and accessing messages using a bicycle and handheld computer]. Available at: http://blasttheory.co.uk/bt/work_rider_spoke.html [Accessed 10 December 2011].

23. Mirjam Schaub and Janet Cardiff (2005) *Janet Cardiff: The Walk Book.* Vienna: Thyssen-Bornemisza Art Contemporary, p. 201.

24. Mark Paterson (2009) Haptic Geographies: Ethnography, Haptic Knowledges and Sensuous Dispositions. *Progress in Human Geography*, 33(6): 768.

25. Tim Ingold (2011) *Being Alive: Essays on Movement, Knowledge and Description.* New York: Routledge, p. 17.

26. Guy Debord [1958] (2006) Preliminary Problems in Constructing a Situation. In K. Knabb, ed. *Situationist International Anthology.* Berkeley: Bureau of Public Secrets, pp. 49–51.

27. Ambient Literature Symposium, 4-5 May 2017, Bristol. The manifesto cards can be accessed here: http://ambientlit.com/wp-content/uploads/2017/05/22-footnotes-for-ambient-literature.pdf and the Ambient Literature project at: http://ambientlit.com/.

28. Janet Cardiff (1999) *The Missing Voice: Case Study B* [locative media: binaural audio walk; London, UK]. Available at: http://cardiffmiller.com/artworks/walks/missing_voice.html [Accessed 10 November 2011].

29. Atom Egoyan (2002) On Janet Cardiff. *Bomb*, 79 (Spring), 60–67.

30. Cosgrove, D. (1999) (p. 7).

31. David Bodenhamer (2010) *The Spatial Humanities: GIS and the Future of Humanities Scholarship.* Indiana University Press.

32. Lorimer, H. and Wylie, J. (2010) LOOP (a geography). *Performance Research,* 15(4), 4–11 (doi:10.1080/13528165.2010.539872).

33. Buschauer, R and Willis, K.S. (2013) *Locative Media: Multidisciplinary Perspectives on Media and Locality.* transcript Verlag. Bielefeld.

34. Joseph Corner (1999).

35. Bodenhamer, D. (2010) (p. xi).

36. Doreen Massey (1994) *Space, Place and Gender.* Cambridge: Polity Press, p. 168.

37. Massey, D. [2005] (2007) *For Space.* London: Sage Publications Ltd., pp. 55, 59.

38. The project was funded by the Dartington Plus, a joint initiative of the Dartington Hall Estate and Dartington College of Arts.

39. Jeffery Sconce (2000) *Haunted Media : Electronic Presence from Telegraphy to Television.* Durham, NC: Duke University Press.

40. Michael, Gallagher (2014) Sounding Ruins: Reflections on the Production of an 'Audio Drift'. *Cultural Geographies*, 21: 3:1.

41. The Hayle Churks app was developed through the PhD studentship 'Designing Landscape Narrative Experience with Locative Media' funded by European Social Fund II grant Grant Number 011200NCO5.

42. See Joshuan Meyrowitz (1985) *No Sense of Place: The Impact of Electronic Media on Social Behavior.* Oxford University Press, p. 38 and Sherry Turkle, S. 2008. Always-On/Always-On-You: The Tethered Self. In James. E. Katz, ed. *Handbook of Mobile Communication Studies.* Cambridge, MA: MIT Press, p. 122.

43. Della Pollock (2005) *Remembering: Oral History Performance.* Palgrave, p. 1.

44. Hayden Lorimer in Toby Butler (2007) Memoryscape: How Audio Walks Can Deepen Our Sense of Place by Integrating Art, Oral History and Cultural Geography. *Geography Compass,* 1(3), 370.

45. The acousmatic musical loop was created by Philip Reeder.The songs were written and performed by Thirty Pounds of Bone, Johny Lamb.

46. Caroline Christov-Bakargiev, Janet Cardiff and George Bures Miller (2012) Künstlergespräch: Cardiff & Miller und Carolyn Christov-Bakargiev [at the exhibition: Janet Cardiff & George Bures Miller. Werke aus der Sammlung Goetz]. Available at: http://www.hausderkunst.de/index.php?id=132&tx_ttnews[tt_news]=1618 [Accessed 16 April 2012].

47. Miller in Christov-Bakargiev et al. (2012).

48. Josephine Reid and Richard Hull. 2011. What Makes Mediascapes Compelling? Insights from the *Riot! 1831* Case-study. In Martin Rieser, ed. *The Mobile Audience: Media Art and Mobile Technologies*. Architecture Technology Culture. New York: Rodopi, pp. 197, 193–204.

49. Alessandro Portelli, 1991. *The Death of Luigi Trastulli and Other Stories: Form and Meaning in Oral History*. Albany, NY: State University of New York Press, 63–65.

50. Reid and Hull (2011:196).

51. James Farman (2012) *Mobile Interface Theory: Embodied Space and Locative Media*. New York: Routledge.

52. Tim Ingold (1993) The Temporality of the Landscape. *World Archaeology*, 25, 159–170.

53. McLucas (2001).

BIBLIOGRAPHY

Alessandro Portelli. (1991) *The Death of Luigi Trastulli and Other Stories: Form and Meaning in Oral History*. Albany, NY: State University of New York Press.

Barbara Bender. (2001) Introduction. In Barbara Bender and Margot Winer, eds. *Contested Landscapes: Movement, Exile and Place*. Oxford: Berg Publishers.

Buschauer, R and Willis, K.S. (2013) *Locative Media: Multidisciplinary Perspectives on Media and Locality*. transcript Verlag. Bielefeld.

Clifford McLucas (2001) *Deep Mapping*. Available at: http://cliffordmclucas.info/deep-mapping.html [Accessed 3 February 2013].

David Bodenhamer (2010) *The Spatial Humanities: GIS and the Future of Humanities Scholarship*. Indiana University Press.

Della Pollock (2005) *Remembering: Oral History Performance*. Palgrave.

Doreen Massey (1994) *Space, Place and Gender*. Cambridge: Polity Press.

Guy Debord [1958] (2006) Preliminary Problems in Constructing a Situation. In K. Knabb, ed. *Situationist International Anthology*. Berkeley: Bureau of Public Secrets.

Hayden Lorimer in Toby Butler (2007) Memoryscape: How Audio Walks Can Deepen Our Sense of Place by Integrating Art, Oral History and Cultural Geography. *Geography Compass,* 1(3): 370.

James Farman (2012) *Mobile Interface Theory: Embodied Space and Locative Media*. New York: Routledge.

Jeffery Sconce (2000) *Haunted Media : Electronic Presence from Telegraphy to Television*. Durham, NC: Duke University Press.

Jeremy Hight 2006. Views From Above: Locative Narrative and the Landscape. *Leonardo Electronic Almanac*, 14(7–8).

Joseph Corner (1999) The Agency of Mapping: Speculation, Critique and Invention. In Dennis Cosgrove, ed. *Mappings*. Critical Views. London: Reaktion, p. 235.

Josephine Reid and Richard Hull. (2011). What Makes Mediascapes Compelling? Insights from the *Riot! 1831* Case-study. In Martin Rieser, ed. *The Mobile Audience: Media Art and Mobile Technologies*. Architecture Technology Culture. New York: Rodopi, pp. 197, 193–204.

Joshuan Meyrowitz. (1985) *No Sense of Place: The Impact of Electronic Media on Social Behavior*. Oxford University Press.

Lorimer, H. and J. Wylie (2010) LOOP (a geography). *Performance Research*, 15(4): 4–11 (doi:10.1080/13528165.2010.539872).

Lucy Lippard (1997) *The Lure of the Local: Senses of Place in a Multicentered Society*. New York: New Press.

Marc Tuters and Kazys Varnelis (2006) Beyond Locative Media: Giving Shape to the Internet of Things. *Leonardo*, 39(4): 359.

Mark Paterson (2009) Haptic Geographies: Ethnography, Haptic Knowledges and Sensuous Dispositions. *Progress in Human Geography*, 33(6): 768.

Michael, Gallagher (2014) Sounding Ruins: Reflections on the Production of an "Audio Drift." *Cultural Geographies*, 21(3): 1.

Mike Pearson (2010) *Site-Specific Performance*. New York: Palgrave Macmillan.

Mike Pearson (2011) Deserted Places, Remote Voices: Performing Landscape. In Steve Daniels et al., eds. *Envisioning Landscapes, Making Worlds: Geography and the Humanities*. New York: Routledge, p. 282.

Mike Pearson and Mike Shanks (1997) Performing a visit: Archaeologies of the contemporary past. *Performance Research*, 2(2): 51.

Misha Myers (2016) *Nobody's Ocean,* Monash Academy of Performing Arts and (2020) The Waves Carve Their Own Desires: The Affects and Agencies of Intramedial Performance. *Research in Drama Education: The Journal of Applied Theatre and Performance* 25(3): 364–377.

Mike Pearson and Mike Shanks (2001) *Theatre/Archaeology*. London: Routledge, pp. 158, 162.

Robert Walton and Jason Milling (2016) *Vanitas,* Available at: http://www.vanitas.online/.

Sherry Turkle, S. (2008) Always-On/Always-On-You: the tethered self. In James. E. Katz, ed. *Handbook of Mobile Communication Studies*. Cambridge, MA: MIT Press.

Tim Ingold (1993) The temporality of the landscape. *World Archaeology*, 25.

Tim Ingold (2011) *Being Alive: Essays on Movement, Knowledge and Description*. New York: Routledge.

William Least Heat-Moon (1999) *PrairyErth: (a deep map)*. Boston: Mariner Books. [Originally published: Boston: Houghton Mifflin, 1991.]

Chapter 14

Unmapping Space (Lines, Smudges, and Stories)

Kimbal Quist Bumstead

> It is a warm spring day in Nishiogikubo. The air is thick, as if a thunderstorm is about to break, punctuated by a gentle breeze. A small group of people are gathered around a long roll of paper, sitting on the floor of a room on the 4th floor of an apartment building. A gentle scent of green tea brewing fills the room, and a soft orange light is cast from fabrics covering the windows. Outside the window, the sound of traffic from one of the main artery roads into central Tokyo rumbles. The people are gathered there to take part in a map-making workshop – an 'Unmapping' workshop to be precise. The maps produced in this workshop are not the kind that will enable someone to locate and find their way – at least not in an objective sense – but rather the kind of map that captures traces of memories, sensory experiences and multiple narratives. These maps are exploratory in nature, both in the sense of exploring selves, but also in the sense of discovering and making connections in an urban landscape.

Unmapping is a term used in computing to mean removing the 'mapping' on a device, such as a hard drive or a MIDI controller. In computing, removing the mapping of a device means removing its programmed usage; it renders the device as a blank slate open for new paths and connections to be made. I use Unmapping as an umbrella term to define a series of projects and workshops which explore spaces and places through the lenses of subjectivity, memory and, at times, fantasy. These projects involve capturing traces of experiences in and of places.

I was not familiar with the computing use of this term when I decided on this word to name my projects. I was looking for a word that hinted at cartography, while conveying ambiguity, and I found that balance in the word

'Unmapping'. Interestingly, my chosen meaning and the computing meaning of unmapping involve almost the exact opposite process, where unmapping a device involves the removal of traces, my projects are about ways of recording traces. There are, however, similarities between these seemingly opposite definitions. A device that is 'mapped' is a physical object with pathways 'written' into it. The possibility to unmap and *re-map* that device indicates that there are multiple ways of using it. A MIDI controller, for example, with a new mapping can enable the buttons on that device to perform different functions, thus by unmapping and remapping it becomes something else. A landscape, on the other hand, if taken as a physical entity or physical landmass, bears traces of those that inhabit it and pass through it. Both in the form of physical marks and impressions as well as the variety of mental impressions of that landscape that are experienced by different eyes and different bodies. Unmapping a landscape in the artistic sense, that I intend through my practice, enables a way to explore these subjective traces. Thus, despite their seemingly opposite processes – that of removal and capture – in both the computing sense and my own definition of the word, 'unmapping' leads to the possibility of new explorations and discoveries.

In this chapter, I propose that the process of making abstract maps can be a way of exploring and capturing subjective and *felt* experiences of places. Using examples from my artistic practice, I discuss mark-making as an embodied practice, the visceral and tactile nature of which can be a tool to explore and capture the complex relationships between body, space, and experience. My interest lies in exploring the ways in which lived experiences *in* and *of* places are imagined, and consequently how *imaginings* can be mapped. I use the term 'imagining' to mean the way in which a place appears to a person when that place is thought about or remembered. Imagining a place does not mean to invent it from nothingness, but rather, as Tim Ingold suggests, imagining is a process that involves being 'in correspondence' with a place via experience.[1] The implication here is that people and the spaces they exist in, and think about, are entangled in an inevitable relationship in which they co-respond. An imagined place could be a real existing place, where a person has physically been, which they revisit through their memories; but it could equally be a place that is dreamt, fantasised about, or envisaged.

The embodied experience of places is inherently subjective, since it is us that live the experience. The emotional attachments we associate with places; how those places feel to us when we are in them is something hard to put into words. A memory of being in a place may become warped or mixed with other memories of other places and representations of that memory can only be what Gaston Bachelard refers to as an '*orientation* towards what is secret without ever being able to tell the secret objectively'.[2] The secret to which he alludes is something internal and intangible; something that cannot quite be explained. I interpret this as the emotional aspect of a memory, how a place is remembered and how it is imagined by the mind's eye. In my Unmapping projects, I work with the notion of the mind's eye as a device for *tuning-in* to those elements of a memory that are 'secret' in Bachelard's sense. I conceptualise the 'mind's eye' as a visual and visceral connection point to a memory; the 'eye' of the imagination as it were – an eye that not only *sees* but also *feels.* The mind's eye is interpreted in a similar sense to the way Proust's narrator in *À la recherche du temps perdu* is transported to a visual and visceral memory from his childhood, a place deep within himself, via the taste of a madeleine dipped into hot tea.[3]

Unmapping is an attempt to communicate the secret of what the mind's eye sees and feels to another person through the process of making maps together. These maps, however, can only ever be an 'orientation' towards felt experience. Unmapping considers a map as a fluid notion, one that is open to interpretation, and relies on the imagination to use it. Imagery and experiences of places past, present, and future are merged, and the relationship between the real and the imagined are in constant dialogue. The maps I discuss in this chapter are not maps that communicate correct or specific information about places that could be universally understood, but rather the process of making them is a way of telling stories. The body plays a central role in the process of drawing these maps; both in terms of a body *in* a space – a body with lived experience, as well as a body *as* a space – as a site in its own right, and as a tool to communicate with.

I will now introduce three pieces of work from the Unmapping project, each of which offers a different approach to participatory mapping. I start with a journey through the various stages of one particular workshop held at Kosaten, a community arts space in Tokyo, in the spring of 2018. The workshop was spread over two consecutive weekends with the first part involving a drawing exercise indoors, and the second involving a walk around the local neighbourhood. Subsequently, I present two projects that involved participatory drawing to map personally significant home spaces.

MARK-MAKING

Imagine that your skin is like a touch-sensitive screen, or the touchpad of a laptop. If I draw a line with my finger onto the back of your hand, you can feel the sensation of my finger against your skin. If you close your eyes you can imagine and visualise a line being created in space. If you concentrate on the sensation, you can sense the direction of the line, the pressure of the point of contact and the speed with which I drew it. With your other hand, holding a stick of graphite you can copy that same gesture onto a sheet of paper. You can imitate the sensation that you feel being drawn on your hand.

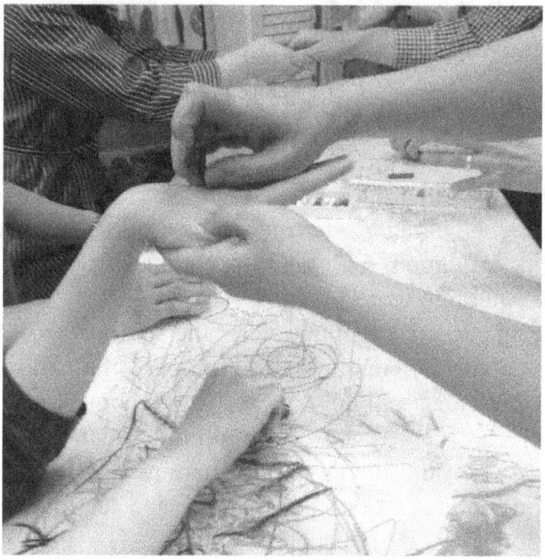

Figure 14.1 Sensory Drawing Workshop at Kosaten, Tokyo, March 2018. Author's Image

> I make another stroke on your hand. This time it is more of a gentle swipe with a curve to it. Once again, you interpret what you feel and transform it into a mark on the paper.
>
> Our bodies and minds are combined into some kind of human-hybrid form of Photoshop, touching and responding. Using different pressures, speeds and strokes, the hand and fingers can apply many different qualities of touch.

A graphite stick has enormous potential for being used to create different types of marks. It can be used on its side, on its edges, and also on its points. It can make sharp crisp lines or be worn down to a smooth rounded point for undefined darkness and softness. It can be jabbed, rotated, swirled, rubbed, flicked, pushed, and rolled. It can make lines, blobs, and smudges. It can produce representative outlines in the form of a sketch, and it can be used to write. The mass and the solidity of the graphite gives it a bodily quality that responds to the way it is held in a person's hand – the resulting marks on the paper respond to the gestures of the body that produced it (Figure 14.1).

> Let's now use this touch sensitive exercise to create a drawing of a place. You are working together with a partner. Both of you have your eyes closed throughout the exercise and you are encouraged not to talk nor explain.

> Think of a place that is personally significant to you, that you can physically imagine yourself being in. With your eyes closed, imagine that you are looking at this place from above as a bird's eye view. Create a rough floorplan of that place in your mind. Now drop yourself down into it, as if you are the little yellow person from Google Maps dropping into a Street View image or a 360° photo. Try to visualise the room that you find yourself in using your mind's eye. See the walls and windows, the furniture, and objects in the room. As you begin to visualise this place from your memory, draw it onto your partner's hand using your hands and fingers as a drawing tool. As you search this mental image, consider any other details that come to mind. What colour was the door? What was the pattern on the curtains? What were the textures in the room? Was it hard, was it soft? Was it hot, was it cold? Can you feel the handrail going up the stairs . . .? Try to think not only about how the place looks and feels but also how it makes you feel. Is there a specific memory or memories of something that happened here? Are there certain objects that have a significance? Your task is to communicate what you are imagining with your mind's eye through touch onto your partner's hand. As you did yourself just a moment ago, your partner translates your touch on their hand with marks on the paper.

As the mind's eye searches and scans a memory, the hand searches along with it, as if ruffling through the pages of an imaginary book, or a finger tracing a route on a road map. The hand becomes a searching tentacle or an endoscope camera, reaching out into an imagined space as it traces what it 'sees' onto the skin of another hand. As the participant takes a journey through their memory with their mind's eye, they become a storyteller, drawing out their story through touch. The person receiving the touch on their hand draws what they feel, as if they were a scribe for their partner's story, drawing it out in map form. Once the map of the first story is completed, the participants switch roles and make another drawing on top of the one that has just been made.

With this exercise, participants are encouraged to tune in to places they have an intimate knowledge of; places they know 'like the back of their hand', rather than places that they have casually passed though. The sensory element of the exercise is geared towards exploring the intimate connections a person has to a place both as a mental image and a bodily experience. The storyteller's hand is the tip of an imaginary paintbrush, painting a representation of an imagined space. This is not a representation in a literal visual sense, but a representation of an immeasurable package of information about what that place *feels* like. The scribe, while tuning in to the sensation may attempt to imagine the space being described, or perhaps create an internal visualisation of the sensation they feel on their hand. But they may of course not experience any visual imagery at all, and perhaps focus purely on the sensation of being touched and the act of mark-making. Describing something through touch is a way of making intangible experiences tangible, both literally speaking as well as visually, albeit in an abstract form.

It is important to note that there are two separate processes of imagining occurring within this exercise, which means that neither of the bodies or minds involved is in control of making the drawing. The storyteller tunes into a memory of an actual place that they know, while the scribe attempts to imagine it. The scribe is unlikely to share the same image of a place nor be able to guess where it is. There have however been some cases when doing this workshop that participants knew each other and had been to a place together and were able to understand which place was being described. For instance, two friends were able to work out that the drawing related to a particular nightclub and another couple worked out that the place was the hall they got married in.[4]

There is no single and correct way to do this exercise. There are no rules about what to convey nor how to convey it. How a touch is intended by the storyteller and how it is read by the scribe is totally dependent on the dynamic and relationship between the two participants. For example, the storyteller may describe a scene; the feeling of being cosy with their lover in a bed on a lazy Sunday afternoon. They may indicate the cosiness with a soft caress against the scribe's hand, and perhaps the physical shape of a door with some hard and straight lines. The scribe might interpret the soft caress as the feeling of floating on water, perhaps on a boat, or perhaps a general expression of a feeling of lightness or calmness, while the hard-line could equally be read as an architectural form, or as a road leading to somewhere or even as a knife cutting through a loaf of bread. In other words, there can be no clear correlation between marks and signified meanings. The aim however is not to produce a clear and readable map but rather that the memory of a place provides a source material for participants to engage in a sensory and bodily experience. In turn, touch is transformed into marks on paper. Those marks tell a story, not only about the place being described but about the moment of togetherness of the participants.

My Unmapping projects depart from literal forms of representation, towards the idea that a drawing is a trace of something that happens. In my workshops, I use the term 'mark-making' rather than 'drawing', as it shifts the focus away from a perceived necessity for a drawing to look like something. It has been my experience as a workshop leader, that if you ask someone to make a drawing, especially adults, they may be embarrassed at their inability to produce an accurate rendition of something. Young children, on the other hand, tend to be freer and will happily produce a page full of expressive marks. My workshops and projects encourage participants to approach mark-making without judgement, as well as encouraging playful styles of mark-making by introducing restrictive measures that make literal representation difficult to achieve. The fact that the drawings are made with eyes closed, while knowing that it is really not important what the drawing looks like, makes the process of drawing less intimidating and also more fun for participants than if they were asked to simply make a drawing of a place from memory. The strangeness of the task and the abstraction involved in the mark-making process enables a tuning-in to emotional and sensory information in a different way than might be possible with words or a more literal representational drawing style. These exercises do capture a *form* of representation, but not a literal one. They represent more embodied aspects of experience and capture traces of a moment in time.

TRACE AND GESTURE

A trace is the material inevitability of something being 'drawn' across a surface or through space, which results in a mark. Such as a snail marking its route with slime or a woodworm boring its way through a piece of wood. In both these cases, the trace which is left behind is an inevitable by-product of their movement through space. In a similar way, a desire path may appear in a city park by repeatedly being walked across, and smudges may build up on a wall when a bike is brought repetitively into a building

Tim Ingold describes a drawn line as the trace of a gesture.[5] A gesture in this sense is an articulation or expression of the body through a space or across a surface.

The qualities of a trace, or mark (i.e. how thick, thin, straight, or meandering it is) is symbolic of the manner in which the gesture was made. A flick of a wrist, for instance, would result in a different mark than a swoop of the hand, or a jab, or a swirl.

The space in which, or onto which, a gesture is produced influences the shape and quality of that gesture. A big swooping motion of an arm would be feasible in an open space, but in a small room, that swooping gesture would be more cautious to avoid hitting things. The human hand is a comparatively small drawing surface in relation to the paper that the scribe is drawing on. As such, a sense of scale needs to be negotiated by the participants to consider how big a mark should be. Take for instance, the touchpad of a laptop, which is comparatively small in relation to the screen. Different speeds and gestures can produce long movements covering the whole distance of the screen, or smaller more detailed nuances.

The hand, as a touch-sensitive surface, is undoubtedly far more complex than the flat surface of a touchpad, not least in terms of nuances of sensitivities but also in terms of its physical shape. Fingers, knuckles, and bones add a physical element to the sensory experience that both the toucher and touched will become aware of. This physicality influences the way in which a hand is touched and also the way that it feels to be touched. A drawing on the back of a hand, therefore to some extent, becomes a drawing of the feeling of a hand.

Sensory drawing is not a one-way relationship, but one that both bodies feel together. The scribe feels a touch on their hand and the storyteller feels the scribe's skin. The dynamic between their two bodies is played out through the point of contact between them, which they both experience. This is not to say that they have the *same* experience, since the emotional experience of touch may be different for different people. Touch can be pleasurable but can also be uncomfortable or could trigger difficult memories. Erin Manning describes the skin as a 'thinking, feeling surface' onto which an 'ethical discourse' is played out between touching bodies.[6] She points out that touching another body comes with a responsibility towards that person, but also that it exposes the body's own vulnerability since a touch cannot be given without also being felt. As such 'touch is singular-plural'.[7]

Sensory drawing is very much a 'singular-plural' activity since the experiences that participants have is simultaneously individual and shared. Participants see and feel in their own ways but share a moment of connection. The multiplicity and complexity of information involved in the process feeds directly into the marks produced on the paper.

Once both participants have switched roles and both have taken turns to draw, the resulting map consists of an abstract bundle of intertwining lines and scribbles (Figure 14.2). The map in its essence contains an experiential interpretation of two separate places as well as capturing an expression of the two bodies that made it. The map spans space and time by simultaneously capturing places beyond the here and now, as well as situating itself in the moment that it was created. To borrow another word from Ingold, the drawing resembles a *meshwork* of lines and marks that intersect and correspond with each other.[8] A meshwork, in Ingold's sense, is an interconnected web of individual and related elements, which serves as a visual metaphor for how everything in the world is entangled with everything else. When one part of the meshwork moves, it has a knock-on effect like a blanket being shaken out. In a sense, the meshwork in an 'Unmap' is a visual expression of how people, places, and experiences are intertwined. The multiple ways in which the marks can be interpreted is suggestive of the multiplicity of ways that places can be experienced, as well as the multiple ways that touch is experienced by the body. An abstract map of this nature might be viewed similarly to an abstract painting, not as a way of communicating a specific message but as a multitude of possibilities.

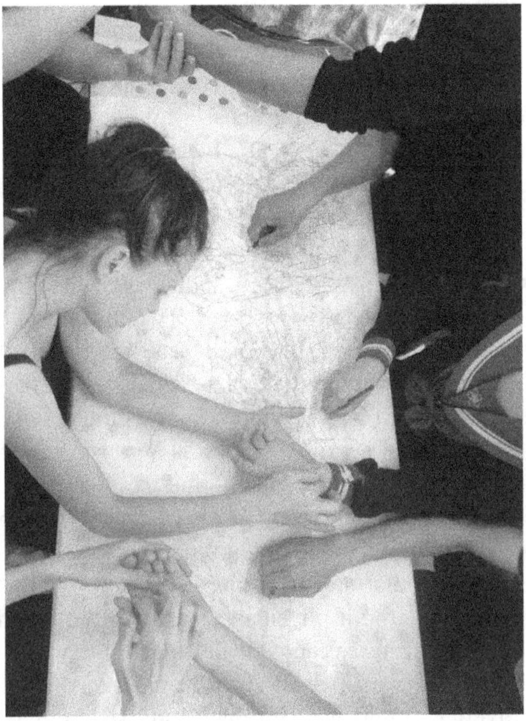

Figure 14.2 A 'Meshwork'. Sensory Drawing Workshop, Kosaten, Tokyo, March 2018. Author's Image

EXPLORATION OF SPACE

Maps in the conventional sense indicate locations of what is already there, such as buildings, landmarks, or routes along which it is possible to travel. An 'Unmap' by contrast, makes itself up as it goes along. Or, using Paul Klee's words, it can be considered as a collection of lines that 'go out for a walk'.[9] There is an implied element of travel with the Unmapping process that functions on multiple levels. First, there is the element of travelling to a place that is not physically here and now, as a sort of time travel to a memory. Next there are the routes taken by hands as they search and trace the surfaces of the skin and the paper. Finally, as I will show in this next section, the Unmap can be used as a tool to navigate a physical journey through an actual place.

Ingold marks a distinction between the different ways that lines can 'travel' and the ways that we can travel along them; 'transport' and 'wayfaring'. 'Transport' emphasises the importance of departure and arrival rather than the process of being on the move. Transport implies that the traveller is taken from point to point in the same way that someone travelling on the London

Underground would do as they move between stations. While sitting on the Tube, the traveller tracks their journey according to pre-determined points indicated on the tube map; arrival at a particular station is an arrival at a point along a line. When travelling in this linear way, we are always either at a point, or between points, until we eventually arrive at our destination. This is comparable to the practice of using waypoints to navigate an unmarked terrain such as the sea. Waypoints are navigational coordinates of latitude and longitude, that when marked onto a map form a join-the-dots style trajectory, that the traveller (or sailor) can move between. In the case of using waypoints, the line between points is not necessarily *drawn on*, as it is between Tube stations on the London Underground map, but rather that line of travel can be imagined by the traveller as the route they traverse across the surface of the planet.

'Wayfaring', on the other hand, is a more fluid notion in which the traveller *is* the line that they travel along. The emphasis here is on the space between departure and arrival, on the journey itself, and the story that unfolds on the way rather than the end goal. 'The wayfarer is continually on the move. More strictly he is his movement'.[10] Ingold exemplifies this with the Rudy Wiebe's research on the Inuit population of Northern Canada, whose travels across snow and ice leave tracks, both physically speaking but also imaginatively. For the Inuit people, each movement made by the body becomes a line; a journey over land is a line etched into the physical territory but also a line etched into the traveller's mind as an intimate and lived knowledge of a place. A landscape in this sense is conceptualised as a mesh of interconnected tracks that a traveller goes along.[11]

In other words, a wayfarer both follows a path, and simultaneously is the path they follow. In this sense, Ingold views the lines made by a wayfarer as equivalent to a storyline, one that tells the story of a person's life, a story which tells itself as it goes along.

The conceptualisation of a line as a form of storytelling brings us back to the use of drawing in Unmapping. Experiences of life become stories, become lines, become traces. Making an Unmap is the visualisation of a story unravelling, while the resulting map, as a by-product of an experience, is symbolic of personal narratives, or indeed storylines.

TAKING A MAP FOR A WALK

While the first part of the Unmapping workshop is about creating maps of places that people know, the second part considers how drawing can be used as a wayfinding tool in a new, perhaps unrelated, place. But before we get to that stage, there is one final step to complete the maps.

By this stage both you and your partner have taken turns to draw maps of places from your memory. On the roll of paper, you will see how your drawings intertwine and how they mingle with the marks made by the pair next to you. Now, with eyes open, the whole group is invited to draw together. You are invited to add any details to the drawing that were pertinent to you while participating in the exercise. Were there any colours, shapes or patterns that mentally appeared or stood out to you? Is there anything that you want to add to the map that you felt unable to communicate through touch? Using coloured pens, markers and crayons add any significant visual details that you wish to embellish. Weave your colourful additions into the meshwork, rather than scribbling randomly. Are any marks already on the paper that you recognise from your own mind's eye journey? Are there any details you wish to add to any particular spot? Once again, there is no correct way to do this, and you are welcome to share as much or as little as you wish (Figure 14.3).

Once these additional colourful details are added, the map is ready to be used as a guide for a walk around the local neighbourhood. Pairs were rearranged into groups of three or four. Each small group chose and cut out a section of the drawing that they wished to use. They could choose any part of the

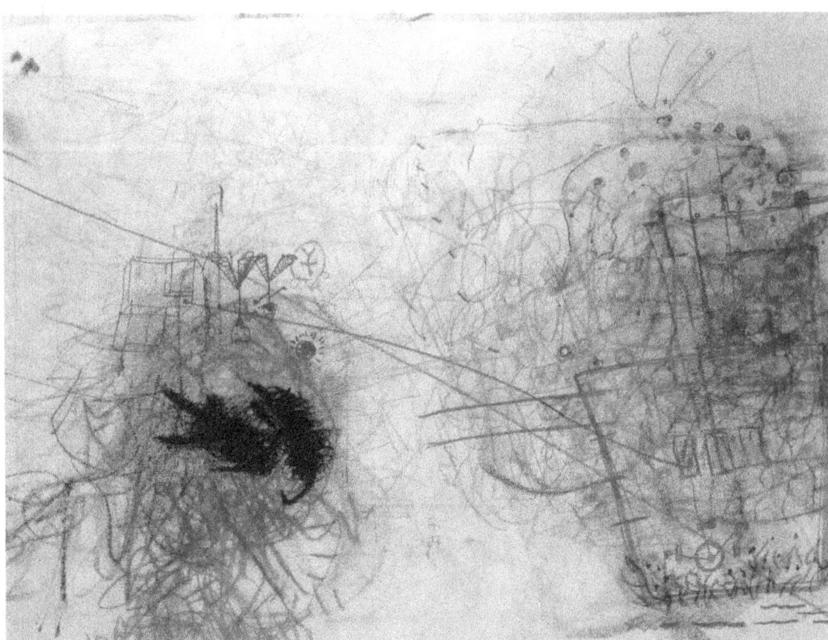

Figure 14.3 An Unmap after the Final drawing Stage of the Workshop, Kosaten, Tokyo, March 2018. Author's Image

drawing, not limited to the parts they had made themselves, and their map section could be whatever shape or size they wished to use. Before setting off on their walks, the groups decided amongst themselves on points of the drawing that they wished to 'visit' – for instance it could be an intriguing mark or a splodge of colour. The map was to be used as a score to help them navigate the streets and to find out what the spots they had identified on paper represented in real life.

In order to use the drawing as a tool to navigate, the abstract marks and lines needed to be interpreted in relation to actual things that participants observed around them. The real landscape had to be abstracted in terms of scale and physical space in order for the map to make any sense. If they were to attempt to follow a wiggling line in the drawing for instance, by matching the same wiggling movement with the path they walk along, then the ordered grid of the Tokyo city streets and buildings would have prevented them from doing so. However, if the scale of the map were to be interpreted as being just one section of a street, or even just a single paving slab, then that same gesture could have been made on a much smaller scale. The map did not necessarily need to be read from an aerial perspective, it could equally be interpreted as a series of interwoven images, or even as a three-dimensional representation. There were no particular rules to determine how the map should be read, it was up to the person reading the map to interpret and lead the way (Figure 14.4).

Figure 14.4 Using an Unmap to Navigate the Local neighbourhood, Tokyo, March 2018. Author's Image

Once the map reader decided that a certain spot on the map had been reached, the group would stop at that location to look around and observe the landscape that they found themselves in. They were asked to consider four questions:

What do you know about this place?
What does this place remind you of?
What do you imagine was here before?
What do you hope/ fear, might be here in the future?

Participants took turns to read the map, tell stories or express thoughts, and to document the process with video and note taking.

The exercise encouraged participants to not be guided by their pre-existing knowledge of how to get somewhere, nor to follow an obvious route, but to see the map as a score to guide their movements. The lines on the map were their paths to follow. The drawing encourages participants to look for details in the landscape that they might not otherwise have seen and to encourage them to stop in places that they might not normally stop. For instance, the map might lead them to stop in the middle of the pavement or next to a particular bin, street sign, or in front of something that had a particular colour that matched up with a particular colour or mark in the drawing. It might encourage unusual behaviours such as slowing down while walking, or crouching under a low tree branch to reach a water hydrant or a patch of grass, if they felt that the drawing told them to do so. Essentially, it encouraged a more playful way of exploring the streets than simply walking through them with an aim to get somewhere.

The strangeness of the task gave a sense of legitimacy to explore the landscape in a way that might normally be considered strange. Participants felt a sense of permission to behave in ways that they would normally they would feel uncomfortable about or embarrassed by.[12] Most of the participants in the second part of the workshop (walking exercise) had also participated in the first (drawing exercise), so they were perhaps already prepared for an unusual way of interpreting space and scale. The engagement of the imagination in the previous exercise fed into their exploration of the neighbourhood, with some participants recounting their own mental images from the mind's eye journey and relating it to things that they saw around them.

The participants in each group consisted of a mix of people who were familiar with the area and those who were not. Some had pre-existing or intimate knowledge of the area, while others brought their own interpretations and associations. During their walks, participants recounted various personal and historical anecdotes, as well as dreamt or imagined scenarios that did or could happen in particular spots.

One participant unexpectedly found himself in front of the primary school he went to, which he had not seen in many years, since he had moved away from the area. The experience of seeing that particular building triggered him to recall a memory of himself being there as a child. A discovery of handmade plastic sculptures in a front garden by one participant triggered a memory of a particular beach where she had spent a lot of time as a child. Another participant, when reaching a footbridge over the main road hoped that it would one day be a garden bridge and dreamt that the road would no longer be there.

This exercise drew inspiration from methodologies used by the Situationist International – specifically the *dérive*, and absurd tasks such as transposing a map of one location to use in another.[13] The Situationists saw subverting the usage of maps and creating new maps as a tool to challenge dominant hierarchies and systems of power.[14] Maps, such as those produced by governments and metropolitan authorities from a birds-eye perspective, uphold an ideological vision of the city as an entity determined by an institutional power rather than as a collection of individual lived experiences.[15] Michel De Certeau argued that the lived city, by contrast is one that can never be fully determined by the system. The pedestrian who inhabits a city and find their own ways through it by taking shortcuts 'plays with spatial organizations'.[16] The pedestrian in this context assumes a subversive agency by being able to walk in their own manner, by finding their own routes.

A dérive, as a more critical form of walking around, is not about succumbing to randomness but rather a total '*insubordination* to habitual influences'.[17] The tools of *a dérive* aim to challenge participants to make unexpected decisions, to go against the flow. This playfulness is at the heart of my Unmapping projects. My intention is not exactly to subvert power structures in the political senses intended by Debord and de Certeau but rather to take a playful approach to exploring and learning about places in unconventional ways from the perspectives of people who are present at the time.

Perhaps the kind of subversion that Unmapping provokes is a reminder for us to connect to the physical world around us and to embrace the multiple ways in which we experience space. The physicality of working with paper maps in an age of smartphones and map apps is an unusual practice in itself. A skill that we tend to forget when we are constantly reminded of where we are, and which way we are facing, by a little blue dot that follows us wherever we go on a map. While looking at a dot on a screen indicating our current location, we can imagine ourselves, physically, as that blue dot.

While on the move, walking through the streets of Nishiogikubo, the group track their own movement in relation to an imagined, fluid, and ever-changing line on a map. There is a constant flickering between deciding and being decided for, between the real and the imagined. These maps could be taken anywhere, because they are maps that become whatever they want to be.

As an exploratory tool, Unmapping sits in a grey area between Ingold's concepts of *wayfaring* and *transport*, refusing to be either, while being both at the same time. In one sense, bodies navigate between pre-determined points on a map, which would connote a form of transport between points which can be explored on arrival. Except that those points on the map are arbitrary and exist only in a relational context – without the map those points do not exist. The route that exists only in the imaginations they are making it up as they go along. Ultimately, Unmapping is a nod in the direction of playfulness, perhaps sending a message that playfulness in everyday life can enable different ways of seeing the world.

AN IMAGINED HOME AND A REMEMBERED HOME

I want to conclude by discussing two different projects that both involved participatory drawing as a way of mapping personally significant domestic spaces. These two projects engage with mark-making in distinct ways – in one case by using drawing as a tool to remember, and in the other a tool draw from the imagination. The first is a project I worked on together with my mother, which focuses on shared memories of the home I grew up in. The second is a performance artwork that used my grandmother's house as a site of imagination, working together with a participating audience.

When remembering a house that we know, we have the capacity to visualise it from the outside, from above as a birds-eye view, as well as imagining ourselves being in it. Likewise if we are to imagine a house that we have never been to, or one that we are trying to recreate from memory, for instance by drawing a picture of it, we can imagine that house as a space in which it is physically possible to be inside. Archaeologist Jo Vergunst describes how the practice of making archaeological drawings is one way of physically imagining the presence of buildings that no longer exist. He points out that drawing a picture of a house from the imagination involves recalling real relations from existing experiences of houses that we already know, which 'quickly precludes any sense of the imagination locked away in the head'.[18] In other words, the imagination is not simply something that is thought about, but something that presupposes the experiences of the body in spaces. The following two projects explore this notion of recreating spaces through drawing and addresses the importance of the body in the processes of 'remembering' and 'imagining'.

'Memories of a Home' is a project I worked on with my mother for an online exhibition (Home Cases, 2020), which focussed on different aspects of being at home during the Covid-19 lockdown in early 2020. Rather than looking at the present situation of being confined within the walls of a house, she in Amsterdam and me in London, we decided instead to focus on a shared memory of a home – the house I grew up in as a child. We each made a drawing of how we remembered the house, marking on any particular details, objects, memories, and emotions that came up through the drawing process and then showed the drawings to each other and discussed them on a Zoom call.[19]

We decided in advance to both focus on the living room of the house as a site to map – rather than a space that was individually significant to us, such as our own bedrooms – since it was a point of common ground. We both drew with our eyes open, and each chose which drawing materials we wanted to use, including the option of collage. The drawing process for this exercise was similar to the mind's eye exercise described earlier. We each visualised the physical space of the house and drew it as we remembered it. The major difference however between this and the sensory drawing was that our eyes were open while drawing and that we drew an image of our *own* memories rather than another person acting as a scribe.

We both engaged with the task quite differently. My drawing focussed on the structural elements of where I thought certain things in the room had been

Figure 14.5 Memories of a home, Drawings on paper, 2020. Author's Image

in the eyes of a child (I was ten-years old when we left), such as the toy shelf, the creepy roots of a monstera plant, and a radiator on which I attempted to melt wax crayons. My mum used a combination of collage and drawing to visualise the interior of the living room, which was filled with various significant objects and items of furniture (Figure 14.5). Each item symbolised a particular story, memory or reflection on her life, and our lives, at that point in time. Her drawing contained a list of words that came to her mind her while drawing and collaging, which signified the key 'themes' in the drawing. These themes encapsulated not only memories and occurrences but also self-reflections, dreams, fears, and desires.

The act of drawing the walls of the house, marking its physical boundaries became a tool to physically imagine being there, within those walls. Drawing created a focal point to think about particular items and details, encouraging us to think about our individual relationships to the house and our associations to the objects within it. The process of thinking about the house, imagining it, and physically representing it with marks on paper acted as a trigger not only for visual memories but also for emotional reflections that were *located* within the room or associated with specific objects.

Our Zoom conversation was an opportunity to speak about what each element in our drawing represented, using the drawing as a tool to guide the conversation. We each worked our way around our drawings, describing the various elements to each other, finding comparisons and differences. It became clear that we not only had different memories of the house itself but also that we had different emotional associations to specific elements within it.

Where the drawing process enabled a way of travelling back to our own memories, the conversation led to reflections about ourselves that were deeper and more personal, perhaps, than if we had simply sat down to talk about our memories of the house.

The second project, 'A walk around my grandmother's house', was a remote live art piece via mobile phone and was part of the Draw to Perform Symposium at [Performance Space] in 2013. My grandmother had always been a somewhat mysterious character to me, having never really known her well. At the time of the performance, she had recently moved to a nursing home and consequently had to leave her house and most of her possessions behind, knowing that she would not be able to take them to the home. I felt moved by this process of letting go of a lifetime worth of things, and by the fact that she took with her only a handful of photographs. I wanted to mark this moment of transition and examine my own relationship to it. After she moved out, I went to her house to make a journey around it. It was a pilgrimage of sorts and a way of saying goodbye to a place that I would be unlikely to ever see again once the house was sold. I wanted to piece together my fragmented memories of having been there as a child and to look at the photographs that she had left behind.

Visitors to the symposium were invited to phone me from the gallery while I explored my grandparents' house. At the gallery, a little table was set up with a pad of paper, pens, and a mobile phone with my phone number as the only contact. Each time someone called me, I would describe to them what I could see at that moment; a description of the room, objects, letters, or photographs that I found. Gallery visitors were invited to put on a blindfold, while they listened to what I was saying over the phone and to draw what I was describing onto the paper in front of them (Figure 14.6). The person drawing on the phone had a similar role to the scribe in the sensory drawing exercise at the start of this chapter, with the difference that they were listening to, and translating from, words and not from touch. The use of the blindfold encouraged participants not to focus too much on what they were drawing but rather to tune in to the descriptions and imagine the place and the described memories for themselves. The resulting drawings became little fragments of the exploration – micro maps of fragmented memories.

As I moved around a room, I described my movement through it and the physical features I could see – *I am standing by the window, I am going through a doorway, the window is divided into hand sized squares.* I tried to give descriptions as simply as possible in ways that could be visualised and drawn, but I also added more elaborate details that could generate visual images in the participant's imagination – *I am holding a ceramic blue elephant adorned with sequins.* Additionally, to visual descriptions, I spoke about my own relationship to certain objects, memories of being in my grandparents' house during brief visits as a child, and my own sense of disconnection to, and general lack of knowledge about my grandparents' lives. For instance, while standing in the kitchen, I remembered once staying with my grandparents, sitting at the table with my brother and the taste of toast with salty butter.

Both of these projects involved reconstructing a place through mark-making and both involved a process of visualising a place. 'Memories of a Home' used the process of drawing as a way to visualise a physical space that my mother and I both knew intimately. The act of making marks on the paper was a way of tuning in to a memory, but also as a way to imagine being there. Whereas in 'A walk around my Grandmother's house', my experience of physically being there became a way for me to recall and tune in to memories. For the participants who were drawing on the other end of the phone, the drawing process was a tool for them to tune in to their own imaginations. Participants drew a place that they did not know based on a description, and thus formed their own visual images of the place that was being described. Their imagined images could have been based on their own experiences of

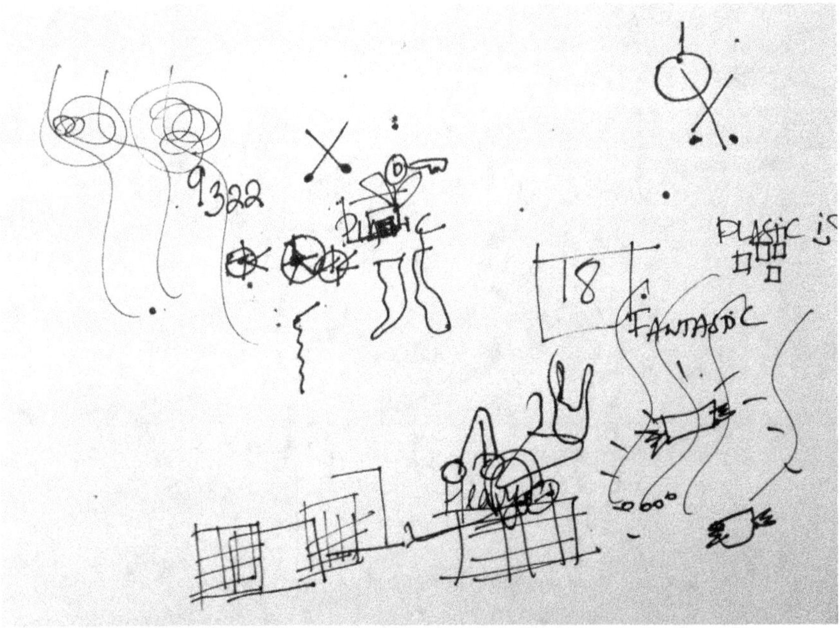

Figure 14.6 Drawing Made by Participant, A Walk Around my Grandmother House, 2013. Author's Image

places they had been in themselves; perhaps to memories of similar rooms, with similar objects and furniture, or perhaps even relating the descriptions to their own memories of visiting grandparents as children. I was surprised when I eventually saw the drawings how closely they corresponded to the reality – one drawing in particular captured the style of a lampshade and armchair almost perfectly (Figure 14.7).

In all my Unmapping projects, the drawing is a by-product and a means of tuning in memories, to allow the process of sharing stories. In the case of the projects described in this chapter, drawing together becomes a way into exploring and discussing the relationship between the body and remembered or imagined spaces. The fact that these are participatory exercises, rather than activities that people do alone, gives these drawings a relational quality. The drawings from 'A walk around my grandmother's house' are not simply depictions of memories, but interpretations of memories through other people's eyes. Likewise 'Memories of a home' was not just a tool to remember a house, but a tool to explore what that house meant to us, both individually and collectively. Equally, in my workshops at Kosaten, the drawing process itself was used as a way for a group of people who did not know each other, to connect and share with each other the place they were in.

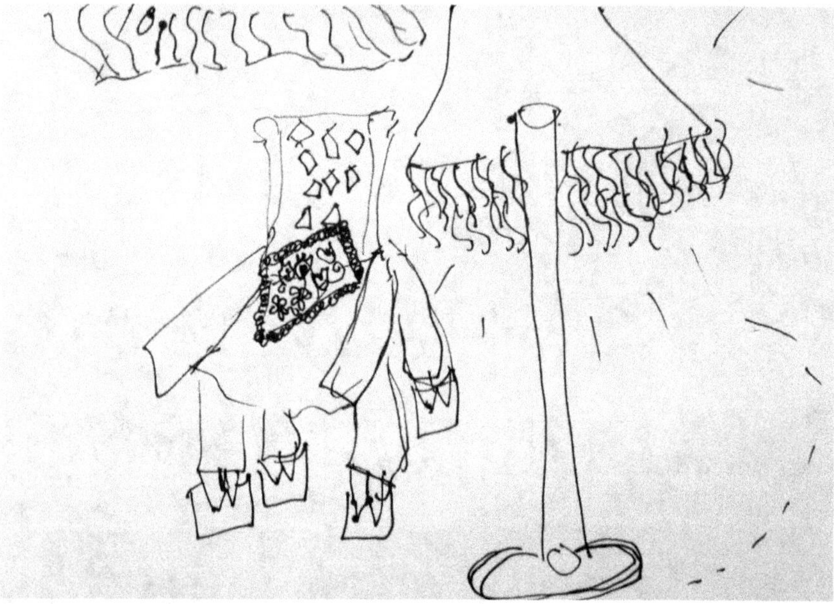

Figure 14.7 Drawing Made by Participant, A Walk Around My Grandmother House, 2013. Author's Image

CONCLUSION

The mapping exercises I have discussed in this chapter exemplify to the utmost extent how maps are subjective. In their subjectivity, they capture the mystery and magic of individual experience. Real encounters in real places may become fused with experiences that are borrowed or extrapolated from elsewhere. Memories warp, and the imagination wanders. The mind's eye becomes a pathway to unexpected discoveries. The maps I have described can be seen as fragments, or vignettes of a bigger whole. They are incomplete, liminal and in process, and as such they are a starting point, or invitation, to mine deeper into memories and experiences relating to places. These maps are fluid and open to interpretation, they are layers to be transposed or perhaps discarded. They may simply be stopping off points to get somewhere else, tools to discover something, perhaps discoveries that are only relevant to the people who made the map.

While these maps may be topographically inaccurate and indecipherable to others, Unmapping as a process is an incredibly valuable tool for exploring personal geographies. However, this does not mean it is a solipsistic activity. In fact it is precisely the opposite. It is a way of being able to share and communicate embodied experiences with other bodies. In other words,

Unmapping in all its formats is an exploratory process rather than a means to produce maps. It highlights the value of subjectivity by acknowledging and validating the multiplicity of experience in places, reminding us that each time we walk down a street, we might see it in a new light.

NOTES

1. Ingold, T. (2016). Introduction in *Imagining Landscapes: Past Present and Future*, edited by Ingold, T. and Janowski, M. Routledge: London. pp. 14.
2. Bachelard, G. (1994). *The Poetics of Space* (Jolas, M. Trans), Beacon Press: Boston, MA (original work published 1958). pp. 13.
3. Proust, M. (1989) *Swann's Way* (1st Vintage International ed). Vintage Books: New York (original work published 1913).
4. Post-workshop feedback from participants two different workshops, Camberwell Arts Festival (2018), Royal Docks Communities Voice (2019).
5. Ingold, T. (2007) *Lines; A brief history*. Routledge: London.
6. Manning, E. (2007) *The Politics of Touch: Sense Movement and Sovereignty*. University of Minnesota Press: Minneapolis, MN, pp. 9–10.
7. Manning, E. (2007). pp. 9.
8. Ingold, T. (2015). *The Life of Lines*. Routledge: London. pp. 3.
9. See Paul Klee (1961:105) in Ingold (2007).
10. Ingold, T. (2007). pp. 78.
11. Ingold, T. (2007). pp. 77.
12. Based on feedback from participants in workshop.
13. See Debord in Pinder, D. (1996). Subverting Cartography: The situationists and maps of the city in *Environment and Planning A*, 28, 405–427.
14. Pinder, D. (1996).
15. de Certeau, M. (1984). *The Practice of Everyday Life* (trans. Steven Rendall). University of California Press: Berkeley, CA.
16. De Certeau 1985 in Ward, G. (2000). *The Certeau Reader*. Blackwell: Oxford. pp. 109.
17. Debord 1981 in Pinder (1996). pp. 417.
18. Vergunst, J. (2016). Seeing Ruins: Imagined and Visible Landscapes in North-East Scotland in *Imagining Landscapes: Past Present and Future*, edited by Ingold, T. and Janowski, M. Routledge: London. pp. 19–20.
19. The full project and video recording can be found here: https://www.unmapping.space/post/home-the-home-to-home-at-home-from-home.

BIBLIOGRAPHY

Bachelard, G. (1994). *The Poetics of Space* (Jolas, M. Trans). Beacon Press: Boston, MA (original work published 1958).

de Certeau, M. (1984). *The Practice of Everyday Life* (trans. Steven Rendall). University of California Press: Berkeley, CA

Ingold, T. (2007). *Lines; A brief history*. Routledge: London.

Ingold, T. (2015). *The Life of Lines*. Routledge: London.

Ingold, T. (2016). Introduction in *Imagining Landscapes: Past Present and Future*, edited by Ingold, T. and Janowski, M. Routledge: London.

Manning, E. (2007) *The Politics of Touch: Sense Movement and Sovereignty*. University of Minnesota Press: Minneapolis, MN.

Pinder, D. (1996). Subverting Cartography: The Situationists and Maps of the City in *Environment and Planning A*, 28, 405–427.

Proust, M. (1989) *Swann's Way* (1st Vintage International ed). Vintage Books: New York (original work published 1913).

Vergunst, J. (2016). Seeing Ruins: Imagined and Visible Landscapes in North-East Scotland in *Imagining Landscapes: Past Present and Future*, edited by Ingold, T. and Janowski, M. Routledge: London.

Ward, G. (2000). *The Certeau Reader*. Blackwell: Oxford.

Part 4

NEW SCOPES, NEW SCALES

Chapter 15

Cartographic Care, or Care-tographies

From London to Hong Kong

Sam Hind

> Every political theory, explicitly or implicitly, contains an account of care.[1]
>
> Joan C. Tronto, 2013

> The promising scenario of a politics of care . . . requires that we think about care in its broadest possible public framework.[2]
>
> Joan C. Tronto, 2013

This chapter looks to several digital mapping platforms – some defunct, others active – that, as I argue, provide a blueprint for an otherwise latent cartographic politics. From 2010 to 2012, 'Sukey' was used by protesters in London to avoid police containments at protest events. It became known as the 'anti-kettling' app. For this brief period, its capacities confounded the police and excited the media. It was to be a new wave of democratic politics. The years since have seen a revanchism of sorts – as police forces throughout the Western world equip themselves with advanced crowd-control technologies. Many of these have taken a cartographic form – with software now capable of tracking the movements (and allegedly inferring the intent) of interested peoples. More recently, in Hong Kong, two similar protest mapping projects – HKmap.live and 103.hk – were launched to help protesters in similar such protests. Opportunities for counter-strategy, thus, remain both possible and necessary.

As a profitable way in, then, I propose to turn to classic feminist texts on 'care'.[3] I do so in order to render legible cartography's latent caring desires, often undervalued in more 'militant' cartographic theories.[4] In its navigational form, mapping involves caring for the lost, disorientated or in danger, allowing one to identify buildings or street names and navigate to bus stops or places of refuge. Maps entail a formalization of spatial relations in order to aid in this navigational care. They are assistive devices, rendered in material form. Yet 'counter-cartographic' approaches – typically 'antagonistic', or subversive – fail to conceptualize how maps engender social relations on their own terms rather than as kinds of 'counter-action'.[5] Articulating this caring practice in more explicit and expressive care-ful terms allows these relationships to be codified and deployed. It builds on, and intensifies, work in feminist Geographical Information Science (GIS) that has considered how mapping projects routinely leave spaces of care 'off the map'.[6] The attempt in this chapter is to consider how this concern for 'care-ful representation' may be further mobilized as 'care-ful practice' – through digital navigation itself. It does so by using protest mapping platforms as a tentative blueprint. I refer to such possible projects as 'care-tographies'. In the following section, I will discuss theories of care, considering how care is a practice, consisting of 'phases' in which care is enacted. Then, I explore how care has been theorized within critical cartography, before discussing how the attendant risks within protest events call for a re-evaluation of care itself. In the second half of the chapter, I detail how protesters 'act care-tographically' during protests in London and Hong Kong, by using mapping apps. Following this, I propose three novel phases of care: the provision of *self-care,* the *resilience* of caring infrastructures, and the *expiration* of care needs. The aim of the chapter, then, is to flesh out the care-tographic coordinates of cartographic care itself.

THEORIES OF CARE

Berenice Fisher and Joan Tronto offer one of the most concrete and comprehensive definition of care. 'On the most general level', they suggest 'that caring be viewed as a *species activity that includes everything that we do to maintain, continue, and repair our "world" so that we can live in it as well as possible.*'[7]

In this, care is cast not as a formal relationship between professional care-giver and receiver, nor as a strictly family relationship between parent and child, but as a more general set of possible relationships throughout the world. This caring sensibility, in Fisher and Tronto's definition, is inherent to all. However, this is not to say that all caring relationships function the same; nor, that care even stands for the same thing. As Tronto suggests, 'caring' entails many things – equally standing for a personal 'burden', a form of love, or a branding exercise.[8] As Victoria Lawson[9] suggests, 'a feminist ethic of care begins from the centrality of care work and care relations to our lives and societies'. Yet, it is '[u]nder neoliberal principles' that 'care is a private affair, occurring in homes and families',[10] with provision supplied by either the traditional family unit or by the market. This common split between care-as-familial and care-as-market-relation denies Fisher and Tronto's more comprehensive definition of care. As Lawson continues, it is in 'the privatization of care [that] we construct certain sorts of people as in need of care – the infirm, the young/elderly, the dependent, the flawed – ignoring the fact that *we, all of us, give and need care*'.[11]

Thus, it is suggested that theories of care should incorporate a significantly broader conception of its operational nature beyond the home, family, and market, to include all aspects of daily and spectacular life. As Joan Tronto[12] has long argued, care should be the 'basis for radical political judgements'.[13] Furthermore, that rather than an ethical stance, care 'is perhaps best thought of as a practice'.[14]

Maria Puig de la Bellacasa, extends this framework even further to argue that 'the idea of care . . . goes beyond moral disposition or a well-intentioned attitude',[15] as well as a form of 'ethically and politically charged practice',[16] to necessarily include the need to 'take care of things'.[17] If we are to understand the world – and our activity in it – as constituted in, through and with material objects, then our understanding of care, and the frameworks necessary to administer care in the world, must include such objects. In other words, de la Bellacasa argues for a thoroughly materialist engagement with care; conceiving it as an ethico-political-material practice.

As part of de la Bellacasa's work, she looks to Bruno Latour's[18] work on 'matters of concern'. Here she finds that Latour, while attentive to the precarious nature of scientific 'facts' and technological 'certainties', does not sufficiently stress the affective nature of matters. While being concerned 'denotes worry and thoughtfulness about an issue',[19] concern itself does not render a great degree of action – ethical or otherwise. Concern, then, is a weak correlate of care. While the act of 'being concerned' might lead to a higher consciousness, it does not connate action or commitment to future action. To address this, de la Bellacasa supposes that Latour's term be adapted to stress 'matters of *care*', demonstrating 'a strong sense of attachment and commitment to something'.[20]

In this more comprehensive conceptualization, Joan Tronto identifies what she calls 'three phases of care'.[21] The first of these is *attentiveness*, or 'caring about'.[22] As she suggests, 'until we care about something, the care process cannot begin. Thus, a constant impulse to return to the details of care processes and structures in life is the starting point of care as a theoretical perspective'.[23] 'To be attentive' as she continues 'requires actual attention to be paid to those who are engaged in the care process'.[24]

Aryn Martin et al. build on Tronto's work, by suggesting that care is a 'selective mode of attention: it circumscribes and cherishes some things, lives, or phenomena as its objects'.[25] Invariably, therefore, 'it excludes others'[26] in the process. Assuming the position of different actors in care relations is critical, therefore, to understanding how this attentiveness operates. Thus, while comprehensive, Fisher and Tronto's definition of care is not flattened and absolute. Care is necessarily selective. As Tronto lays out:

> The public housing debate looks quite different to someone living in substandard housing who has to cope with that situation (which affects all other aspects of life: how to keep and prepare food, how to protect property, how to arrive safely home from school, etc.) than to an economist who focuses solely on 'market forces.' Thus, a shift occurs in what counts as 'knowledge' in making philosophical and political judgements.[27]

Equally, in public protests, these care relations also start to look radically different; depending on which actors are brought into focus.

The second of these is *responsibility*, or the act of 'taking care of'.[28] 'Care requires that humans . . . take responsibility for one another',[29] and 'involves the recognition that one can act to address . . . unmet needs'.[30] As Joan Tronto writes, '[d]emocratic politics should centre upon assigning responsibilities of care, and for ensuring that democratic citizens are as capable as possible of participating in this assignment of responsibilities'.[31]

Over the last thirty years, these responsibilities have demonstrably shifted. In the West, this has invariably involved the shrinking of state responsibilities for forms of social care and the expansion of market responsibilities to carry out these same functions. In Hong Kong, a similar shift is captured in the economic policy of 'positive non-interventionism', guiding the territory since British rule, until more recent Beijing-led conflict.[32]

The third is *competence*. This 'requires that the actual care-giving work be done',[33] as well as being able to evaluate the degree to which care can be given sensitively. This demands taking into account the nature of the care-work, or care relations, as well as the intended recipients of the care-work or relations: 'the practice of caring for someone else's children requires some different competencies than caring for one's own children . . . Knowing how to negotiate [this] is part of the caring practice of being a good nanny, which is different from the practice of being a good mother'.[34]

In essence, '[i]t involves physical work, and almost always requires that care-givers come in contact with the objects of care'.[35] As Tronto points out, however, the gifting of money rarely constitutes an act of care-giving itself:

> As feminist economists have long noted, there is a great deal of work that goes into converting a pay check, or other kind of money, into the satisfying of human needs. That we quickly equate...the provision of money with the satisfaction of needs points to the undervaluing of care-giving in our society.[36]

The final phase of care is *responsiveness*, or 'care-receiving'.[37] Care, therefore, is not afforded and exercised without acknowledgement or understanding of how the care recipient responds or reacts to the care process. This does not necessarily take a linear form, however, with responsiveness considered only after attention, responsibility, and competence. As Martin et al. argue, 'prior to securing a thing to care for, a person [or thing] must have the capacity or willingness to respond, to be called into action, to be hailed by that object or phenomenon'.[38] Recognizing, and securing, this 'capacity to respond' is the responsibility of all involved in the care relationship formed; those on all sides must be 'willing' to engage.

As Tronto suggests:

> It is important to include care-receiving as an element of the caring process because it provides the only way to know that caring needs have actually been met . . . But perceptions of needs can be wrong. Even if the perception of a need is correct, how the care-givers choose to meet the need can cause new problems.[39]

This is perhaps the most appropriate moment at which to bring care directly into conversation with maps, mapping, and navigation. Think, for a second, that you are lost; either driving to a relative's house, hiking alone, or cycling home from a night out in a new city. On each occasion – for the purposes of this argument – you find yourself without a typical navigational device that would usually ensure your safe arrival. You decide to find the nearest person who might be able to assist. With a flurry of confident hand gestures, reassuring head nods, and concise instructions from an affable passer-by, you head on to your presumed destination.

But in each of these hypothetical cases the otherwise willing, and seemingly more knowledgeable individual, mishears your destination. With your misplaced confidence you bound on, replaying the hand gestures, impersonating the head nods, and repeating the instructions. Now further away from your intended destination, lost in a labyrinthine suburb, on a barren moorland, or along a never-ending city street, you realize that despite their best intentions, your navigational assistant has failed to properly 'perceive your needs' (navigate to point B), 'causing new problems' (you are now even more hopelessly lost!) along the way.

In short, navigational instructions – issued to someone who is lost – are a lifeline. Considered in relation to the theories of care discussed in this section, navigation itself can, I argue, be considered as a caring practice not unlike typical acts of care. Thus, if we are to take seriously Tronto's suggestion that care should not be limited to professional and/or familial relationships, then it is worth considering how other such practices – including navigation – can be thought of as caring acts (or not). In the next section, I look to how maps, mapping, and care have been conceptualized in critical cartography thus far.

CAREFUL REPRESENTATION

The concept and practice of care is already well-explored throughout critical cartography, most notably through gender. Agnieszka Leszczynski and Sarah Elwood[40] suggest that gender matters 'because the introduction and pervasiveness of emergent spatial information technologies, and the things we do with them, have material consequences'. These 'emergent spatial information technologies', as they call them, include new mobile-mapping platforms with the potential to geolocate users. Two, now defunct, apps – WhereTheLadies .at and Girls Around Me – mined Foursquare for 'check-ins from users with female-sounding first names' with the latter going one step further 'by linking those check-ins with the women's Facebook profiles'.[41] As Leszczynski and Elwood emphasize, '[m]asculinst values can be encoded into technologies in explicit ways, particularly in instances where male privilege drives the very design and conceptualization of the end product'.[42] Far from rudimentary dating apps, each of these platforms 'promote[d] and enable[d] potentially predatory behaviour and encourage[d] unsolicited advances to women' with the possibility of 'reaching the level of sexual harassment'.[43] It is these material consequences that Leszczynski and Elwood acutely identify as effects of spatial media design and programming.

Furthermore, that gender 'is a significant axis along which difference is (re)produced through the design of new spatial media themselves, the ways in which they encode space, and the ways in which they presuppose and reify normative gendered and sexual subjectivities'.[44] Monica Stephens also examines evidence of the gendered nature of OpenStreetMap (OSM), through an 'examination of the amenities that have been proposed and approved as features on the map'.[45] As Stephens[46] explains, 'amenities are features that provide a service or facility for map users' such as a convenience store or a local hospital. As OSM is a 'wiki-style' platform edited by an active community, users can 'propose features and vote to approve what will appear as "map features" . . . that will be rendered on the basemap'.[47] But as Stephens[48] explains, while 'OSM users approved features to delineate between a restaurant, pub, bar, biergarten, nightclub, stripclub, swingerclub and brothel', proposals for similar distinctions between spaces of care, such as 'childcare' were voted down by OSM contributors for fear of ambiguity.

In essence, users believed that the amenity was sufficiently similar to already-existing features such as 'kindergarten', such that 'spaces of care and nurture that are associated with feminized skills garner less attention than the facilities where women are commodified (strip clubs, brothels, etc.) and therefore do not obtain the votes necessary to become features'[49]. The result is that the 'lack of childcare features on the map adversely affects mothers as women are still primarily responsible for childcare and the lack of these services on the map can reduce their access to urban opportunities'.[50]

The power of the OSM platform resides in its ability to prescribe a particular world. It is inscribed in the lines, labels, and amenity hierarchies that govern its design and is wielded by OSM editors acting as gatekeepers of 'reasonable', 'necessary', and 'functional' cartographic knowledge, rendering a mapping interface devoid of amenities and spaces typically required for, used by, and of value to, care-givers. If these 'emergent spatial information technologies', as Leszczynski and Elwood remind us, have 'material consequences' then the outcome of such as state of affairs is an absence of all four phases of Tronto's care: a lack of attentiveness (to varying forms, and spaces of childcare), responsibility (of a mapping platform to provide cartographic information on such), competence (in the delivery, and sensitivity of care-giving), and responsiveness (to the needs of children). In each phase, from Stephens' analysis, the platform fails to deliver a care-ful solution.

In the next section, I consider how the circumstances of particular protests, where mapping apps have been used, requalify the relationship between risk and care.

RISKY SITUATIONS

The taking of risks ordinarily implies the abandoning of care, attention, and responsibility. Yet, I argue that in some protest events, risk-taking is necessary in order to exercise care relations. As such, rather than the antonym of care, risk becomes a simultaneous and complimentary force. In non-'A to B' demonstrations, this risk-taking might ordinarily involve ensuring the mobility of fellow activists, to avoid police containment or 'kettling'.[51] It is this mobility that, in turn, ensures the safety of the protest participants. It is through the provision of what I refer to as 'navigational knowledge' – that is, map-based information relevant to the specific moments within a protest event – that this safety is secured. Only by offering this information *live*, rather than in advance, can such safety be assured, as such protest events are often fast-moving and unpredictable.

Although risk has been variously theorized, here I look to the work of Louise Amoore,[52] Ben Anderson,[53] and Rob Shields,[54] who have contributed to conceptualizing risk as the calculation of possible futures. In this form, risk becomes an orientation towards the not-yet-happened. It is at once a technology, a quality, and a calculation through which particular forms of value are assigned to a future event. Yet this future event cannot be known entirely, or perhaps, not even identified in the first place. It is elusive. This uncertainty – of what possibly lies ahead – is the essence of this conceptualization of risk.

Yet there are different modes of risk, and Amoore suggests a 'possibilistic logic' is replacing probabilism as the more dominant force in contemporary life. Rather than seeking to avoid future risky events, this possibilistic logic actively works across its terrain to simulate, model, and manage it, acting:

> . . . not strictly to *prevent* the playing out of a particular course of events on the basis of past data tracked forward into probable futures but to *preempt* an unfolding and emergent event in relation to an array of possible projected futures.[55]

The logic of possibility strikes up a peculiar relationship between the past, present, and the future – working across the terrain of all three at once. It does so through the deployment of yet-more novel technologies throughout the 'diverse worlds of risk management consulting, computer science, commercial logistics, and data visualization'[56] as well as, of course, global security – whether in state or private forms. Indeed, it is within these non-governmental worlds – as much as within state research departments themselves – that we have seen the growth of such technologies and strategies. As Nathaniel O'Grady[57] suggests, '[e]ngendering anticipatory forms of governance requires new temporal arrangements to coordinate the calculative practices by which . . . risk is made sense of'. Rob Shields provides a conceptual guide to the relations drawn between the past, present, and the future in this possibilistic mode, suggesting that 'risk is always more than concrete danger and calculations of probability because of the importance of perception and understanding as ingredients in risk assessment'.[58] As such, calculating risk (in the possibilistic sense invoked by Amoore) involves taking into account both future actions and other less concrete elements.

Appropriately, with this shift from a probabilistic to a possibilistic logic, we see the emergence of a new quality of care. While Fisher and Tronto's four phases still apply, they do not consider the presence of risk as a threat to care itself. More precisely, they consider risk as a manageable state *inherent to* care – especially so within the kinds of protest events I discuss here. In other words, care must necessarily entail an orientation towards risk and risky situations. In the remainder of this chapter, I will explore how this relationship manifested in the use of (a) the Sukey platform in London from 2010–2012 and (b) the HKmap.live and 103.hk platforms in Hong Kong from 2019 to 2020. I end by proposing four new phases of care: the provision of *self-care*, the *resilience* of caring infrastructures, and the *expiration* of care needs.

ACTING CARE-TOGRAPHICALLY IN LONDON

Monica Stephens' case of OSM demonstrates a considerable lack of care at the heart of digital, spatial media, echoed across digital media more broadly.[59] In this section, I detail the first counter-example, the likes of which can, I suggest, provide a blueprint for acting 'care-tographically'.

Sukey was a digital platform designed to help protesters navigate during student and anti-austerity demonstrations in the United Kingdom.[60] It was launched in December 2010 as a Google Maps 'mash-up' now viewed over 244,000 times,[61] re-launched as a web application in January 2011, re-designed as a more comprehensive platform in October 2012, before being retired not long after. During the early years of the austerity era in the United Kingdom (2010–2015), and the first term of a new Conservative–Liberal Democrat coalition, numerous mobile, volatile, and unpredictable protest events were organized in London and other cities around the country.

Many of these demonstrations were hosted by student groups to protest against the rise in higher education tuition fees. Although typically A to B demonstrations, these events often culminated in far less routed and spatio-temporally predictable circumstances. As a way of preventing widespread disruption, the police would begin kettling. Once contained, protesters were often held for hours, without access to food, water, or the use of a bathroom.[62] On release, their details would be taken, despite no public order crimes having been committed. The speed to which the police resorted to such tactics was unprecedented; defining the physical brutality of these early austerity years.

The Sukey platform was launched as an 'anti-kettling' platform designed to ensure protesters could avoid such containments – not necessarily as a point of principle, but as a necessary, practical fix to a self-evident problem. The kettle is an indiscriminate manoeuvre. Unlike other tactics carried out by riot police during protest events, the containment is neither (a) defensive, (b) passive, nor (c) selective. It is not performed to defend a particular location (a building, square, statue), nor does it facilitate general policing duties during an A to B demonstration (i.e. lining a route), or, entail the careful selection of individuals judged to have committed a crime (i.e. assault, damage to private property). Instead, the containment is (a) offensive, (b) active, and (c) collective. That is to say, it involves the forward, choreographed, mobile movement of a mass of riot officers, entails deliberate and intended corralling of individuals into a designated, impermeable space, and is aimed at preventing the movement of proximal activists, en masse.

The aim of the platform was to provide live navigational updates on police containments. Messages were routinely sent by users of the platform and participants in the demonstration, providing Sukey with up-to-date information on the whereabouts of police officers, riot vans, and mounted police. Once verified, these messages were either distributed back to users via a Sukey Twitter account, or, more radically, rendered cartographically on a refreshable digital map of the protest environment. Junctions blocked by riot police would be represented by red lines, with those freely accessible mapped in green. These cartographic signs became the navigational lifeblood of many protesters on the ground, ensuring they stayed clear from potential police containments; free to continue protesting.

What was unique about the platform was its ability to provide navigational updates to protesters, while they were on the move. As such, it became responsible, re-active, and sometimes even pre-emptive of otherwise dangerous situations. Although maps have long been used in demonstrations: as navigational tools, in information leaflets, and as planning and organizational documents, their use as on-the-fly navigational prompts, responsive to, and cognizant of, possible threats was unheard of until the Sukey platform. It became a kind of 'anticipatory technology' – able to assist users in pre-empting future situations.

With the rise of various protest movements in recent times, from Black Lives Matter to the Climate Strikes, the Sukey platform – although long since gone – provides a radical blueprint for a latent cartographic politics. Thus, there is a need to understand how care might be exercised through digital tools; embedded in digital life. As such I outline here, how the Sukey platform engendered a form of care, working through Fisher and Tronto's four phases.

First, it ensured participants were *attentive* to the collective needs of other protesters beyond that of immediate friends or comrades. Sukey, considered care-tographically, entailed the construction of a broad, connective network demonstrating Bennett and Segerberg's logic of 'connective action'.[63] Such networks, as Bennett and Segerberg suggest, 'operate importantly through the organizational processes of social media', eschewing 'strong organizational control or the symbolic construction of a united "we"'.[64] This attention, conditional of this connective capacity, was cast far beyond organizational affiliation, not only augmenting but entirely replacing other mechanisms designed to ensure the safety of protesters during demonstrations (i.e. buddy systems or blocs). This is patently not care-as-familial, nor care-as-market-relation, but care-as-connective-practice.

Second, it routed a form of care *responsibilities* through both a digital device (smartphone) and a digital platform (map app). In so doing, these responsibilities were rendered visually, textually and cartographically, and the 'unmet needs' of protesters, materialized. Without a routing of these responsibilities through a platform capable of processing 'on-the-fly' needs, these responsibilities remained unassigned. The needs of fellow protesters, furthermore, remained pre-determined, fixed and assured *before* the protest event rather than dynamically addressed *during* it. The platform allowed new risks to be verified and mapped – such as the formation of a police line across a particular junction. Here is why risk is not the antonym of care: it openly worked across the terrain of the former, in order to instantiate the latter.

Third, it was able to evaluate the *competence* of care-giving by virtue of whether protesters had been contained or not. As Sukey was primarily an 'anti-kettling' platform it served a singular, obvious purpose: to prevent the containment of protesters. Its on-the-ground success, therefore, was measured by the degree to which activists remained 'un-kettled' and free to move through the city streets. In other words, to continue protesting. This competency, therefore, was judged not through a singular, bi-directional care-giver > care-receiver > care-giver relationship, but through a multi-directional, plural formation; a 'many-to-many' care-tography.

Then, lastly, it became possible to engender, foster, and actualize a *perceptive capacity* in demonstration participants – that is, 'care-receivers' – that police containments represented the most dangerous, possible threat to their continued right to protest. It enabled possible protesters to engage with, and understand, what became to be the most significant threat to bodily safety during demonstrations at that time: police containments. Much of this involved communicating 'how to spot a containment', 'how to avoid a containment' and, if necessary, 'how to survive a containment'. The platform therefore enabled care-receivers to be (a) open to being understood as such and (b) also, simultaneously, act as care-givers.

The platform, needless to say, was not without its problems. As with any blueprint, the reality was somewhat messier. In many ways, the version of events presented above is an idealized view of the platform. In reality, only some participants were attentive towards fellow protesters in a way the platform demanded. Furthermore, that these care needs were not always met, dependent as they were on a successful circulation through the Sukey network, and reliant upon qualification and verification of risk reports sent to the team. In addition, that due to the uneven relationship between protesters and attendant police forces – with all the latter's labour, equipment, legal recourse, and organizational resource – this desire to remain free, mobile, and disruptive was unrealized. Lastly, the capacity to perceive threats was, and continues to be, dependent on the successful translation of activist pedagogies into concrete navigational knowledges. Nonetheless, the platform provided a glimpse of what a care-tographic project might look like.

In the next two sections, I want to focus on two protest mapping projects used during protests in Hong Kong from 2019 to 2020.

ACTING CARE-TOGRAPHICALLY IN HONG KONG

Both HKmap.live, a dynamic, online mapping platform; and 103.hk, a static alternative, provide a fascinating update to Sukey both (a) advancing the digital protest map in new ways and (b) demonstrating a new example of care-tographic practice. I argue that not only did both projects activate four phases of care – attentiveness, responsibility, competence, and responsiveness – for protesters within Hong Kong but also three further phases or aspects of protest care-tography, not necessarily cultivated through the Sukey platform from 2010 to 2012. I refer to these as the provision of *self-care,* the *resilience* of caring infrastructures, and the *expiration* of care needs.

In 2019, Hong Kong once again became the site of protest, five years after the 'Umbrella Revolution' in 2014. Not unlike the student protests in London, Tin-yuet Ting characterizes these most recent protests as 'wildcat' actions, in which 'digitally savvy citizens engage . . . in largely ad hoc networked forms of pop-up protests'.[65] The adoption of the Bruce Lee quote 'be water, my friend' by activists is a distillation of the autonomous ethic that pervades Hong Kong itself, as well as abstract tactical advice.[66] In 2019, activists eschewed the more sedate occupational tactics of 2014, choosing instead to 'rise up simultaneously in multiple locations', drawing 'upon mobile social media to coordinate . . . operations and . . . avoid police detection'.[67] Following the passing of a new national security law in June 2020,[68] protesters once again returned to the streets of Hong Kong, a year on from the 2019 protests against a contentious extradition law. The new law gives China increased powers to intervene in Hong Kong affairs, threatening the relative autonomy the territory has had since British handover in 1997, under the 'one country, two systems' rule.

Unlike Sukey, which used a Twitter account to simultaneously issue updates, HKmap.live uses a telegram channel (@HKmaplive) to push map links and message updates to its 106,869 subscribers.[69] Users can navigate directly to HKmap.live in any web browser or download the HKmap.live app on an Android device. In a much publicised account, the app was removed by Apple from the App Store, citing its use 'in ways that endanger law enforcement and residents in Hong Kong' suggesting that the app had 'been used to target and ambush police' and 'threaten public safety'.[70] 103.hk also use a telegram channel (@RealTimeMapHK), cross-linked in the HKmap.live channel, to push map links and message updates to its 27,784 subscribers.[71] Instead of a dynamic map, however, 103.hk posted fresh, static digital map images up to every 15 minutes during major events,[72] both to its website, and into their dedicated telegram channel.

To begin with, HKmap.live were distinctly aware of the specificity of their project, and the particular ways in which it provided situated, cartographic assistance to protesters in Hong Kong. In a series of tweets on 12 June 2020, a year after its original deployment, HKmap.live mentioned that they had been receiving inquiries to make '#HKmap available to other part [*sic*] of the world'.[73] Their careful response was three-fold. First, they identified the 'security implications' of exporting their open-source project to other locations.[74] Second, they emphasised the 'very specific logic' of the project, mentioning both the 'local knowledge' and 'high population density' of Hong Kong that made the project possible.[75] Then, third, that it had taken considerable effort to keep the project running for ten months while also 'working full-time irl [in real life] to put . . . bread on the table'.[76] As such, the makers of HKmap.live

Figure 15.1 HKmap.live (left) as Shown in the Android App, in the Aftermath of Protests against the New National Security Law. 103.hk (right) produced Maps as shown in the @RealTimeMapHK Telegram Channel, in the midst of the Same Protests Earlier in the Day. Author's Image.

were attentive both towards their actual and possible users (security implications, etc.), the environment in which they worked (the densely populated, and unique territory of Hong Kong),[77] as well as the necessary labour required to make the project both efficient and effective (juggling activism and paid employment, etc.). In other words, HKmap.live – with sentiments echoed by 103.hk – lend themselves to a more thorough analysis of the care practices it engendered through the mapping projects.

Both the maps attempted to cultivate a similar 'connective action'[78] to Sukey, *attentive* to the collective assistance required by protesters. However, they took two paths to achieve this: one producing dynamic maps, the other generating static maps. In the case of 103.hk, this attentiveness was largely enabled through so-called runners – on-the-ground data collectors charged with reporting on the current situation. Maintaining a core group of reliable runners was important for generating accurate information, as well as ensuring information was collected according to an agreed protocol. But setting up and maintaining a reliable network of contributors was difficult, as 103.hk contend: '[r]unners need training and experience' and are 'NOT plug-and-play'.[79] Similar to reservations had by the Sukey developers, the 103.hk team was 'quite skeptical of crowd-sourcing all information indiscriminately',[80] instead desiring a systematic approach reliant on a smaller number of dedicated runners.

Both projects show the critical value of routing care *responsibilities* through a digital device, using bespoke mapping platforms. In the 'city without ground',[81] the risks were patently different, adding to the list of responsibilities. Neither HKmap.live nor 103.hk were strictly, or only 'anti-kettling' maps, per se, in

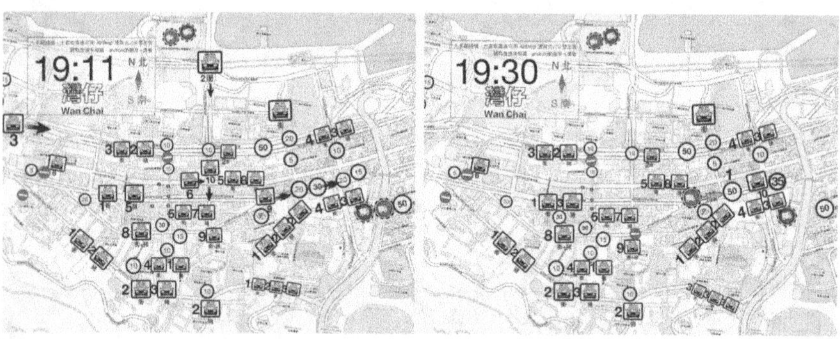

Figure 15.2 103.hk Updates on the @RealTimeMapHK Telegram Channel, 19 Minutes Apart, from 19.11 to 19.30 on 1 July 2020 in the Wan Chai District of Hong Kong. Note the heavy police presence, as well as a water cannon (white van in red circle) to the right of the image at 19.11. Protester numbers are estimated in the purple circles. In the 19.30 map, the water cannon is now moving at speed along Hennessy Road, towards a group of fifty protesters. A number of roads continue to be blocked in the area. Author's Image.

that their primary aim was to prevent protesters from being kettled. Instead, both were more broadly, anti-police or anti-arrest platforms intended to facilitate the on-going mobility and safety of protesters in fast-moving and rapidly-changing situations from protests in Admiralty outside the Legislative Council Complex, to those in Sheung Shui in the New Territories, and from occupations of Chek Lap Kok Airport to the Hong Kong Polytechnic University.[82] A whole range of police tactics were deployed in the Hong Kong protests – from roving riot police units equipped with batons and round shields, to water cannon, road blocks, and the liberal use of both pepper spray and tear gas. Organized mob attacks on protesters, such as those on 21 June 2020 in Yuen Long,[83] only added to the manifold, assumed responsibilities of the mapping teams.

The 103.hk team was clearly aware of the limitations of their work and the outstanding issues that hampered their *competence* in assisting people during the protest events. These issues manifested in various ways – from runner knowledge, to reporting protocols, and from areal mapping issues to bandwidth restrictions. While identification of these issues was easily done, the comparative 'success' of 103.hk's efforts was more difficult to gauge – not least because of the shear breadth of possible situations encountered by protesters (thus synthesised by runners and mapped by the respective teams) but also because of the manifold tactical aims, to match the plurality of risks: preventing arrest, restricting police movement, combatting pepper spray or water cannon,[84] neutralizing tear gas,[85] disabling possible surveillance infrastructure,[86] and maintaining a semblance of protest itself. Here, an evaluation of the competence of care-giving through the respective map projects was overwhelmingly difficult – at least with respect to ensuring the on-going safety of protesters during particular events, ordinarily wary of surveillance by the Hong Kong authorities.[87]

The provision of both dynamic and static maps enabled the delivery of near real time information *responsive* to on-the-ground developments. This dual emphasis provided at least two options for receiving tactical information, via two maps, and multiple apps/browsers/channels. More specific suggestions offered by 103.hk demonstrate reflexivity in the operation, enabled by the accumulation of situated, operational knowledge by protesters. For instance, due to the sheer volume of people in attendance at major demonstrations, Wi-Fi access in central Hong Kong districts (such as Admiralty) would become severely strained. In order to circumvent these recurring issues, 103 .hk suggested that uses should utilize the 'AirDrop' function on their mobile devices to pass on the static maps, as this 'can be an effective way to ripple information';[88] ensuring protesters had up-to-date maps. This echoed similar efforts during the 2014 Hong Kong protests, during which the mesh networking app 'FireChat' was used by protesters to maintain communications in crowded areas.[89]

'THE MAP IS NOT AS IMPORTANT AS YOU ARE'

In addition, the use of protest maps during the Hong Kong protests engendered some novel aspects of care-tographic practice that require further elaboration. Here I suggest that these constitute three further phases of care, beyond those identified by Fisher and Tronto.[90]

First, the provision of *self-care*. The Hong Kong mapping projects demonstrated the need to develop caring strategies for runners. As well as receiving appropriate training to hone their data collection (i.e. the work to be done), they were also given advice on how to maintain their *own safety*. This required runners to pay attention to their *own provision* of care, while providing it care-tographically to protesters, too. 103.HK runners were asked to memorize reporting protocols, such as the right mark-up terms for representing crowd density (Y for a normal, 'loose' crowd or YYY for 'shoulder-packed') or movement speeds (> for 'slow/congested flow' or >>> for 'running speed').[91] They were also provided with a list of things to bring to the protest (areal print-outs, a powerbank, umbrella, 'water and nourishment' etc.), and initial setup requirements ('go to your chosen zone', 'locate barricades/supply stations/first aid stations', etc.), to aid their safety in the field. This reflexive provision can perhaps best be summarized by a final word of advice given by 103.hk. Ultimately, '[t]he map is not as important as you are'.[92]

Second, the *resilience* of caring infrastructures. The projects also demonstrated a need to develop a sustainable framework that could both support and streamline reporting. Here, 103.hk recognised that the vocabulary they had developed, to ensure reporters could send information in a formalized manner, was insufficient. As they note, '[i]f you are building a system for collecting input[s], and it is missing some of the elements, there are elements of the world that you are not seeing'.[93] As mentioned in relation to Sukey, matching the 'swarming reality' of the protest with the 'clean, clear and categorized features'[94] of the map is a difficult task – only made easier over time if one tweaks and adjusts reporting protocols. Any extra time spent, for instance, (a) asking for contextual information (police direction of travel, etc.) or (b) interpreting information for translating to a map-ready form ('how dense *is* the crowd, here?') slows the care-tographic process down. Building a robust reporting and translation workflow was identified by 103.hk as critical to ensuring the long-term sustainability of the project – allowing care to be practiced through the provision of cartographic information in any situation, no matter how volatile, unpredictable, or confusing.

Third, the *expiration* of care needs. A final recurrent issue referenced by 103.hk was the inability to recognise when map annotations were no longer relevant. While the process of relaying information to add *onto* the map had been formalized, with a workable solution for streamlining translation to the best of the 103.hk team's abilities, no solution had been found to report what runners *couldn't* see or *stopped* seeing. That is, to *remove* events from the map. As they argued, '[i]t is often impractical for them to download the map and comment on what is out-dated, and most of the time only our most experienced runners have the presence-of-mind to do that'.[95] In short, when to know when the caring – in relation to specific incidents – can stop, with attention redirected to other emerging situations where care might be needed. Although clearly a critical component of mapping protests – not least because of limited resources, and the need to parse time-sensitive information – the 103.hk team found no workable solution to codify the reporting of incidents no longer requiring their care-tographic assistance.

To summarise, the Hong Kong protest mapping projects, HK.map and 103.hk, provide further evidence for how theories of care might be applied to cartographic practice. While Sukey demonstrated the feasibility of delivering digital forms of care through a dynamic, mobile mapping app; HK.map and 103.hk evidence the additional ways in which care is enacted, cartographically. In these examples, there is ample evidence that engenders the phases of care Fisher and Tronto[96] explicate, of attentiveness, responsibility, competence, and responsiveness. However, documentation produced by 103.hk, and other evidence from HK.map, demonstrate other such phases, or care-related concerns. Adding these aspects to Fisher and Tronto's[97] original phases, rather than unnecessarily complicating their abstractions, provides three further dynamics evidenced during the development of these two protest mapping projects during the latest wave of protests in Hong Kong. As such, they provide an enhanced framework through which to surface, interrogate, and evaluate, care-tographic practice as encountered in the wild.

CONCLUSION

This chapter has sought to explore a 'care-tographic' landscape, rendered from, and through, various digital mapping platforms. It does so in order to recast the possibility of a digital, cartographic politics forged through a new mobilization of the feminist concept of care. This extends the representational consideration of 'spaces of care' within critical cartography and GIS to consider the navigational properties of a 'care-ful practice'. Following Fisher and Tronto,[98] I consider this care not as an emotional force, or moral principle, but as a pragmatically oriented interrelation between thought and action.

But this interweaving of careful thought and action in cartographic politics is nothing new. Maps, mapping, and navigation have always involved a care-ful sensibility, despite often being used, or created, carelessly.[99] Nonetheless, it has been a latent dynamic, waiting to be surfaced. Fisher and Tronto's[100] four phases of care – attention, responsibility, competence, and capacity – provide a productive infrastructure to thinking both abstractly *and* concretely about how care-tographic projects may play out in the future. Additional insights from protests in Hong Kong – nearly ten years after student protests in London during which Sukey was launched – suggest the presence of other phases, or concerns, that provide a fuller picture of how maps enable caring practices. I have referred to these as the provision of self-care, the resilience of caring infrastructures, and the expiration of care needs. Yet many more situations, beyond protest events themselves, offer the possibility of thinking about care-tographies: from the mapping of informal settlements, to the operation of maritime rescues. While in the former, mapping is often uncritically seen as 'empowering', in the latter, map apps have (falsely) claimed the ability to help identify stranded migrants.[101] In both, there are clearly care-ful dynamics at play, but these remain underexplored, even by well-meaning participants. Thus, this chapter is an attempt to 're-animate' a careful logic that underlies all counter-mapping projects: a desire to care for lost, disorientated, or in danger. Rendering this logic legible is the first step.

NOTES

1. Joan C. Tronto, *Caring Democracy: Markets, Equality, and Justice* (New York: New York University Press, 2013), p. 25.
2. Joan C. Tronto, *Moral Boundaries: A Political Argument for an Ethic of Care* (New York: Routledge, 1993), p. 172.
3. Berenice Fisher and Joan C. Tronto, 'Toward a Feminist Theory of Caring', in *Circles of Care: Work and Identity in Women's Lives*, ed. by Emily Abel and Margaret Nelson (Albany: University of New York Press, 1991), pp. 35–62; Joan C. Tronto, *Moral Boundaries*.
4. Counter Cartographies Collective, Craig Dalton and Liz Mason-Deese, 'Counter (Mapping) Actions: Mapping as Militant Research', *ACME: An International E-Journal for Critical Geographies*, 11.3 (2012), 439–66.
5. Sam Hind, 'Disruptive Cartographies: Manoeuvres, Risk and Navigation', *PhD Thesis* (2017), 1–315 (p. 84).
6. Monica Stephens, 'Gender and the GeoWeb: Divisions in the Production of User-Generated Cartographic Information', *GeoJournal*, 78.6 (2013), 981–96; Agnieszka Leszczynski and Sarah Elwood, 'Feminist Geographies of New Spatial Media: Feminist Geographies of New Spatial Media', *The Canadian Geographer*, 59.1 (2015), 12–28.
7. Fisher and Tronto, p. 40.
8. Joan C. Tronto, *Caring Democracy*, p. x.

9. Victoria Lawson, 'Geographies of Care and Responsibility', *Annals of the Association of American Geographers*, 97.1 (2007), 1–11 (p. 3).

10. Lawson, p. 3.

11. Lawson, p. 3.

12. Joan C. Tronto, 'Care as a Basis for Radical Political Judgements', *Hypatia*, 10.2 (1995), 141–49 (p. 141).

13. See the 'Pirate Care' project for how care is conceived as radical practice: https://syllabus.pirate.care/ or The Care Collective, *The Care Manifesto: The Politics of Interdependence* (London: Verso, 2020).

14. Joan C. Tronto, *Moral Boundaries*, p. 108.

15. Maria Puig de la Bellacasa, 'Matters of Care in Technoscience: Assembling Neglected Things', *Social Studies of Science*, 41.1 (2011), 85–106 (p. 86).

16. de la Bellacasa, p. 90.

17. de la Bellacasa, p. 90.

18. Bruno Latour, 'Why Has Critique Run out of Steam? From Matters of Fact to Matters of Concern', *Critical Inquiry*, 30.2 (2004), 225–48 (p. 200).

19. de la Bellacasa, p. 89.

20. de la Bellacasa, pp. 89–90.

21. Joan C. Tronto, p. 144.

22. Joan C. Tronto, *Moral Boundaries*, p. 106.

23. Joan C. Tronto, p. 145.

24. Joan C. Tronto, p. 145.

25. Ayrn Martin, Natasha Myers and Ana Viseu, 'The Politics of Care in Technoscience', *Social Studies of Science*, 45.5 (2015), 625–41 (p. 627).

26. Martin, Myers and Viseu, p. 627.

27. Joan C. Tronto, p. 145.

28. Joan C. Tronto, *Moral Boundaries*, p. 106.

29. Joan C. Tronto, p. 145.

30. Joan C. Tronto, *Moral Boundaries*, p. 106.

31. Joan C. Tronto, *Caring Democracy*, p. 7.

32. Catherine R. Schenk, 'Negotiating Positive Non-interventionism: Regulating Hong Kong's Finance Companies, 1976-1986', *The China Quarterly*, 230 (2017), 348–70; Lau Siu-kai and Kuan Hsin-chi, 'Public Attitude toward Laissez Faire in Hong Kong', *Asian Survey*, 30.8 (1990), 766–81; Elizabeth Wong, 'Hong Kong's Move Away From Non-interventionism Should be Judged on Its Results, Not Intentions', *South China Morning Post*, September 18, 2015, https://www.scmp.com/news/hong-kong/politics/article/1859163/hong-kongs-move-away-non-interventionism-should-be-judged; Richard Wong, 'Capitalist and Socialist Rent-seekers Eroding Positive Non-interventionism', *South China Morning Post*, November 3, 2015, https://www.scmp.com/business/markets/article/1875210/capitalist-and-socialist-rent-seekers-eroding-positive-non.

33. Joan C. Tronto, *Caring Democracy*, p. 22.

34. Joan C. Tronto, *Caring Democracy*, p. 23.

35. Joan C. Tronto, *Moral Boundaries*, p. 107.

36. Joan C. Tronto, *Moral Boundaries*, p. 107.

37. Joan C. Tronto, *Moral Boundaries*, p. 107.
38. Martin, Myers and Viseu, p. 635.
39. Joan C. Tronto, *Moral Boundaries*, p. 108.
40. Leszczynski and Elwood, p. 13.
41. Leszczynski and Elwood, p. 18.
42. Leszczynski and Elwood, p. 18.
43. Leszczynski and Elwood, p. 19.
44. Leszczynski and Elwood, p. 13.
45. Stephens, p. 989.
46. Stephens, p. 989.
47. Stephens, p. 989.
48. Stephens, p. 990.
49. Stephens, p. 991.
50. Stephens, p. 991.
51. Scott Sørli, 'A Short History of Kettling', *Scapegoat*, 2012, 1–44.
52. Louise Amoore, 'Data Derivatives: On the Emergence of a Security Risk Calculus for Our Times', *Theory, Culture & Society*, 28.6 (2011), 24–43; Louise Amoore, *The Politics of Possibility: Risk and Security Beyond Probability* (London: Duke University Press, 2013).
53. Ben Anderson, 'Preemption, Precaution, Preparedness: Anticipatory Action and Future Geographies', *Progress in Human Geography*, 34.6 (2010), 777–98.
54. Rob Shields, *The Virtual* (London: Routledge, 2003).
55. Louise Amoore, p. 9.
56. Louise Amoore, n.p.
57. Nathaniel O'Grady, 'A Politics of Redeployment: Malleable Technologies and the Localisation of Anticipatory Calculation', in *Algorithmic Life: Calculative Devices in the Age of Big Data* (London: Routledge, 2016), pp. 72–86 (p. 81).
58. Shields, p. 185.
59. Karin Schwiter and Jennifer Steiner, 'Geographies of Care Work: The Commodification of Care, Digital Care Futures and Alternative Caring Visions', *Geography Compass*, 14.12 (2020), 1–16.
60. Patrick Kingsley, 'Inside the Anti-Kettling HQ', *Guardian*, February 3, 2011, https://www.theguardian.com/uk/2011/feb/02/inside-anti-kettling-hq.
61. The original mash-up map is still online: https://www.google.com/maps/d/viewer?mid=1vyznzbdE7u-jknxBzfcn4FI29T0&msa=0&ll=51.501049%2C-0.12707199999999474&z=16. My own records show twice the total number of views (244,000 rather than 116,000).
62. Dan Hancox, *Kettled Youth: Summer of Unrest* (London: Vintage Digital, 2011).
63. W. Lance Bennett and Alexandra Segerberg, 'The Logic of Connective Action: Digital Media and the Personalization of Contentious Politics', *Information, Communication & Society*, 15.5 (2012), 739–68.
64. Bennett and Segerberg, p. 748.
65. Tin-yuet Ting, 'From 'Be water' to 'Be fire': Nascent Smart Mobs and Networked Protests in Hong Kong', *Social Movement Studies*, 19.3 (2020), 362–68 (p. 362).

66. Antony Dapiran, 'Be Water!': Seven Tactics That Are Winning Hong Kong's Democracy Revolution', *New Statesman*, August 1, 2019, https://www.newstatesman.com/world/2019/08/be-water-seven-tactics-are-winning-hong-kongs-democracy-revolution.

67. Ting, p. 363.

68. Lily Kuo and Verna Yu, 'Controversial Hong Kong National Security Law Comes into Effect', *Guardian*, June 30, 2020, https://www.theguardian.com/world/2020/jun/30/controversial-hong-kong-national-security-law-comes-into-effect.

69. Correct as of June 29 2020.

70. Sam Byford, 'Apple Removes App Used in Hong Kong Protests After Pressure from China', *The Verge*, October 10, 2019, https://www.theverge.com/2019/10/10/20907596/apple-hong-kong-protests-app-removed-china.

71. Correct as of June 29 2020.

72. 103.hk, 'Real Time Map', *103.hk*, accessed June 29, 2020, https://103.hk/en/map/#map.

73. HKmap.live, 'Have Been Receiving Lots Inquiry Regarding Making #HKmap Available to Other Part of the World. Sadly It's Not Possible At the Moment [. . .]', *Twitter*, June 12, 2020, https://twitter.com/hkmaplive/status/1271518304109080576.

74. HKmap.live, n.p.

75. HKmap.live, '2. HKmap Have Some Very Specific Logic That Adapt to HK, and It Works Well (Sorta) Due to Our Local Knowledge and the High Population Density in HK [. . .]', *Twitter*, June 12, 2020, https://twitter.com/hkmaplive/status/1271518305744850944.

76. HKmap.live, '3. The Movement in HK Just Reached the 1 Year Mark, and Our Team Kept HKmap Up and Running 24x7 for 10 Mo While Working Full-Time IRL to Put the Bread on the Table [. . .]', *Twitter*, June 12, 2020, https://twitter.com/hkmaplive/status/1271518307292557313.

77. Special thanks go to one individual who helped in my understanding of this unique geography, and socio-cultural history. Any errors within are my own.

78. Bennett and Segerberg, p. 748.

79. 103.hk, 'Difficulties With Communications', *103.hk*, accessed June 29, 2020, https://103.hk/en/map/refl/#difficulties-with-communications.

80. 103.hk, n.p.

81. Adam Frampton, Jonathan D. Solomon and Clara Wong, *Cities without Ground: A Hong Kong Guidebook* (San Francisco, CA: ORO Editions, 2012); Clancy Wilmott, 'Cities without Ground: A Hong Kong Guidebook (review)', *Cartographica*, 48.3 (2013), 254–55.

82. Su Xinqi, Kimmy Chung, Alvin Lum and Christy Leung, 'Hong Kong Border Town of Sheung Shui Rocked by Protest Violence and Chaos before Police Finally Clear Streets at Night', *South China Morning Post*, July 13, 2019, https://www.scmp.com/news/hong-kong/politics/article/3018473/sheung-shui-readies-itself-protests-hongkongers-take; Danny Lee and Zoe Low, 'Hong Kong Airport Occupied by Hundreds of Protesters amid Tightened Security', *South China Morning Post*, August 9, 2010, https://www.scmp.com/news/hong-kong/transport/article/3022074/security-increased-hong-kong-international-airport-ahead.

83. Christy Leung and Victor Ting, 'Hong Kong Police Chief Defends Officers Arriving 35 Minutes After First Reports of Yuen Long Mob Violence Against Protesters and MTR Passengers', *South China Morning Post*, July 22, 2019, https://www.scmp.com/news/hong-kong/law-and-crime/article/3019657/hong-kong-police-chief-defends-officers-arriving-35.

84. Chloe Whiteaker, 'The Essential Tool for Hong Kong Protesters? An Umbrella', *Bloomberg*, September 20, 2019, https://www.bloomberg.com/graphics/2019-hong-kong-protesters-umbrellas/.

85. Taylor Mahlandt, 'Water on the Smoke: The Brilliantly Simple Hong Kong Method for Fighting Tear Gas', *Slate*, June 17, 2019, https://slate.com/news-and-politics/2019/06/hong-kong-tear-gas-water-bottle.html.

86. Yuan Yang, 'Why Hong Kong's Protesters Fear the City's 'Smart Lamp Posts'', *Financial Times*, January 8, 2020, https://www.ft.com/content/f0300b66-30dd-11ea-9703-eea0cae3f0de

87. Trey Smith, 'In Hong Kong, Protesters Fight to Stay Anonymous', *The Verge*, October 22, 2019, https://www.theverge.com/2019/10/22/20926585/hong-kong-china-protest-mask-umbrella-anonymous-surveillance; Blake Schmidt, 'Hong Kong Police Already Have AI Tech That Can Recognize Faces', *Bloomberg*, October 22, 2019, https://www.bloomberg.com/news/articles/2019-10-22/hong-kong-police-already-have-ai-tech-that-can-recognize-faces.

88. 103.hk, 'Real Time Map', n.p.

89. Archie Bland, 'FireChat – The Messaging App That's Powering the Hong Kong Protests', *Guardian*, September 29, 2014, https://www.theguardian.com/world/2014/sep/29/firechat-messaging-app-powering-hong-kong-protests; James Ash, *Phase Media: Space, Time and the Politics of Smart Objects* (London: Bloomsbury, 2018).

90. Fisher and Tronto.

91. 103.hk, 'Procedures', *103.hk*, accessed July 1, 2020, https://103.hk/en/map/obs/#procedures.

92. 103.hk, n.p.

93. 103.hk, 'Closing: Recommendations / Requests / Room for Collaboration', *103.hk*, accessed July 1 2020, https://103.hk/en/map/refl/#closing-recommendations-requests-room-for-collaboration.

94. Sam Hind, 'Between Capture and Addition: The Ontogenesis of Cartographic Calculation', *Political Geography,* 78 (2020), 1–9 (p. 6).

95. 103.hk, 'Remaining Problems', *103.hk*, accessed July 1, 2020, https://103.hk/en/map/refl/#remaining-problems.

96. Fisher and Tronto.

97. Fisher and Tronto.

98. Fisher and Tronto.

99. Timothy Amos, 'Fighting the Taboo Cycle: Google Map Protests and Buraku Human Rights Activism in Historical Perspective', *Japanese Studies*, 35.3 (2015); Alex Hern, 'Fitness Tracking App Strava Gives Away Location of Secret US Army Bases', *Guardian*, January 28, 2018, https://www.theguardian.com/world/2018/jan/28/fitness-tracking-app-gives-away-location-of-secret-us-army-bases.

100. Fisher and Tronto.

101. Jiri Panek and Rostislav Netek, 'Collaborative Mapping and Digital Participation: A Tool for Local Empowerment in Developing Countries', *Information*, 10.8 (2019); Alex Hern, 'Refugee Rescue App Pulled from App Store After it is Outed as Fake', *Guardian*, June 21, 2016, https://www.theguardian.com/technology/2016/jun/21/refugee-rescue-i-sea-app-pulled-app-store-outed-as-fake.

BIBLIOGRAPHY

Adam Frampton, Jonathan D. Solomon and Clara Wong, *Cities without Ground: A Hong Kong Guidebook* (San Francisco, CA: ORO Editions, 2012).

Agnieszka Leszczynski and Sarah Elwood, 'Feminist Geographies of New Spatial Media: Feminist Geographies of New Spatial Media', *The Canadian Geographer*, 59.1 (2015), 12–28.

Alex Hern, 'Fitness Tracking App Strava Gives Away Location of Secret US Army Bases,' *Guardian*, January 28, 2018, https://www.theguardian.com/world/2018/jan/28/fitness-tracking-app-gives-away-location-of-secret-us-army-bases.

Alex Hern, 'Refugee Rescue App Pulled from App Store After it is Outed as Fake,' *Guardian*, June 21, 2016, https://www.theguardian.com/technology/2016/jun/21/refugee-rescue-i-sea-app-pulled-app-store-outed-as-fake.

Antony Dapiran, 'Be Water!': Seven Tactics That Are Winning Hong Kong's Democracy Revolution,' *New Statesman*, August 1, 2019, https://www.newstatesman.com/world/2019/08/be-water-seven-tactics-are-winning-hong-kongs-democracy-revolution.

Archie Bland, 'FireChat – The Messaging App That's Powering the Hong Kong Protests,' *Guardian*, September 29, 2014, https://www.theguardian.com/world/2014/sep/29/firechat-messaging-app-powering-hong-kong-protests.

Ayrn Martin, Natasha Myers and Ana Viseu, 'The Politics of Care in Technoscience', *Social Studies of Science*, 45.5 (2015), 625–41 (p. 627).

Ben Anderson, 'Preemption, Precaution, Preparedness: Anticipatory Action and Future Geographies', *Progress in Human Geography*, 34.6 (2010), 777–98.

Berenice Fisher and Joan C. Tronto, 'Toward a Feminist Theory of Caring', in *Circles of Care: Work and Identity in Women's Lives*, ed. by Emily Abel and Margaret Nelson (Albany: University of New York Press, 1991), pp. 35–62.

Blake Schmidt, 'Hong Kong Police Already Have AI Tech That Can Recognize Faces,' *Bloomberg*, October 22, 2019, https://www.bloomberg.com/news/articles/2019-10-22/hong-kong-police-already-have-ai-tech-that-can-recognize-faces.

Bruno Latour, 'Why Has Critique Run out of Steam? From Matters of Fact to Matters of Concern', *Critical Inquiry*, 30.2 (2004), 225–48.

Catherine R. Schenk, 'Negotiating Positive Non-interventionism: Regulating Hong Kong's Finance Companies, 1976-1986', *The China Quarterly*, 230 (2017), 348–70.

Chloe Whiteaker, 'The Essential Tool for Hong Kong Protesters? An Umbrella,' *Bloomberg*, September 20, 2019, https://www.bloomberg.com/graphics/2019-hong-kong-protesters-umbrellas/.

Christy Leung and Victor Ting, 'Hong Kong Police Chief Defends Officers Arriving 35 Minutes After First Reports of Yuen Long Mob Violence Against Protesters and MTR Passengers,' *South China Morning Post*, July 22, 2019, https://www.scmp.com/news/hong-kong/law-and-crime/article/3019657/hong-kong-police-chief-defends-officers-arriving-35.

Clancy Wilmott, 'Cities without Ground: A Hong Kong Guidebook (review)', *Cartographica*, 48.3 (2013), 254–55.

Counter Cartographies Collective, Craig Dalton and Liz Mason-Deese, 'Counter (Mapping) Actions: Mapping as Militant Research', *ACME: An International E-Journal for Critical Geographies*, 11.3 (2012), 439–66.

Dan Hancox, *Kettled Youth: Summer of Unrest* (London: Vintage Digital, 2011).

Danny Lee and Zoe Low, 'Hong Kong Airport Occupied by Hundreds of Protesters amid Tightened Security,' *South China Morning Post*, August 9, 2010, https://www.scmp.com/news/hong-kong/transport/article/3022074/security-increased-hong-kong-international-airport-ahead.

Elizabeth Wong, 'Hong Kong's Move Away From Non-interventionism Should be Judged on Its Results, Not Intentions,' *South China Morning Post*, September 18, 2015, https://www.scmp.com/news/hong-kong/politics/article/1859163/hong-kongs-move-away-non-interventionism-should-be-judged.

James Ash, *Phase Media: Space, Time and the Politics of Smart Objects* (London: Bloomsbury, 2018).

Jiri Panek and Rostislav Netek, 'Collaborative Mapping and Digital Participation: A Tool for Local Empowerment in Developing Countries', *Information*, 10.8 (2019).

Joan C. Tronto, *Moral Boundaries: A Political Argument for an Ethic of Care* (New York: Routledge, 1993).

Joan C. Tronto, 'Care as a Basis for Radical Political Judgements', *Hypatia*, 10.2 (1995), 141–49 (p. 141).

Joan C. Tronto, *Caring Democracy: Markets, Equality, and Justice* (New York: New York University Press, 2013).

Karin Schwiter and Jennifer Steiner, 'Geographies of Care Work: The Commodification of Care, Digital Care Futures and Alternative Caring Visions', *Geography Compass*, 14.12 (2020), 1–16.

Lau Siu-kai and Kuan Hsin-chi, 'Public Attitude toward Laissez Faire in Hong Kong', *Asian Survey*, 30.8 (1990), 766–81.

Lily Kuo and Verna Yu, 'Controversial Hong Kong National Security Law Comes into Effect,' *Guardian*, June 30, 2020, https://www.theguardian.com/world/2020/jun/30/controversial-hong-kong-national-security-law-comes-into-effect.

Louise Amoore, 'Data Derivatives: On the Emergence of a Security Risk Calculus for Our Times', *Theory, Culture & Society*, 28.6 (2011), 24–43.

Louise Amoore, *The Politics of Possibility: Risk and Security Beyond Probability* (London: Duke University Press, 2013).

Maria Puig de la Bellacasa, 'Matters of Care in Technoscience: Assembling Neglected Things', *Social Studies of Science*, 41.1 (2011), 85–106.

Monica Stephens, 'Gender and the GeoWeb: Divisions in the Production of User-Generated Cartographic Information', *GeoJournal*, 78.6 (2013), 981–96.

Nathaniel O'Grady, 'A Politics of Redeployment: Malleable Technologies and the Localisation of Anticipatory Calculation', in *Algorithmic Life: Calculative Devices in the Age of Big Data* (London: Routledge, 2016), pp. 72–86.

Patrick Kingsley, 'Inside the Anti-Kettling HQ', *Guardian*, February 3, 2011, https://www.theguardian.com/uk/2011/feb/02/inside-anti-kettling-hq.

Richard Wong, 'Capitalist and Socialist Rent-seekers Eroding Positive Non-interventionism,' *South China Morning Post*, November 3, 2015, https://www.scmp.com/business/markets/article/1875210/capitalist-and-socialist-rent-seekers-eroding-positive-non.

Rob Shields, *The Virtual* (London: Routledge, 2003).

Sam Byford, 'Apple Removes App Used in Hong Kong Protests After Pressure from China,' *The Verge*, October 10, 2019, https://www.theverge.com/2019/10/10/20907596/apple-hong-kong-protests-app-removed-china.

Sam Hind, 'Disruptive Cartographies: Manoeuvres, Risk and Navigation', *PhD Thesis* (2017), 1–315 (p. 84).

Sam Hind, 'Between Capture and Addition: The Ontogenesis of Cartographic Calculation', *Political Geography,* 78 (2020), 1–9 (p. 6).

Scott Sørli, 'A Short History of Kettling', *Scapegoat*, 2012, 1–44.

Su Xinqi, Kimmy Chung, Alvin Lum and Christy Leung, 'Hong Kong Border Town of Sheung Shui Rocked by Protest Violence and Chaos before Police Finally Clear Streets at Night,' *South China Morning Post*, July 13, 2019, https://www.scmp.com/news/hong-kong/politics/article/3018473/sheung-shui-readies-itself-protests-hongkongers-take.

Taylor Mahlandt, 'Water on the Smoke: The Brilliantly Simple Hong Kong Method for Fighting Tear Gas,' *Slate*, June 17, 2019, https://slate.com/news-and-politics/2019/06/hong-kong-tear-gas-water-bottle.html.

Timothy Amos, 'Fighting the Taboo Cycle: Google Map Protests and Buraku Human Rights Activism in Historical Perspective', *Japanese Studies*, 35.3 (2015).

Tin-yuet Ting, 'From 'Be water' to 'Be fire': Nascent Smart Mobs and Networked Protests in Hong Kong', *Social Movement Studies*, 19.3 (2020), 362–68 (p. 362).

Trey Smith, 'In Hong Kong, Protesters Fight to Stay Anonymous,' *The Verge*, October 22, 2019, https://www.theverge.com/2019/10/22/20926585/hong-kong-china-protest-mask-umbrella-anonymous-surveillance.

Victoria Lawson, 'Geographies of Care and Responsibility', *Annals of the Association of American Geographers*, 97.1 (2007), 1–11.

W. Lance Bennett and Alexandra Segerberg, 'The Logic of Connective Action: Digital Media and the Personalization of Contentious Politics', *Information, Communication & Society*, 15.5 (2012), 739–68.

Yuan Yang, 'Why Hong Kong's Protesters Fear the City's 'Smart Lamp Posts',' *Financial Times*, January 8, 2020, https://www.ft.com/content/f0300b66-30dd-11ea-9703-eea0cae3f0de.

Chapter 16

Mapshop

Learning to Map, Mapping to Learn

Emily Barrett and Matthew W. Wilson

One of the premises of critical cartography and critical GIS is that mapmaking and GIS is worthy of study due to the possibilities for intervention – that *how* we know matters for what we might *do*. This sensibility towards mapmaking hit an inflection point in the late 1990s, as the GIS & Society debates reached fever pitch[1]. Now, as we enter the third decade of these discussions, what might be the new possibilities for intervening with maps and GIS? What are the new capacities for resonate and responsible mapmaking? In an attention-economy, how might we cultivate care differently?

In what follows, we overview Mapshop, an initiative at the University of Kentucky that attempts to leverage the technical resources and the expertise of students and faculty in the Department of Geography to both support community partners with mapping and visualisation needs as well as offer training opportunities for campus and community.

Of course, the scholarship in community geography recognises that there are no one-size-fits-all approaches to doing engaged work with mapmaking and GIS. Therefore, we do not suggest that our efforts will necessarily have similar results, effects, or affects. Our work begins and ends with a recognition of the unique histories and on-going struggles in our communities, noting that we might only ever scratch the surface, and our work is inconclusive, and perhaps suggestive at best.

We situate Mapshop within a longer history of the university and its relationship to the city. The college that would later become the University of Kentucky was carved out of Transylvania University in 1878, a university that was established in Lexington, Virginia (later Kentucky), in the late eighteenth century. The support of the Morrill Act of 1862 funded a federal land grant to establish the A&M College as part of Kentucky University (as Transylvania University was named at the time) in 1865. The stories of the

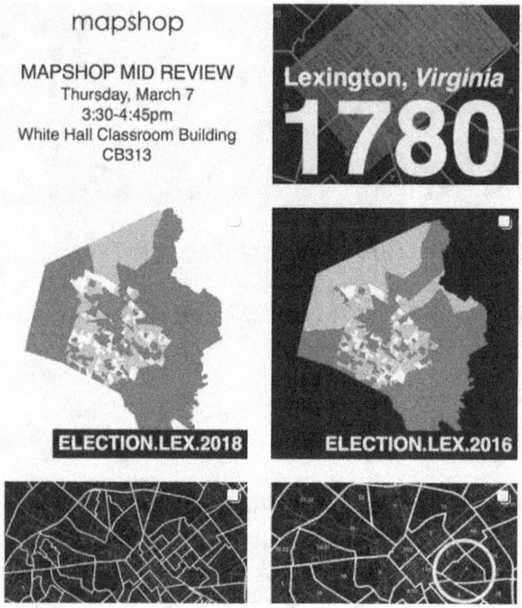

Figure 16.1 Authors' Image.

histories of University of Kentucky and Transylvania University are inextricably tied to the stories of settler colonialism in the Bluegrass and the emergence of Lexington as a frontier city for a new nation.

The importance of these universities in Lexington cannot be emphasised enough. Along with the Bluegrass Community and Technical College, these institutions are the largest landholders in the city – stabilising, creating, and generating new speculation on real estate in a city that has had an urban growth boundary since 1958 (the first of such boundaries in the United States). Lexington has a population of over 300,000 and a combined statistical area of nearly 900,000. The University of Kentucky student population is just over 30,000 and the campus is one of the few flagship public universities with all liberal, fine and performing arts, sciences, professional, and applied colleges on a single, urban campus. The opportunities (and risks) associated with university–community relations are ever-present.

UNIVERSITY–COMMUNITY PARTNERING

Geographers in North America have a well-established history of engaged scholarship. Notably, William Bunge's *Fitzgerald: Geography of a Revolution* (1971) encapsulates a collaborative and boots-on-the-ground study of a

Figure 16.2 Authors' Image.

community's struggle for justice. By focusing on 'concrete peculiarities', Bunge advocated for geographers to be 'engaged in the radically democratic project of providing pedagogical resources to enable suppressed and exploited communities to manage for themselves, to facilitate flourishing geographical lives'. Although dismissed by some of his contemporaries[2] and later rightfully revisited and critiqued by his research collaborators,[3] Bunge's ideas were later echoed by David Harvey's call for a people's geography: a geography that would depict, analyse, and understand the world not as an abstract ideal, 'but as it really is'.[4]

However, it was not until the 1990s, with the rise of the GIS & Society debates, that geographers began to more seriously and systematically partake in engaged and participatory scholarship as a tool to leverage the power and resources of universities to address community concerns, social inequalities, and environmental injustices.[5] Considering GIS as a social, political, and power-laden process rather than an analytical tool solely designed to represent and analyse spatial relationships, initial participatory efforts attempted to invert power dynamics, were context and issue-driven, and emphasised community involvement in the creation and/or use of geographical information.[6]

Public Participation GIS (PPGIS), for instance, sought to address growing concerns over the prevalence of state-produced spatial data and the use of persuasive visualisations to inform public policy.[7] Developed within the field of planning, PPGIS initially emphasised ways in which it could make 'decision-making tools available and accessible to all those with a stake in official decisions'.[8] Under this approach, universities often acted as intermediaries or facilitators providing community members with access to expensive technology and training, student labour, and a variety of other institutional resources to support their engagement with public policymaking.[9] In exchange, community members provided academics and students with opportunities to gain invaluable contextual knowledge and apply their research both within and beyond the classroom. PPGIS approaches have diversified, covering topics from food insecurity, to housing, to transportation, to natural resource management.[10] As such, they are often resistant to clear categorisation. However, PPGIS can broadly be considered as the confluence of social and participatory activities, community advocacy and technology – all grounded and contextualised in specific places.[11]

An alternative approach, Participatory GIS (PGIS), re-evaluated and more critically examined the definition of public inherent to PPGIS projects. Apprehensive over representation, PGIS focused more explicitly on participatory methods to engage marginalised populations, especially in the developing world.[12] Although sometimes collapsed into PPGIS, PGIS draws its methods, context, and understanding more specifically from community-integrated GIS and counter-mapping.[13] By enabling the participation of underrepresented populations, and considering them as experts in their own right, PGIS has provided avenues to map alternative views and to articulate different, even contradictory, narratives of the same problems from different positions of power.[14]

Despite the successes of both PPGIS and PGIS (shortened to P/PGIS), familiar barriers persist, and new challenges have emerged. For example, although P/PGIS have diffused the means of participation, particularly in terms of access to cheaper and a greater diversity of geospatial technologies, 'at the bottom of the digital divide relatively little has changed'.[15] P/PGIS is subsequently in danger of 'becoming a numbers game', whereby a reliance on ever-changing technologies prioritises the participation of a greater volume of technology adept users.[16] Community-based and grassroot organisations, for example, despite a desire to participate in geographic and GIS-based inquiries, continue to lack capacity, in terms of time, skills, and financial resources.[17]

Community geographies (CG), a more recent approach to participatory university–community collaborations, seeks to address some of these constraints. Ideally, CG programs develop sustained and reciprocal relationships between university and community partners, negotiate collaborative knowledge production and shared power, flexibly respond to a variety of community priorities, and leverage the assets of universities and communities to bridge the spatial digital divide. While CG continues to implement the P/PGIS ethos of confronting existing power structures to address community concerns, it has a broadened focus including the use of a multiplicity of methods, not all dependent upon the use of computer technology. Examples of these methods include sketch mapping, community-input surveys and focus groups, participatory radio shows and film-making, as well as the use of local archives and oral narratives.[18] Similarly, the structure, scale, scope, and duration of CG projects vary, often even within the same CG program or institution. Given greater autonomy to determine the limits and expectations of participation, community partners actively and more collaboratively delineate the research objectives, processes, and outcomes.[19]

CG has highlighted a series of new pressures facing university–community collaborations. Most prominently, CG challenges community-engaged scholarship to be more fully integrated into academic disciplines and institutions.[20] Eric Sheppard, as president of the American Association of Geographers, highlighted that '"elite universities" rhetoric talks the talk of public scholarship without walking the walk: It remains too often a pro bono activity to be undertaken in addition to everything else'.[21] A more integrated CG advocates for changes to the traditional academic reward system, flexible teaching schedules and evaluations, as well as updated student learning outcomes, a recognition of community partners as peers in the peer-review process, and more formalised professional and financial support.[22] Without integration and institutional support, CG can struggle to materialise and be sustained given the extent of time necessary to establish reciprocal partnerships and given the difficulty in evaluating and communicating some of the more intangible, yet invaluable, impacts of CG projects. Finally, CG has also drawn attention to the ways in which the politics of research becomes increasingly complex as community geographers remain in place and as the boundaries between scholar and community blur.[23]

ENGAGEMENT AND ATTENTION

Mapshop was established in 2015, growing alongside these debates and experiments with P/PGIS and CG. Originally offered as a course in the Department of Geography (GEO509) since 2012, Mapshop now operates both a connected course and supports undergraduate student research assistantships. As the campus is located in the heart of downtown Lexington, the opportunities for greater reciprocity with the neighbours and neighbourhoods near the university meant that the course for university–community partnerships was insufficient. By sponsoring research assistantships with undergraduate students, we were able to expand these partnerships out of the academic semester.

Mapshop has provided an opportunity to pause and reflect upon the ways in which maps and mapmaking might intervene in, as well as disrupt hegemonic narratives about Lexington, as expressed by undergraduate students and new faculty, as well as challenge long-standing tensions created by the university: what might be generalised as 'town–gown' relations. We advance a post-representational viewpoint in Mapshop. In doing so, we draw upon Kitchin, Perkins, and Dodge (2009) to suggest that mapmaking is not about providing a view upon a world; instead, mapmaking is worlding.[24] Mapmaking, as art, 'does not reproduce the visible, but makes visible', to borrow an oft-cited quote of Paul Klee. Mapshop is therefore an invitation to experiment with engagement and the fashioning of attention – about how to produce greater care for things of urgency. Here, we are inspired by renewed efforts in psychogeography, map art, and radical cartography.[25]

However, the space of potential engagement and attention is a crowded one.[26] For non- profits and for-profits as well as activists, students, faculty, and neighbours, the ability to capture the attention of those who would be impacted by such methods is increasingly challenged. New forms of media and new platforms for interaction have fragmented any simple notion of 'the public' and 'the community'. Engagement and attention are subjected to algorithmic sorting and herd-mentality. Far from letting this numb us to getting involved, Mapshop instead experiments with multiple forms of engagement and attention capture.

Figure 16.3 Authors' Image.

One of these experiments was to fully alter the format of the connected course (GEO509). Taught as a lecture and lab course, students were presented with projects imagined by community partners and asked to choose which projects to work toward – with mixed results. For example, community partners would put forward projects asking students to map locations of community gardens, analyse the spatial relationships around waste management, or create a reference map for a service-providing organisation. The effort and responsibility to design the projects required months of meetings with community partners in advance of the course, leaving the instructor with the task of managing expectations such that the community partner would ideally receive something of direct value. Some of these projects would meet the needs of community organisations; others came up short.

Regardless, any engagement with the community partner began and ended inside the course – leaving the partner without the crucial follow-up as their needs changed. After running a version of this course for four years, it was felt that students were missing the most important part of this partnership – design – and that community partners were generally treated as clients and not as experts in their own domain.

In 2015, the course was redesigned using a studio model. The students were presented with a first assignment – to understand the issues that shape Lexington, using what is available from the American Community Survey.[27] This first assignment then, by week four, emerges into a series of concerns driven by the students' explorations (e.g. housing affordability, food insecurity, transportation access, educational attainment, class and racial segregation).

Community partners are brought in as experts to discuss these issues and provide feedback on student research projects. Lectures and technical demonstrations are given as needed, and not pre-planned. Instead, the instructor and students work, week by week, to better understand through mapmaking the impacts of these issues on our neighbours and neighbourhoods.

In that same year, a research assistantship was created with support from the College of Arts and Sciences. These assistantships run the entire academic year, allowing the mapping efforts of the connected course to expand into other projects with community partners. A project submission form was created, allowing community organisations and non-profits to contact Mapshop with mapmaking ideas or questions for spatial analysis. We have found the submission form to be an important, if still underutilised, aspect of our support for community partners, decentring the course and the schedules of students in the process of requesting mapping expertise. Unbounding these efforts from the classroom puts us one step closer toward the idea of the land-grant public university – one that leverages the expertise of the university in support of our communities.

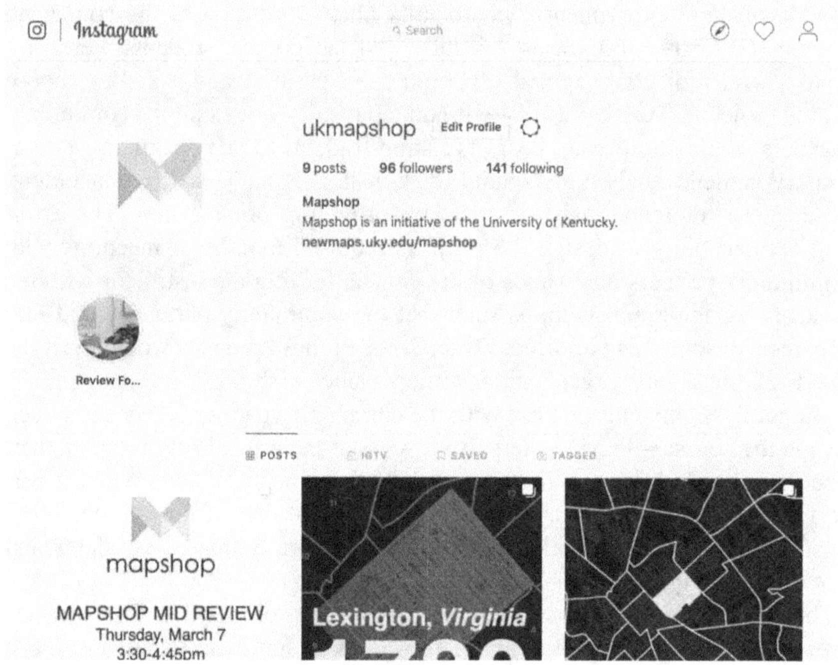

Figure 16.4 Authors' Image.

In the fall of 2019, we created an Instagram account with the intent to modularise some of our Mapshop research efforts into a format for shorter attention spans. This has provided an enjoyable way to connect with members of our communities, while providing Mapshop with an intermediate outlet for impactful ideas and representations. Here, we can try on different forms of analysis and a different approach to map stories.

IMPLICATIONS

While Mapshop continues to experiment with engagement, we also encounter a series of ambiguous and often conflicting implications. For example, as we work to create, curate, and analyse data around a multitude of topics within Lexington, we are also increasingly viewed as a consultancy service for our partners. Of course, for the fashioning of attention, this means that Mapshop is given opportunities to direct the gaze of our partners – to draw focus to the concerns and injustices that we believe require attention. Some examples of this work include highlighting the ways in which the city's legacy of a racially divided urban morphology[28] continues to influence the contours and

Figure 16.5 Authors' Image.

effects of contemporary development, visualising the missing data points that while collected, remain inaccessible to the public, and demonstrating inequities in city services.

However, being viewed as consultants also works to legitimise the authority of the university as a knowledge producer and re-establishes problematic forms of objectivity and neutral science that further black box urban data processes and decision-making. This is evident, for instance, in project requests that seek to legitimise the anecdotal stories of community members with the facts of data. Although these data are undeniably relevant and important, we attempt to resist participating in a digital culture where there are many maps but *few stories being told through them.*[29]

We face similar concerns over reconstituting the city as a laboratory; a playground for teaching and learning. The framing of Lexington neighbourhoods as opportunities for project-based learning (for students of GEO509) or for gaining hands-on experience (for assistants), or to situate case studies (for academics), for instance, can all be seen as extractive processes that once again privilege and locate the university as separate from its neighbours. This is particularly pertinent as institutions of higher education throughout North America and Europe face neoliberal pressures to provide students-customers with greater opportunities for applied learning experiences and to provide philanthropic benefactors with socially relevant research with broader impacts. Programmes, like Mapshop, always have the potential to be co-opted for the interests of the university. As we work, therefore, we are cognizant of the ways in which we can direct the attention of the university to reflect on its own footprint as well as the shadows that footprint casts on its neighbours.

At the same time, Mapshop intentionally works to transform the role of community members from participants in student-led projects to community experts directing our attention and critiquing the results. In this way, we follow calls within P/PGIS, CG, and citizen science literatures to recast community members as more than bodies for data collection, but as analysts, peers, and community experts in their own right.[30] In doing so, we seek to destabilise the casting of the university as the sole voice of authoritative knowledge. This does not mean pedestalling the voice of community rather it requires listening, with an intent to reset our own rhythms of project design and implementation.

Navigating these changing relationships with our partners and our students requires its own form of care and attention. For instance, in a previous iteration of the Mapshop connected course, a student group became frustrated that their community partner was not responding to their emails, and further, that this lack of response was going to impact the students' ability to finish the project and the course. In these moments, it becomes important to work towards an understanding alongside our students that community partners are not beholden to the academic semester. They are not employees nor students of the university. The rhythms of partners' work are not the same as the rhythms of the course. For them, there is no such thing as a 'final grade'. Some students are sought after by community partners following their time with Mapshop, to continue the relationship and the work. Altering this dynamic has been instrumental to lengthening the rhythms of engagement, while also providing both students and partners with a more realistic sense of this relationship.

Treating community partners as experts, as well as moving away from forms of partnerships that treat partners as clients receiving a Mapshop 'service', we believe, creates greater opportunities for reciprocity. In particular, we judge the reciprocity of our partnerships by the ability of community experts to decline to engage. Not only is the space of potential engagement crowded but the demands to engage are increasingly pervasive, requiring organisations and individuals to devote ever more resources, more time, and already limited capacity to potentially unrelated project aims.[31] As academics face increasing expectations to make their research more directly relevant and socially engaged, it is pertinent to remember the demands that such calls place on communities themselves. Situated within an elite institution, with a series of potential resources to leverage, we recognise the ability of community experts to decline requests to engage as a metric of the flexibility and resiliency of our partnerships.

To resist slipping into some of the endless ambiguities of these thoughts, we attempt to cultivate within ourselves and our students an intense engagement with and responsibility to place. Mapshop is more than creating maps within the detached confines of a university classroom or computer lab. It involves an investment in the study area, its community members, and the peculiarities of its concerns. The experiences that students have through engaging with Mapshop are designed to be intensive, collaborative, and unfinished. They are an entry to further engagement. For some, that continued engagement may not be situated within Lexington. Yet, Mapshop aims to instil independent and ethical mapmaking practices within all of its students and assistants so that while at the University of Kentucky our engagement extends beyond simple play and experimentation to scratch (however suggestive or inconclusive) at issues of social justice and inequality. We hope that even as students leave the university, they will take these practices with them.

CONCLUSIONS

Mapshop will continue to seek support from the university in order to provide student assistantships while supporting our community partners with spatial data analyses and revisualisations. We continue to build community-based projects into the design of our introductory and advanced mapmaking

Figure 16.6 Authors' Image.

coursework, to best prepare students with both technical expertise and more substantial knowledge about the issues that impact our neighbours and neighbourhoods in Lexington.

However, the work of Mapshop is incomplete. This is perhaps one of the more difficult lessons of learning to map and mapping to learn. The GISciences have taught generations of students how to design a research project, collect and analyse data, and find ways to represent the findings from that research. However, what might it mean if the project is one that attempts to address social and environmental injustices? How might these projects be designed and implemented? When does one report findings? Instead, the experiences assembled by Mapshop have no easy end-points. There are only more data, more community concerns, more meetings to attend. To pull these struggles out of the confines of the classroom, the syllabus is the real objective. To fashion novel and imminent forms of inquiry that keeps students and instructors on their toes is the goal. Learning to map is not enough. In Mapshop, on our best days, we prefer to map in order to learn.

NOTES

1. Nadine Schuurman, "Trouble in the Heartland: GIS and Its Critics in the 1990s," *Progress in Human Geography* 24, no. 4 (2000): 569–90.

2. Lewis. "Review of Fitzgerald: Geography of a Revolution." Ley. "Review of Fitzgerald: Geography of a Revolution."

3. Gwendolyn C. Warren, Cindi Katz, and Nik Heynen, "Myths, Cults, Memories, and Revisions in Radical Geographic History," in *Spatial Histories of Radical Geography* (John Wiley & Sons, Ltd, 2019), 59–85, https://doi.org/10.1002/9781119404781.ch2; Rich Heyman, "'Who's Going to Man the Factories and Be the Sexual Slaves If We All Get PhDs?' Democratizing Knowledge Production, Pedagogy, and the Detroit Geographical Expedition and Institute," *Antipode* 39, no. 1 (February 2007): 99–120, https://doi.org/10.1111/j.1467-8330.2007.00508.x.

4. David Harvey, "On the History and Present Condition of Geography: An Historical Materialist Manifesto," *The Professional Geographer* 36, no. 1 (February 1984): 1–11, https://doi.org/10.1111/j.0033-0124.1984.00001.x.

5. John Pickles, *Ground Truth: The Social Implications of Geographic Information Systems* (Guilford Press, 1995); Nadine Cato Schuurman, "Critical GIS: Theorizing an Emerging Science" (University of British Columbia, 2000), https://doi.org/10.14288/1.0089782.

6. Christine E. Dunn, "Participatory GIS — a People's GIS?," *Progress in Human Geography* 31, no. 5 (October 2007): 616–37, https://doi.org/10.1177/0309132507081493; Eric Sheppard, "GIS and Society: Towards a Research Agenda," *Cartography and Geographic Information Systems* 22, no. 1 (January 1995): 5–16, https://doi.org/10.1559/152304095782540555.

7. Nancy J. Obermeyer, "The Evolution of Public Participation GIS.," *Cartography and Geographic Information Systems* 25, no. 2 (1998): 65; Renee Sieber, "Public Participation Geographic Information Systems: A Literature Review and Framework," *Annals of the Association of American Geographers* 96, no. 3 (September 2006): 491–507, https://doi.org/10.1111/j.1467-8306.2006.00702.x.

8. Schroeder. "Report on Public Participation GIS Workshop".

9. Mordechai Haklay and Carolina Tobón, "Usability Evaluation and PPGIS: Towards a User-Centred Design Approach," *International Journal of Geographical Information Science* 17, no. 6 (September 2003): 577–92, https://doi.org/10.1080/1365881031000114107; Sieber, "Public Participation Geographic Information Systems."

10. Amber J. Boll-Bosse and Katherine B. Hankins, "'These Maps Talk for Us:' Participatory Action Mapping as Civic Engagement Practice," *The Professional Geographer* 70, no. 2 (April 3, 2018): 319–26, https://doi.org/10.1080/00330124.2017.1366788; Sarah Elwood, "Beyond Cooptation or Resistance: Urban Spatial Politics, Community Organizations, and GIS-Based Spatial Narratives," *Annals of the Association of American Geographers* 96, no. 2 (June 2006): 323–41, https://doi.org/10.1111/j.1467-8306.2006.00480.x; William J. Craig, Trevor M. Harris, and Daniel Weiner, *Community Participation and Geographical Information Systems* (CRC Press, 2002); Dunn, "Participatory GIS — a People's GIS?"

11. Melinda Laituri, "The Issue of Access: An Assessment Guide for Evaluating Public Participation Geographic Information Science Case Studies," *URISA Journal* 15 (2003): 8.

12. Bandana Kar et al., "Public Participation GIS and Participatory GIS in the Era of GeoWeb," *The Cartographic Journal* 54, no. 4 (2016): 296–99; Sieber, "Public Participation Geographic Information Systems."

13. For community-integrated-GIS see: Craig, Harris, and Weiner, *Community participation and geographic information systems.* For counter-mapping see: Peluso. "Whose Woods Are These? Counter-Mapping Forest Territories in Kalimantan, INDONESIA." For more information on the distinctions see: Dunn, "Participatory GIS — a People's GIS?"; Sieber, "Public Participation Geographic Information Systems: A Literature Review and Framework."

14. Kar et al., "Public Participation GIS and Participatory GIS in the Era of GeoWeb."

15. Sarah Elwood, "Negotiating Knowledge Production: The Everyday Inclusions, Exclusions, and Contradictions of Participatory GIS Research*," *The Professional Geographer* 58, no. 2 (May 2006): 197–208, https://doi.org/10.1111/j.1467-9272.2006.00526.x.

16. Kar et al., "Public Participation GIS and Participatory GIS in the Era of GeoWeb."

17. Jonnell A. Robinson, Daniel Block, and Amanda Rees, "Community Geography: Addressing Barriers in Public Participation GIS," *The Cartographic Journal* 54, no. 1 (January 2, 2017): 5–13, https://doi.org/10.1080/00087041.2016.1244322.

18. Heather Davis, "Using Radio to Promote Community Engagement around Well-Being issues"; Lyall, "Participatory Film-making"; Torres, "Sketch Mapping and Emotional Attachment in Restoration"

19. Robinson, Block, and Rees, "Community Geography."

20. Jonnell A. Robinson and Timothy L. Hawthorne, "Making Space for Community-Engaged Scholarship in Geography," *The Professional Geographer* 70, no. 2 (April 3, 2018): 277–83, https://doi.org/10.1080/00330124.2017.1366775.

21. Eric Sheppard, "Public Geographies," *Resident's Column. American Association of Geographers.*, 2013, http://news.aag.org/2013/01/public-geographies/.

22. Robinson and Hawthorne, "Making Space for Community-Engaged Scholarship in Geography."

23. Robinson, Block, and Rees, "Community Geography."

24. Martin Dodge, Rob Kitchin, and Chris Perkins, *Rethinking Maps: New Frontiers in Cartographic Theory* (Routledge, 2009).

25. Bhagat and Mogel, "An Atlas of Radical Cartography"; Kanarinka, "Art-Machines, Body-Ovens and Map-Recipes: Entries for a Psychogeographic Dictionary."; Wood, "Map Art."

26. Matthew W. Wilson, "Paying Attention, Digital Media, and Community-Based Critical GIS," *Cultural Geographies* 22, no. 1 (January 2015): 177–91, https://doi.org/10.1177/1474474014539249.

27. The American Community Survey (ACS) is a continuous and sample-based demographic survey administered by the United States Census Bureau. Supplementing the decennial census, ACS data is released annually and intends to provide more up-to-date estimates of demographic trends. Beginning in 2005, the ACS tracks variables such as income, employment, poverty, housing characteristics, ancestry, language proficiency and migration. The ACS is often used by public and non-profit entities for a variety of strategic planning efforts such as allocating funding, identifying at-risk populations, community advocating and much more. There is some debate over whether the ACS is constitutional. For more see: https://www.census.gov/programs-surveys/acs.

28. Richard H. Schein, "Urban Form and Racial Order," *Urban Geography* 33, no. 7 (October 2012): 942–60, https://doi.org/10.2747/0272-3638.33.7.942.

29. Wilson, Syllabus for GEO509: Community Mapshop. Available at: https://wilsonism.github.io/geo509/syllabus.html#i-course-description.

30. Jessica Breen et al., "Mapping Grassroots: Geodata and the Structure of Community-Led Open Environmental Science," *ACME: An International E-Journal for Critical Geographies,* 14, no. 3 (2015): 849–73.

31. Wilson, "Paying Attention, Digital Media, and Community-Based Critical GIS."

BIBLIOGRAPHY

Bhagat, Alexis, and Lize Mogel. "An Atlas of Radical Cartography." *Journal of Aesthetics and Protest Press*, 2008.

Boll-Bosse, Amber J., and Katherine B. Hankins. "'These Maps Talk for Us:' Participatory Action Mapping as Civic Engagement Practice." *The Professional*

Geographer 70, no. 2 (2018): 319–26. https://doi.org/10.1080/00330124.2017.1366788.

Breen, Jessica, Dosemagen, Shannon, Warren, Jeffery, and Lippincott, Matthew. "Mapping Grassroots: Geodata and the Structure of Community-Led Open Environmental Science." *ACME: An International E-Journal for Critical Geographies,* 14, no. 3 (2015): 849–73.

Bunge, William. *Fitzgerald: Geography of a Revolution.* Edited by Andrew Herod, Melissa Wright, Nik Heynen, and Trevor Barnes. Athens: University of Georgia Press, 2011.

Craig, William J., Trevor M. Harris, and Daniel Weiner. *Community Participation and Geographical Information Systems.* CRC Press, 2002.

Davis, Heather. "Using Radio to Promote Community Engagement around Well-Being Issues". Presented at Community Geography Workshop, 2019: http://www.communitymappinglab.org/commgeog19.html.

Dodge, Martin, Rob Kitchin, and Chris Perkins. *Rethinking Maps: New Frontiers in Cartographic Theory.* Routledge, 2009.

Dunn, Christine E. "Participatory GIS — a People's GIS?" *Progress in Human Geography* 31, no. 5 (2007): 616–37. https://doi.org/10.1177/0309132507081493.

Elwood, Sarah. "Beyond Cooptation or Resistance: Urban Spatial Politics, Community Organizations, and GIS-Based Spatial Narratives." *Annals of the Association of American Geographers* 96, no. 2 (2006): 323–41. https://doi.org/10.1111/j.1467-8306.2006.00480.x.

———. "Negotiating Knowledge Production: The Everyday Inclusions, Exclusions, and Contradictions of Participatory GIS Research*." *The Professional Geographer* 58, no. 2 (2006): 197–208. https://doi.org/10.1111/j.1467-9272.2006.00526.x.

Haklay, Mordechai, and Carolina Tobón. "Usability Evaluation and PPGIS: Towards a User-Centred Design Approach." *International Journal of Geographical Information Science* 17, no. 6 (September 2003): 577–92. https://doi.org/10.1080/1365881031000114107.

Harvey, David. "On The History And Present Condition Of Geography: An Historical Materialist Manifesto." *The Professional Geographer* 36, no. 1 (1984): 1–11. https://doi.org/10.1111/j.0033-0124.1984.00001.x.

Heyman, Rich. "'Who's Going to Man the Factories and Be the Sexual Slaves If We All Get PhDs?' Democratizing Knowledge Production, Pedagogy, and the Detroit Geographical Expedition and Institute." *Antipode* 39, no. 1 (2007): 99–120. https://doi.org/10.1111/j.1467-8330.2007.00508.x.

Kar, Bandana, Renee Sieber, Muki Haklay, and Rina Ghose. "Public Participation GIS and Participatory GIS in the Era of GeoWeb." *The Cartographic Journal* 54, no. 4 (2016): 296–99.

Lanarinka. "Art-Machines, Body-Ovens and Map-Recipes: Entries for a Psychogeographic Dictionary." Cartographic Perspectives 53 (2006): 24–40.

Laituri, Melinda. "The Issue of Access: An Assessment Guide for Evaluating Public Participation Geographic Information Science Case Studies." *URISA Journal* 15 (2003): 8.

Lewis, Peirce. "Review of Fitzgerald: Geography of a Revolution." *Annals of the Association of American Geographers* 63, no. 1 (1973): 131–32.

Ley, David. "Review of Fitzgerald: Geography of a Revolution." *Annals of the Association of American Geographers* 63, no. 1 (1973): 133–35.

Lyall, Angus. "Participatory Film-making". Presented at Community Geography Workshop, 2019: http://www.communitymappinglab.org/commgeog19.html.

Obermeyer, Nancy J. "The Evolution of Public Participation GIS." *Cartography and Geographic Information Systems* 25, no. 2 (1998): 65.

Peluso, Nancy Lee. "Whose woods are these? Counter-mapping forest territories in Kalimantan, Indonesia." *Antipode* 27, no. 4 (1995): 383–406. https://doi.org/10.1111/j.1467-8330.1995.tb00286.x.

Pickles, John. *Ground Truth: The Social Implications of Geographic Information Systems*. Guilford Press, 1995.

Robinson, Jonnell A., Daniel Block, and Amanda Rees. "Community Geography: Addressing Barriers in Public Participation GIS." *The Cartographic Journal* 54, no. 1 (2017): 5–13. https://doi.org/10.1080/00087041.2016.1244322.

Robinson, Jonnell A., and Timothy L. Hawthorne. "Making Space for Community-Engaged Scholarship in Geography." *The Professional Geographer* 70, no. 2 (2018): 277–83. https://doi.org/10.1080/00330124.2017.1366775.

Schein, Richard H. "Urban Form and Racial Order." *Urban Geography* 33, no. 7 (2012): 942–60. https://doi.org/10.2747/0272-3638.33.7.942.

Schroeder, Paul. Report on Public Participation GIS Workshop. In Harris, T. and Weiner, D., GIS and Society: The Social Implications of how People, Space, and Environment are Represented in GIS, Scientific Report for the Initiative 19 Specialist Meeting, 2–5 March, Koinonia Retreat Center, South Haven, Minnesota, National Center for Geographic Information and Analysis Report (1996): 96–97.

Schuurman, Nadine. "Critical GIS : Theorizing an Emerging Science." University of British Columbia, 2000. https://doi.org/10.14288/1.0089782.

———. "Trouble in the Heartland: GIS and Its Critics in the 1990s." *Progress in Human Geography* 24, no. 4 (2000): 569–90.

Sheppard, Eric. "GIS and Society: Towards a Research Agenda." *Cartography and Geographic Information Systems* 22, no. 1 (1995): 5–16. https://doi.org/10.1559/152304095782540555.

———. "Public Geographies." *Resident's Column. American Association of Geographers.*, 2013. http://news.aag.org/2013/01/public-geographies/.

Sieber, Renee. "Public Participation Geographic Information Systems: A Literature Review and Framework." *Annals of the Association of American Geographers* 96, no. 3 (2006): 491–507. https://doi.org/10.1111/j.1467-8306.2006.00702.x.

Torres, Hannah. 2019. "Sketch Mapping and Emotional Attachment in Restoration" Presented at Community Geography Workshop, 2019: http://www.communitymappinglab.org/commgeog19.html.

Warren, Gwendolyn C., Cindi Katz, and Nik Heynen. "Myths, Cults, Memories, and Revisions in Radical Geographic History." In *Spatial Histories of Radical*

Geography, 59–85. John Wiley & Sons, Ltd, 2019. https://doi.org/10.1002/9781119404781.ch2.

Wilson, Matthew W. "Paying Attention, Digital Media, and Community-Based Critical GIS." *Cultural Geographies* 22, no. 1 (2015): 177–91. https://doi.org/10.1177/1474474014539249.

Wood, Denis. "Map Art." *Cartographic Perspectives* 53 (2006): 5–14.

All unaccredited images were taken of the Mapshop Mid Review, held on 7 March 2019, with several guests in attendance, including the vice mayor of Lexington and two members of the Lexington City Council. Source: Authors.

Chapter 17

Map Orkney Month

Imagining Archaeological Mappings

Daniel H. J. Lee

During March 2015, a new map of Orkney was created. This counter-map[1] was made by residents and visitors from their everyday journeys, favourite walks, island tours, encounters, significant places and objects. The map is an unfamiliar Orkney, revealed through the experience and creativity of its inhabitants and contributions from outside. Participants were asked to carry a handheld Global Positioning System (GPS), or use their smartphone, to map their day and record a 'site' of their choice; a kind of countywide archaeological walkover survey. Map Orkney Month was as much about process, an archaeology of the event itself and the drawing out of alternative heritage rather than polished cartography and recording archaeology. This 'counter' position poses a challenge to archaeological practice and power structures, the generation and dissemination of archaeological knowledge, and the way we engage with and give agency to others as a discipline.[2]

This chapter will introduce this project and its processes by dipping into a selection of the contributions (shared here as quotes and associated images; please read these in sequence along with the main text). Full details of the project and how it unfolded can be found on the Public Archaeology 2015 blog.[3] The story told here draws more thematically across the Map Orkney Month assemblage and proposes new modes of archaeological cartography. Map Orkney Month was archaeological in conception, but multi-disciplinary in its outlook, bringing together elements of archaeology, geography, cartography, and arts practice. For me, this sort of project exposes the importance of archaeology of the present within the sub-field of Contemporary Archaeology.[4] Indeed, archaeology of the contemporary world *should,* and can only be, all of these things and more.

The politics of archaeological cartography are then briefly discussed in order to contextualise the innovative approach employed here. Archaeologists and the discipline of archaeology have tended to use cartography – in a multitude of guises – in highly codified and power-loaded ways.[5] Part of the approach used here, as will become clear, aims to break down these power structures and experiment with the democratisation of archaeology[6] and explore the use of 'Alternative Mappings' in the discipline,[7] a theme central to this book. To push this even further, imaginative sites[8] were included in the project as a direct challenge to common dualities such as subjectivity/objectivity and fact/fiction, the role of imagination and the real in the generation of knowledge and narrative, and the use of memory in archaeology.[9]

In this way, the *imagining* of archaeological mappings, as stated in the chapter title, prompts archaeologists and others to think about (archaeological) mapping in a new way. And at the same time asks us to *imagine* the Map Orkney Month assemblage without the usual cartographic signposts. The individual contributions are 'maps' in themselves. As such, this chapter contains no overall map – only fragments will be shared – leaving the rest to your imagination (figure 17.1).

Figure 17.1 Rowena Baker, Kirkwall Bay 15/03/15: 'My mapping day started with feeding the hens and helping my daughter revise for her National Fives. We both then went to Hatston Slip for rowing practice with the Orkney Rowing Club, rowing a traditional Fair Isle Yole and getting a "Viking Eye" view of the Orkney landscape. Afterwards, back home and more revision, followed by a walk down the field to the shore from where we can see the buoy marking the site of the Royal Oak (sunk by a German U-boat in 1939 during WWII). While walking along the shore and within sight of the Royal Oak buoy I found part of a poppy wreath that may have been laid down on the wreck on the 14th October to commemorate the loss of 883 young lives'. Rowena Baker / Map Orkney Month.

PROJECT BACKGROUND

Map Orkney Month was part of Public Archaeology 2015 (PA2015), an online project aimed at engaging the public in archaeology and archaeological themes in a creative and innovative way.[10] While all projects are based in the United Kingdom, its online presence has given it global outreach. Six archaeologists and six non-archaeologists each contributed a month-long project during the year. In March 2015, I embarked upon an archaeological public mapping project. In doing so, Map Orkney Month responded directly to one of the central themes behind the Public Archaeology project – that ultimately archaeology will and should also be undertaken by non- archaeologists, breaking down the control and authority of the discipline and its role in translating the past and present. As several archaeologists have recently suggested or discussed, we are all archaeologists now,[11] but what does this mean in terms of participatory mapping in archaeology? As PA2015 curator James Dixon suggests, the overall project 'will be realised by archaeologists inspiring non-archaeologists to action'.[12] While initially the case for Map Orkney Month, I think in the end, the non-archaeologists have provided much of the inspiration to archaeology, and certainly to me. The project took this idea in a certain direction, by not only handing complete control to the mappers but also dispensing with an explicit interest in the past. Some of the mappers were also archaeologists as they (we) too are part of the island community (figure 17.2).[13]

Figure 17.2 Norna Sinclair, Stromness 12/03/15: 'A VERY windy start to the day and an invigorating walk to the viewpoint on top of Brinkie's Brae, looking over Stromness to Hoy. I continued on round the west shore to my special groatie buckie beach (small cowrie shell, the finding of which brings good luck!). Then on past the Point 'o' Ness and through Stromness. Later a quick "nip tae the toon" via Waulkmill Bay, Orphir'. Norna Sinclair / Map Orkney Month.

The role of 'archaeology' within Map Orkney Month was loose, and most of the 'archaeological' aspect emerged from participants as they operated under a perceived archaeology banner (i.e. Public *Archaeology* 2015, *archaeological* survey). Guidelines for participants suggested that sites could be heritage or non-heritage related, the *archaeology* was left up to them. The result was that

Figure 17.3 Rosey Priestman and Brendan Colvert, Stove, Sanday 01 & 13/03/15: 'Forming the Pearl. In the introduction to "The Diary of Patrick Fea of Stove, Orkney 1766–1796" Bill Hewison writes how the Vikings saw the island of Sanday as shaped like a lobster stretching its arms towards Norway with Stove clamped in its tail. (WS Hewison, (ed.). *The Diary of Patrick Fea of Stove, Orkney 1766-1796*. (East Linton: Tuckwell Press, 1997), 11.) I have always remembered this differently. Stove is the pearl the lobster holds in its claw. Gardemeles, the farm on the sands, became Stove sometime in the 1500s. The Bay grew and the farm was rebuilt, each time further back from the sea. Old stones emerge from the sand. Clamped or held like a treasure. We have mapped the present Stove, the model farm and its burnt outbuildings, and walked around the perfect circle of the pearl in the sands of the bay'. Rosey Priestman and Brendan Colvert / Map Orkney.

many of the project outcomes were non-archaeological, but were *inspired* by archaeology, albeit in non-traditional and more-than-representational ways. For example, mapping everyday journeys with GPS is nothing new, but Map Orkney Month facilitated new and unique explorations; creating new networks of people, places, and things – some of the main themes for archaeological studies of the past, present, and future (figures 17.3, 17.4, and 17.5).

The project takes the idea of an archaeological walkover survey and unfolds it. Such surveys are used to assess the archaeological resource within a given area by walking, mapping, photography, and written description. Walkover surveys are commonly used in developer- funded and research-led archaeology and aim to provide an objective account of heritage assets, although in the latter more experimental and phenomenological practice is common.[14] In the case of Map Orkney Month, the walkover methodology provides a 'snapshot' survey that draws together a wide range of sites where their perceived archaeological 'value' is of little significance. Of interest here is the use of GPS and walking in archaeological fieldwork practice, capturing the way heritage is woven into everyday life, and in particular the application of participatory mapping in archaeology.[15] The aim is to move on from the 'this is what we do, come and have a go if you like' approach used in some community archaeology projects to one where the participant is an active and creative agent.[16] In particular, participatory mapping could be a powerful tool for investigating an archaeology of the present that has political and social meaning; an accessible and richer archaeology that allows everyone to meaningfully contribute. In this way, Map Orkney Month resulted in a quite different group of sites and interactions (figure 17.6).

MAP ORKNEY MONTH

In the lead up to March 2015, a call for participants was made using local media (BBC Radio Orkney, The Orcadian newspaper, posters, and web content) with the aim of finding a group of participants, hopefully spread throughout the archipelago. Once the group was assembled and a timetable worked out, GPS devices were posted out to islands or delivered to houses (there were up to seven GPS devices circulating to allow for postage, travel, and downloading time; figure 17.7). The majority of people borrowed a small GPS receiver, but some used a GPS app on their smart phones and others drew sketch maps. I wanted to keep the technical side of things down to a minimum for participants and not let this put people off the project. I had considered using open access online mapping (e.g. Google My Maps) but decided to cut out the emphasis on technology where possible and explore using a loaning system, trust, and the postal service. Despite this, the project

Figure 17.4 Forming the Pearl, Stove Bay, Sanday; Walking a Perfect 133m Diameter Circle in the Sands of the Bay. Rosey Priestman and Brendan Colvert / Map Orkney.

was still too 'techy' for some, and the majority of people had never used a GPS before (figure 17.8).

I was to act as the facilitator (or curator) for the project but did not guide or even ask the thirty participants what they were planning. The mapping and content was very much in their hands. The only guidelines were that participants had a single day (midnight to midnight) to map and they were asked to record a 'site' during their travels. An information sheet was provided showing basic GPS operation tips, mapping with a smartphone, and a short practical GPS workshop was occasionally held on the doorstep. On the sheet, some pointers were given, such as the following:

- You could forget about the GPS and use it to passively record your day, or take a deliberate walk to a certain place/site or even draw-by-walking – it's up to you.
- Don't worry about taking the same route more than once.

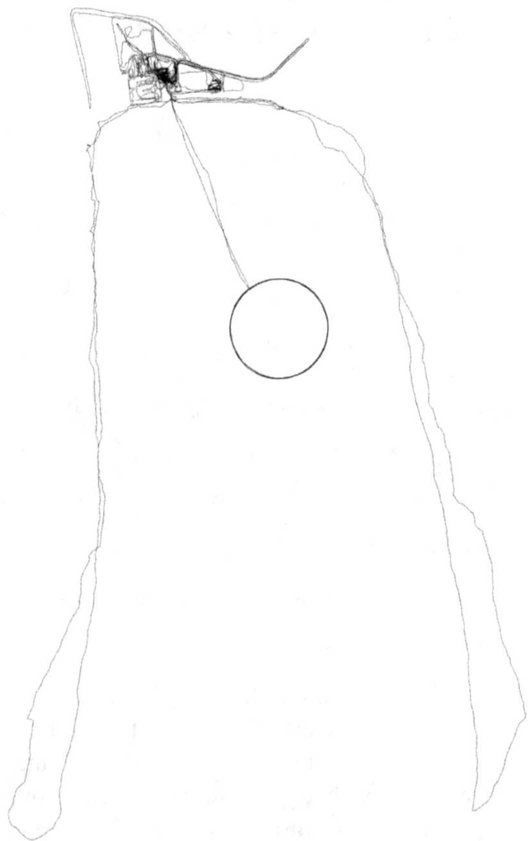

Figure 17.5 Mapping the Pearl, Stove Bay, Sanday; Digital Map from GPS Tracks. Author's Image.

- Why not take a new route or go somewhere you have been meaning to visit for a while and record it?
- Your site doesn't have to be archaeological or heritage related. It will *become* a site through you recording it.
- Why not collaborate with others?

Participants were encouraged to mark a waypoint[17] at their site (although not essential) and record it using photography, sound recordings, or video. I asked for a small piece of text describing the day, but again this was not essential (figure 17.9).

Mapping was undertaken on all the main permanently inhabited islands in the archipelago, and on most days during March, often with multiple

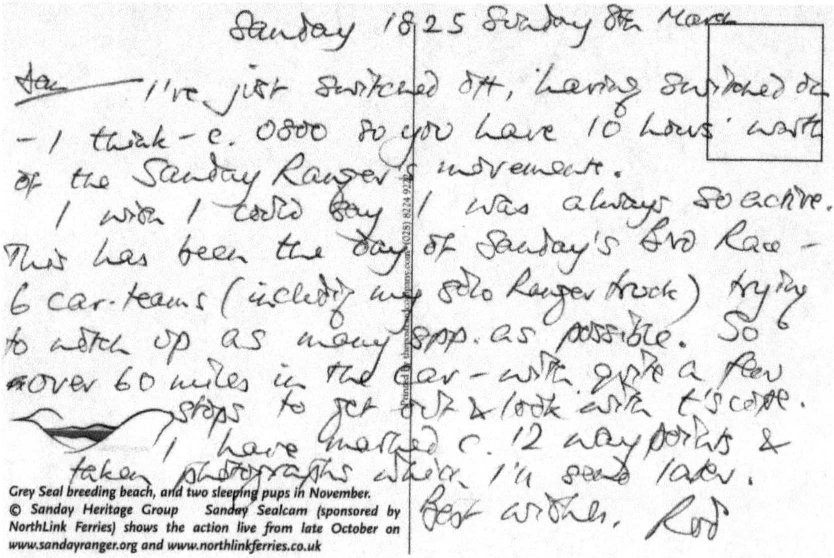

Sanday 1825 Sunday 8th March

Dan — I've just switched off, having switched on — I think — c. 0800 so you have 10 hours' worth of the Sanday Ranger's movements.

I wish I could say I was always so active. This has been the day of Sanday's Bird Race — 6 car-teams (including my solo Ranger truck) trying to notch up as many spp. as possible. So over 60 miles in the car — with quite a few stops to get out & look with t'scope.

I have marked c. 12 waypoints & taken photographs which I'll send later.

Best wishes, Rod

Grey Seal breeding beach, and two sleeping pups in November.
© Sanday Heritage Group Sanday Sealcam (sponsored by NorthLink Ferries) shows the action live from late October on www.sandayranger.org and www.northlinkferries.co.uk

Figure 17.6 Rod Thorne 08/03/15. Postcard from Sanday with a Summary of the Day. Author's Image.

contributions per day. The repertoire of the traditional walkover survey – that of walking – was expanded to include all modes of transport: islander plane, ferry, rowing boat, bus, car, push bike, and of course walking.

Despite what many initially thought, however, it was not about covering ground, charting large areas or recording everything. The emphasis (at least from my point of view) was on everyday journeys, less familiar places, stories, and creating – or drawing out – heritage sites through enacting or choreographing the project. Participants were free to interpret this in their own way, with some recording part of the day or a short walk, some recording numerous

Figure 17.7 The Materiality of Map Orkney Month behind the Scenes. Author's Image.

Figure 17.8 Mark Cook, Kirkwall 07/03/15: 'A typical day in the taxi never knowing where my journey will take me and who will be my traveling companion. Sometimes they are regulars and we have a few minutes to blether and catch up, other times it's someone I've not met before, and like speed dating on wheels I have a limited time to find out about their story!' Author's Image.

sites, and others keen to show the archaeological highlights of their island. For example, island wide tours of archaeological sites and viewpoints were undertaken on Sanday, Stronsay, and Flotta. Some took recording everyday journeys literally and stuck to their normal routine, for example, travelling to work or going to the gym. For some, the journey formed most of the contribution, for example, a photo essay of the ferry trip from Westray to Kirkwall. Others chose a day that they knew something different was happening, and some undertook choreographed mini-projects, for example, mapping Stove farm on Sanday and drawing a perfect circle in the bay (see above), introducing the story of Peedie Pete on Hoy and visiting red phone boxes in the West Mainland on a bicycle (see figures 17.15 and 17.16). For me, the most powerful experience of the

Figure 17.9 Jo Inkster, Rousay 01/03/15: 'A typical Sunday on the farm for this time of year. Cattle feeding duties followed by a wet and windy hack out on my favourite horse Storm. Rode out to the Westside of Rousay and my Waypoint picture is taken looking out over Quandale (site of General Burrough's Clearances) towards the Mainland. The rest of my day was spent with more cattle feeding, a quick dog walk and sometime in the workshop'. Jo Inkster / Map Orkney Month.

project was giving away the control; posting out the technical equipment and basic know-how, not asking what participants were planning, and waiting for the GPS to be posted back or data emailed. The results from this brand of public archaeology were experimental and unexpected. Some of the most exciting contributions arrived as a surprise in the project email account after people mapped on their smartphones, for example, a trip out to North Ronaldsay 'between planes' from The Mainland (figure 17.10).

A particular highlight of Map Orkney Month was a live 'phone box conference' around the West Mainland. Ian Garman cycles around Orkney, and other places in the United Kingdom, to find red telephone boxes and see if they still work. He live tweets the number and waits for people to call to participant in a 'phone box conference' (Twitter: @phoneboxconf). The point is not the phone boxes themselves, but the ability to hold conversations in them. His contribution to Map Orkney Month provided an intriguing snapshot of this technology as it fades away unnoticed; mapping obsolescence (figure 17.11).

Figure 17.10 Helga Tulloch, North Ronaldsay 04/03/15: 'Isabella and I went out between planes to feed the sheep at Cruesbreck and hens at Verracott, pick up a dehumidifier and managed to fit in a walk round the West Beach and pancakes at Purtabreck'. Helga Tulloch / Map Orkney Month.

Figure 17.11 Ian Garman, West Mainland 08/03/15: 'Sunday 8th March was West Mainland's turn for a phone box conference. While the GPS would track where I went, I was keen to share success and failure in real-time on Twitter (although the paradox is I need a mobile phone signal to send a tweet.). Who called? Six people in total: from Orkney, mainland Scotland, England and Wales. For once, most of the callers were folk I didn't already know. So two successful conferences from the last two working red phone boxes I could find in West Mainland. Who knows how long even they will last?' Ian Garman / Map Orkney Month; Digital map: Author.

IMAGINARY CONTRIBUTIONS

Imaginary sites, as introduced above, provided an additional innovative angle to Map Orkney Month. The aim was to push archaeological counter-mapping further and destabilise archaeological authority, tradition, and power structures – further disrupting the relationship between the territory and the map (see the Introduction to this volume for a wider discussion). Imaginary sites could be made by people within Orkney or elsewhere, negating the need to actually be within the county or physically enact their journeys or visit their sites. In other words, an imaginary site can be anything you want it to be, as long as it was 'within' or had some connection to Orkney. The call for imaginary contributions extended the scope for participating in the project from elsewhere. Imaginary contributions were received from throughout the United Kingdom and Bergen in Norway, as well as some from within Orkney itself.

In one multi-authored contribution, other team members from the Public Archaeology 2015 project joined forces to follow an imaginary tour of Orkney devised by James Dixon. The group mapped a collective journey having arrived at Kirkwall airport, following a route west to Skara Brae calling at sites on the way. The 'mapping', however, was situated in the landscape around which they lived. The results were an intriguing blend of imaginings and juxtapositions where central London and other places in the United Kingdom were transposed onto the rural Orcadian landscape.[18]

Inspired by this, Lara Band and Dave Webb undertook a particularly detailed mapping disjuncture in London, translating the imagined Orkney tour into the city and disrupting the map and territory yet further. They mapped the population of the Orkney archipelago (c. 21,000) onto an equivalent area of population in East London – approximately 1.5 by 2.5 km – centered on Kirkwall Place (Kirkwall is the capital of Orkney) and re-created the tour. They tried to find the same sites in the urban landscape, having never visited Orkney and not knowing what the original sites looked like (figure 17.12). This pyschogeographical tour took in city versions of James Dixon's prompts, translating and blurring the rural onto the urban.[19]

Back in Orkney, Rebecca Marr took us on an imaginary journey around the islands using photographs by photographer Gunnie Moberg held by the Orkney Library and Archive (figure 17.13).

Finally, Mark Cook shared an old photograph that played an intriguing part in causing him to live in Orkney (figure 17.14).

For some, imaginary contributions may be a step too far, but I would argue that these interpretations of the Map Orkney Month brief were amongst the most exciting and opened up possibilities for a truly multi-vocal cartography. They demonstrate the potential for an archaeology in and of the present[20] that

Figure 17.12 Lara Band and Dave Webb 01/03/15: '. . . emerging on the northern side we headed westwards and, nearing the coast we reached the site of four Great Rings. Turning south we searched for Skara Brae in Temple Street and though the settlement did not take the form we were expecting, imagining and hoping for, by its angularity and simplicity of form we did believe we'd found it. Though we'd found Rings in several places before, the actual Ring of Brodgar should have been somewhere near the corner of Old Bethnal Green Road and St Jude Road. In fact, it appeared on Middleton Green'. Lara Band and Dave Webb / Map Orkney Month.

is both archaeological and non-archaeological; archaeological cartography that is critical, multi-disciplinary, and creative. As such, Map Orkney Month was an archaeology with, and of, the participants as they went to familiar places, made detours, constructed new stories, imaginings, and connections, and found other networks and material relationships. As usual, the most interesting results from any project are those that are unexpected, and Map Orkney Month had results like this every day.

Imaginary sites blur the distinction between reality and non-reality, fact and fiction, and past and present as the Map Orkney Month archive moves into the future. They play on our remoteness to the past and how we interpret it, even if we are there in direct contact with the material. It exposes the role of imagination, memory work, and storytelling involved in many aspects of archaeological interpretations.

ARCHAEOLOGICAL CARTOGRAPHY

Archaeologists make maps in a variety of ways from a bewildering array of primary and secondary data, based on antiquarian beginnings and grounded in the modernist project. Maps and plans are still hand-drawn on excavation projects, despite the increased use of digital technology, such as laser scanning and photogrammetry. Archaeologists routinely make maps of sites, landscapes, and regions at a range of scales using various sources: hand-drawn plans, measured survey, geophysical and landscape survey, aerial

Figure 17.13 Rebecca Marr, Kirkwall 04/03/15: When, after my commute from Stromness, I arrive at Orkney Library and Archives in Kirkwall the journey begins. Travelling in my work-room I can cover astonishing distances; Papay and North Ronaldsay before tea break, Hoy and Wyre after lunch. Visiting places fleetingly or sometimes lingering longer, I do this through the photography of Gunnie Moberg (Gunnie Moberg Archive, accessed 04/10/20 https://gunniemobergarchive.wordpress.com/). I decided to map the photographs in Gunnie Moberg's first publication stone Built' published in 1979 by Stromness Books and Prints"; (which happens to be where my physical GPS mapped journey began; Gunnie Moberg Archive, "Stone Built", accessed 04/10/20). Gunnie Moberg: 'A bucht, Mainland', Gunnie Moberg Archive, Orkney Library & Archive, reproduced with permission.

Figure 17.14 Mark Cook, view of Scapa Flow (via Wei Ha Wei, China) 07/03/15: 'My photo is a large panoramic print that approximately 100 years old. We were given it as a present nearly 20 years ago and told it was Scapa Flow in Orkney. We had for many years wanted to visit Orkney, and when we finally did we brought the picture with us to find the location, and quickly confirmed it was not around Orkney after all and also noticed it was inscribed "Wei Ha Wei, China". Nevertheless, we loved Orkney and 9 months later had moved here. The picture, therefore, is an imaginary view of Scapa Flow from Houton Tower which I visited on the way home on Saturday'. Mark Cook / Map Orkney Month.

Figure 17.15 Sian Thomas, Graemsay 11/03/15: 'My mapping day on Graemsay dawned with an average wind speed of about 47mph with gusts about 60 mph. But, undaunted, I donned waterproofs and wellies, with GPS firmly in a pocket and set off first to feed my hens at Sandside. I could barely stand up and the hens were getting blown about, so no photo opportunity there. The stone hen house is part of the old farm buildings and gets some shelter from the wind. The 5ft garden dyke that leads to the buildings also helps, especially as I'm quite short! But as soon as I get away from any shelter I'm nearly blown over. Not the time for a walk along the shore yet then'. Sian Thomas / Map Orkney Month.

photographs, GPS surveys, satellite imagery, etc. Archaeologists, along with artists,[21] were quick to adopt new GPS technology in the mid-1990s and new media now forms a sub-discipline(s) of archaeology (e.g. digital archaeology).[22] Yet, cartographic traditions in archaeology generally remain highly codified. They use a code and visual language specific to archaeology (e.g. drawing conventions, symbols, colours, scales, etc.) and the skills involved in 'reading' these expose the often privileged nature of the discipline, as seen for example in the hierarchy of certain site personnel and organisations.[23]

Karen O'Rourke has classified artistic walking projects as 'top-down', where participant walks are structured and controlled by the artist, or 'bottom-up', where the participant is the active agent[24] (although interestingly the *idea* for the project can still come from above.[25] Bottom-up projects in archaeology are instigated by the participants or community, and often

Figure 17.16 Fran Flett-Hollinrake: Custodian of St Magnus Cathedral, Kirkwall 18/03/15: 'Mid- morning I get a visit from our friend Paz who is staying with us for a week. He was with us in 1999 when we saw the total eclipse in Devon, and he has come up to see the eclipse in Orkney. At the same time, Sophie the apprentice stonemason comes in; the three of us climb the cathedral tower and wind the clock. The clock is 100 years old this year, and is proper clockwork that needs to be wound every day. There are three parts to wind and it takes about 5 minutes to do the job. After that, we climb to the very top of the tower and go out onto the parapet to see Kirkwall laid out below in the sunshine. The cruise ship Marco Polo can be seen berthed at Hatston Pier. After a long, busy day (during which over 500 visitors have come through the cathedral doors), I wait for the bells to chime five o'clock, then I lock the big front door and head for home'. Fran Flett-Hollinrake / Map Orkney Month.

facilitated by archaeologists). Almost all mapping in archaeology, be it professional or participant focused, is top-down. For example, students and volunteers are readily trained in drawing plans and maps, and taught archaeological recording practices (e.g. undergraduate field schools and Scotland's Rural Past project, where volunteers are trained in 'archaeological code'). The process of planning a large open area excavation is often collaborative and relatively egalitarian, but the process and results are highly structured and ultimately conform to agreed standards, objectives, and outcomes (e.g. professional field manuals and Chartered Institute for Archaeologists guidelines). Archaeological maps and plans remain highly structured and codified; and as such, they possess, create, and reinforce cartographic authority.[26]

All this is fine, and hard to deny in a professional and academic discipline. The result, however, is that archaeology has experimented much less with giving away control and authority and exploring bottom-up approaches to survey and mapping. Notable exceptions would include Scotland's Urban Past Project where less traditional outcomes such as film are now being foregrounded. In this way, we should strive to free the mapping process and mapmakers from traditional expectations and outputs. Does mapping-as-process even need to include maps?

Professional archaeology is becoming increasingly good at training others in archaeological practice, but rarely does it give out the authority to 'do what you like' with that knowledge and filtering or authenticating the results. In addition, can we really free ourselves from the military rooted top-down way of seeing inherent in cartography, GPS technology,[27] and archaeological traditions? The digital revolution has opened up cartography to the masses (e.g. Google Earth, Google Sketch Up, Smartphone GPS),[28] and in the same way the public are starting to bypass archaeologists and get on with archaeology themselves. Map Orkney Month may not achieve all of this, but it certainly exposes some of these issues for debate (figure 17.15).

Shanks and Pearson's deep maps provide inspiration and comparison for innovative mapping projects in archaeology and beyond.[29] Deep maps attempt

> to record and represent the grain and patina of place through juxtapositions and interpretations of the historical and the contemporary, the political and the poetic, the factual and the fictional, the discursive and the sensual; the conflation of oral testimony, anthology, memoir, biography, natural history and everything you might ever want to say about a place.[30]

Map Orkney Month provides all of this at once, but as a 'snapshot', situating all these pasts very much in the present. This compression of time, through lived experience, everyday journeys, and the choreography of the mapping process, provides a collective interpretation of what Orkney is *now* rather than a rehearsed and researched historical account, even though many archaeological sites were included.

My journey in innovative mapping developed during a recent archaeological residency at Papay Gyro Nights contemporary art festival[31] on the island of Papa Westray (Papay), Orkney involving myself and Antonia Thomas. Here, as archaeologist in residence, I mapped artists and festival spaces, creating archaeologies *with* rather than *of* the festival, and followed these spaces and materials into the future.[32] This project facilitated a creative engagement with our own practice and allowed us to push and subvert archaeological mapping practices (e.g. non-representational archaeologies). I have also undertaken participatory counter-mapping where alternative sites were recorded at iconic Neolithic sites, such as the Ring of Brodgar and Ness of Brodgar in Orkney (for example, carrier bags, car tyres, and archaeological tools).[33] Here the focus was on ephemera, and transitory sites and materials, using GPS as a tool for exploration rather than facilitating an explicit interest in the Neolithic.

From these projects, I would add to O'Rourke's distinctions of 'top-down' and 'bottom-up' participatory mapping discussed above,[34] to include 'passive' and 'active' mapping with GPS. Passive mapping records movements, such as journeys, activities, and events as they happen, while active mapping involves deliberate wayfaring by walking and consciously drawing with the body in space. Drawing by walking, and the digital lines left by using a GPS (perceived in the field through praxis or in digital reality on the laptop screen and printed page), were not considered by anthropologist Tim Ingold in his biography of the line[35] but are central modes of practice in all of my mapping projects. My approach on Papay foregrounded a more-than-representational archaeology that focused on practice and process, rather than – or as equal to – outcomes. This is an archaeology that does not follow traditional practice (e.g. maps that show and could be used to locate, sites, or finds)[36] and results in unexpected outcomes.

MATERIALITY

Map Orkney Month engaged materials, archaeological, and non-archaeological, in numerous ways and generated new assemblages in the process. A small archive was created, of digital data (GPS and GIS data, photographs and the blog) along with artefacts such as postcards, notes on slips of paper, reused jiffy bags, worn zip seal bags used to carry GPS receivers, and the GPS units themselves. Much of this material was circulated amongst fellow participants via me; an exchange of materials between those who had briefly or never met. The materiality of the project is therefore the relationship between humans, non-humans, and material (things, places, sites).

Other material connections were made between participants, beyond the materiality of the project. Groatie Buckie shells were found and not found in Stromness and Egilsay (it is the *finding*, not the objects themselves that brings good luck). Sound recordings made by Fran Hollinrake exposed the workings of the cathedral clock in delicate detail, followed by chimes and locking the large door: stone steps, brass cogs, timber door, iron lock, bronze bell (figure 17.16).[37]

Perhaps the most ephemeral but powerful object was the weathered fragment of a poppy wreath found by Rowena Baker near the Royal Oak washed up like other jetsam on the shore: in plastic we shall not forget.

CONCLUSIONS

Map Orkney Month was an unusual mapping project for archaeology. First, it was process led and allowed the project to create itself with little or no guidance. The original aim was to make a map, but in the end, the finished map was not the most important part and the contributions stood so powerfully for themselves. For this reason, this chapter does not contain a 'final' map,[38] leaving this to your imagination.[39] Instead, certain connections and material objects created or found during the project tell this story for me. For example, Rod Thorne's postcard, the groatie buckie shell, the winding of the cathedral clock, and the fragment of commemorative wreath all make a new map of their own, individually and combined. Rather than being one map, therefore, Map Orkney Month, is a multitude of 'maps'.

Second, participants were active agents who mapped 'passively' or 'actively' as they chose. In doing so, the project was bottom-up, as nearly everything was left to the mappers, although the original conception and curation was mine.[40] Significantly, this approach seeks to destabilise archaeological power structures and critique knowledge generation within the discipline.

Third, the project was not explicitly archaeological, but combined archaeology, geography, cartography, and arts practice, which I would argue is one of its strengths. This brand of archaeological cartography focuses on an archaeology of the present, of surfaces (physical and metaphorical), without an explicit interest in the past or the need to record archaeology for archaeology's sake. All sites recorded within Map Orkney Month have meaning and were 'generated' or 'drawn into the foreground' in some unique way by the participants and the project. Map Orkney Month became a collaborative mapping performance by all those who took part, orchestrated by the places they visited and the journeys that they took.[41]

More significantly, feedback from participants suggests that Map Orkney Month has led people to think about archaeology in a new way (i.e. it is not all about digging holes or the past), learn how to use GPS, explore the potential of their smartphone, discover GIS, think about space in a new way, and most importantly find new threads of inspiration that they are now continuing to follow.

In a way, the 'idea of the new map' was just a way of channelling an alternative archaeology project. By giving away the control, archaeologists can learn from non-archaeologists' approaches which are replete with creativity and connections that may not necessarily have been made. Above all, Map Orkney Month challenged people to make new connections with familiar places; as one participant said to me, 'you are changing the way people think about space, which is really hard to do'.[42]

ACKNOWLEDGEMENTS

Thanks to Mike Duggan and the Livingmaps Review team for including Map Orkney Month in this book. Thanks originally to Phil Cohen for inviting me to contribute to Livingmaps Review journal, Waypoints editor Barbara Brayshay and for inclusion in the very first issue! This chapter was originally published in Livingmaps Review No 1 (2016) and is republished here with minor revision. Many thanks to James Dixon for the inspiration that is Public Archaeology 2015 and fellow team members, what a year! I'm indebted to all of the participants of Map Orkney Month, for their creativity and generosity, without whom there would have been no project. Particular thanks here to Rowena Baker, Norna Sinclair, Rosey Priestman, Brendan Colvert, Rod Thorne, Mark Cook, Jo Inkster, Helga Tulloch, Ian Garman, Tonje Birkeland, Lara Band, Dave Webb, Rebecca Marr, Sian Thomas, Fran Flett-Hollinrake, and other imaginary contributors who are quoted or their contribution referenced in this chapter. Thanks to Orkney Library and Archive for permission to reproduce 'A bucht, Mainland' by Gunnie Moberg. Thanks to Antonia Thomas and Barbara Brayshay for commenting on drafts of the original paper in Livingmaps Review. All errors and omissions are my own.

NOTES

1. Counter-mapping, in this context, is an alternative to the traditional practices of mapping landscapes and sites in archaeology.

2. These arguments have been further developed in Daniel HJ Lee, 'Experimental mapping in archaeology: process, practice and archaeologies of the moment', *Re-Mapping Archaeology: Critical Perspectives, Alternative Mappings,* edited by Mark Gillings, Piraye Hacıgüzeller and Gary Lock (London: Routledge, 2018), 43–176.

3. Daniel Lee[a] 'Map Orkney Month blog posts' accessed 04/10/20 https://publicarchaeology2015.wordpress.com/category/map-orkney-month.

4. Paul Graves-Brown, Rodney Harrison and Angela Piccini, 'Introduction', in *The Oxford Handbook of the Archaeology of the Contemporary World*, edited by Paul Graves-Brown, Rodney Harrison and Angela Piccini (Oxford: oxford University Press, 2013) 1–26.

5. For a discussion of maps as objects, their co-production and their role in generating and maintaining archaeological knowledge see Chris Whitmore, 'The world on a flat surface: maps from the archaeology of Greece and beyond', in *Representing the Past: Archaeology through Text and Image*, edited by Sheila Bonde and Stephen Houston (Oxford: Oxbow Books, 2013), 127–152. For recent discussions of maps in archaeology see Mark Gillings, Piraye Hacıgüzeller, and Gary Lock (eds), *Re-Mapping Archaeology: Critical Perspectives, Alternative Mappings* (London: Routledge, 2018), Daniel HJ Lee, 'Experimental mapping in archaeology: process, practice and archaeologies of the moment', *Re-Mapping Archaeology: Critical Perspectives, Alternative Mappings,* edited by Mark Gillings, Piraye Hacıgüzeller and Gary Lock (London: Routledge, 2018), 43–176, and Piraye Hacigüzeller[a], 'Collaborative mapping in the age of ubiquitous internet: an archaeological perspective', *Digital Classics Online*, BD. 3(2) (2017): 5–16. For creative mapping in archaeology see Daniel HJ Lee, 'Experimental mapping in archaeology: process, practice

and archaeologies of the moment', *Re-Mapping Archaeology: Critical Perspectives, Alternative Mappings,* edited by Mark Gillings, Piraye Hacıgüzeller and Gary Lock (London: Routledge, 2018), 43–176, and Daniel Lee and Jasper Coppes, 'Flux Tower', *Livingmaps Review* no. 7 (2019).

6. Cornelius Holtorf, *Archaeology is a Brand! The meaning of Archaeology in Contemporary Popular Culture* (Oxford: Archaeopress, 2007), 119; on democratisation through digital archaeology and the internet, see Lorna Richardson 'A digital public archaeology?', *Papers from the Institute of Archaeology* (23)1: 10 (2013): 1–12.

7. Piraye Hacigüzeller has reconceptualized archaeological maps as performances, specifically promoting alternative mapping practices, which provide alternatives to representational modes of thinking. This approach will 'help free archaeologists from map-related epistemological anxieties and let them experiment while collecting cartographic information and creating maps', while exploring the 'political and ethical consequences' of their mapping, see Piraye Hacigüzeller[b], 'Archaeological (digital) maps as performances: towards alternative mappings', *Norwegian Archaeological Review* 50(2) (2017): 166; For case studies in alternative archaeological mapping see Daniel H.J. Lee, 'Experimental mapping in archaeology: process, practice and archaeologies of the moment', *Re-Mapping Archaeology: Critical Perspectives, Alternative Mappings,* edited by Mark Gillings, Piraye Hacıgüzeller and Gary Lock (London: Routledge, 2018), 43–176.

8. Imaginary sites are: fictional, exist but were not actually visited, provide disjuncture in time and space/territory, or those that make links through the actions or creativity of the mapper.

9. The aim here is to break down dichotomies, but by stating them so boldly they are often reinforced. Because so little critique of archaeological mapping has so far been undertaken it is important to start the debate here at the beginning, and then move onto the more subtle ways maps and maps are made, used and reproduced in archaeology.

10. Public Archaeology 2015 project blog: https://publicarchaeology2015.wordpress.com accessed 04/10/20.

11. Cornelius Holtorf (2005) and Michael Shanks (2012) have both suggested that 'we are all archaeologists now' in the context of archaeology, popular culture and the archaeological imagination. Cornelius Holtorf has more recently asked the question 'are we all archaeologists now' in Cornelius Holtorf, 'Introduction', in Cornelius Holtorf (ed) Forum: 'Are we all archaeologists now?', *Journal of Contemporary Archaeology* 2(2) (2016): 217–219. The consensus, from a range of archaeologists and non-archaeologists, is mixed and demonstrates that this statement is far from a simple matter. Map Orkney Month suggests, perhaps, that there is some 'archaeologist' within all of us, and that 'non-archaeologists' can draw out contemporary archaeologies in new and different ways. Moreover, are trained archaeologists *all* 'archaeologist', or are there other parts to the whole? I certainly do not restrain myself from wandering into other disciplines, but at the same time argue that even my more experimental cross-disciplinary work and mappings *are* part of archaeology, in a drive to explore innovative forms of practice *within* rather than outside the discipline.

This process aims to invite others in, and even let them take the lead in some circumstances, rather than viewing alternative power and control structures as a threat.

12. James Dixon, 'Project Details', Public Archaeology 2015 blog, accessed 04/10/20 https://publicarchaeology2015.wordpress.com/.

13. This is something that archaeologists and other professionals very easily forget.

14. For experimental practice see Oscar Aldred, *An Archaeology of Movement: A Methodological Study* (PhD thesis, University of Iceland, 2015); Daniel Lee and Antonia Thomas, 'Archaeologies in between: walking (in)visible lines', *Flettverk* #2 (SCENE), (2015): 4–13. For phenomenology, or experiential approaches, see recent discussions by Joanna Brück, 'Experiencing the past? The development of a phenomenological archaeology in British prehistory', *Archaeological Dialogues* 12(1) (2005): 45–72; Sue Hamilton, Ruth Whitehouse, Keri Brown, Pamela Combes, Edward Herring, and Mike Saeger-Thomas, 'Phenomenology in practice: towards a methodology for a "subjective" approach', *European Journal of Archaeology* 9(1) (2006): 31–71 and Chris Tilley 'Phenomenological approaches to landscape archaeology', in *Handbook of Landscape Archaeology*, edited by Bruno David and Julian Thomas (Walnut Creek, CA: Left Coast Press, 2008), 271–276.

15. For the potential of collaborative mapping in archaeology, see Piraye Hacigüzeller[a], 'Collaborative mapping in the age of ubiquitous internet: An archaeological perspective', *Digital Classics Online*, BD. 3(2) (2017): 5–16.

16. Ideas employed by James Dixon, 'Buildings archaeology without the recording', Presentation given at European Association of Archaeologists conference, Session LV6 Creative Archaeologies, Glasgow 05/09/15.

17. A waypoint is a point on the earth's surface recorded with coordinates using a GPS device.

18. Daniel Lee[b], 'Map Orkney Month Week 4 . . .' 2015, accessed 04/10/20, https://publicarchaeology2015.wordpress.com/2015/03/29/map-orkney-month-week-4-winding-the- clock-drawing-stromness-and-imaginary-journeys-from-elsewhere/.

19. Daniel Lee[c], 'Imaginary Tours of Orkney from Elsewhere . . .' 2015, accessed 04/10/20, https://publicarchaeology2015.wordpress.com/2015/03/31/imaginary-tours-of-orkney-from- elsewhere-mapping-archipelagos-in-east-london/.

20. Rodney Harrison, 'Surface assemblages: towards an archaeology in and of the present', *Archaeological Dialogues* 18(2) (2011): 141–161.

21. Karen O'Rourke, '*Walking and mapping: Artists as Cartographers*' (Cambridge, MA: Massachusetts Institute of Technology Press, 2013), xvii.

22. Lorna Richardson, 'A digital public archaeology?' *Papers from the Institute of Archaeology* (23)1: 10 (2013): 1–12.

23. See Matt Edgeworth, '*Acts of Discovery: An Ethnography of Archaeological Field Practice*' (eBook published by Matt Edgeworth, 2006) for an ethnography of archaeological field practice, specifically around the act of discovery.

24. Karen O'Rourke, '*Walking and Mapping: Artists as Cartographers*' (Cambridge, MA: Massachusetts Institute of Technology Press, 2013), xviii.

25. Sarah Casey and Gerry Davies 'Lines of engagement: drawing walking tracking', *Journal of Visual Art Practice* 14(1) (2015): 72–83.

26. Gavin Lucas, *Critical Approaches to Fieldwork: Contemporary and Historical Archaeological Practice* (London: Routledge 2001); Gavin Lucas, *Understanding the Archaeological Record* (Cambridge: Cambridge University Press, 2012).

27. However, much GPS technology is democratised, we can never get away from its routes in military technology, thus it is ultimately about power and control; it could be switched off or (re)scrambled at any time.

28. Jeremy W Crampton and John Krygier, 'An introduction to critical cartography', *ACME: An International E-Journal for Critical Geographies* 4(1) (2006): 11–33.

29. For example their early deep mapping around the abandoned farmstead of Esgair Fraith, near Lampeter, Wales see Mike Pearson and Michael Shanks, *Theatre/Archaeology* (London: Routledge, 2001), 154–162, and Pearson's explorations of his home village Hibaldstow, North Lincolnshire in Mike Pearson, *In Comes I: Performance, Memory and Landscape* (Exeter: University of Exeter Press, 2006), and theatre making in Cardiff in Mike Pearson, *Marking Time: Performance, Archaeology and the City* (Exeter: University of Exeter Press, 2014).

30. Mike Pearson and Michael Shanks, *Theatre/Archaeology* (London: Routledge, 2001), 64–65.

31. http://www.papaygyronights.papawestray.org/ accessed 05/11/15.

32. See Daniel Lee and Antonia Thomas, 'Archaeologists-in-residence at Papay Gyro Nights: experience, expectations and folklore-in-the-making', Paper presented at *Contemporary and Historical Archaeology in Theory* (CHAT) Conference 2013, UCL London 09/11/13. Available at, accessed 04/10/20, https://archaeologistsinresidence.wordpress.com/2013/11/16/archaeologists-in-residence-at-papay-gyro-nights-experience-expectations-and-folklore-in-the-making/.

33. See for example Daniel Lee 'Kirkwall Grammar School S1 rapid counter mapping . . .' accessed 04/10/20 https://archaeologistsinresidence.wordpress.com/2015/06/01/kirkwall-grammar-school-s1-rapid-counter-mapping-of-the-ring-of-brodgar/.

34. Karen O'Rourke, "*Walking and Mapping: Artists as Cartographers*" (Cambridge, MA: Massachusetts Institute of Technology Press, 2013), xviii.

35. Tim Ingold, *Lines: A Brief History* (London: Routledge, 2007).

36. See Daniel Lee 'Collaborative drawings', accessed 04/10/20, https://archaeologistsinresidence.wordpress.com/2014/04/13/collaborative-drawings/.

37. Listen to 'Winding the clock at St magnus cathedral', accessed 04/10/20 https://soundcloud.com/public-archaeology-2015/winding-the-clock-at-st-magnus-cathedral.

38. Or dots on a map – the typical archaeological anti-phenomenological visualisation.

39. Maps as process, practices and mappings, see Rob Kitchin, Michael Dodge, 'Rethinking maps', *Progress in Human Geography* 31(3) (2007): 331–344.

40. See Sarah Casey and Gerry Davies 'Lines of engagement: drawing walking tracking' *Journal of Visual Art Practice* 14(1) (2015): 72 for discussion of the walked line used to 'embody and exchange information and creative experience' and Karen O'Rourke, '*Walking and Mapping: Artists as Cartographers*' (Cambridge, MA:

Massachusetts Institute of Technology Press, 2013), xviii bottom-up participant-led mapping in an arts context.

41. For discussion of archaeological maps as performances see Piraye Hacigüzeller[b], 'Archaeological (digital) maps as performances: towards alternative mappings', *Norwegian Archaeological Review* 50(2) (2017), for explorations in performing sites and landscapes see Daniel Lee and Jasper Coppes, 'Flux Tower', *Livingmaps Review* no. 7 (2019).

42. Listen to 'The end of the day at St Magnus Cathedral', accessed 04/10/20, https://soundcloud.com/public-archaeology-2015/the-end-of-the-day-at-st-magnus-cathedral.

BIBLIOGRAPHY

Aldred, Oscar RO, "*An archaeology of Movement: A Methodological Study*" (PhD thesis, University of Iceland, 2015).

Brück, Joanna, "Experiencing the past? The development of a phenomenological archaeology in British prehistory", *Archaeological Dialogues* 12(1) (2005): 45–72.

Casey, Sarah and Davies, Gerry, "Lines of engagement: drawing walking tracking", *Journal of Visual Art Practice* 14(1) (2015): 72–83.

Crampton, Jeremy W and Krygier, John, "An introduction to critical cartography", *ACME: An International E-Journal for Critical Geographies* 4(1) (2006): 11–33.

Dixon, James, "Project Details", Public Archaeology 2015 blog, accessed 04/10/20 https://publicarchaeology2015.wordpress.com/project-details/.

Dixon, James, "Buildings archaeology without the recording", Presentation given at European Association of Archaeologists conference, Session LV6 Creative Archaeologies, Glasgow 05/09/15.

Edgeworth, Matt, *Acts of Discovery: An Ethnography of Archaeological Field Practice* (eBook published by M. Edgeworth, 2006).

Gillings, Mark, Hacıgüzeller, Piraye and Lock, Gary (eds), *Re-Mapping Archaeology: Critical Perspectives, Alternative Mappings* (London: Routledge, 2018).

Graves-Brown, Paul, Harrison, Rodney, and Piccini, Angela, "Introduction", *The Oxford Handbook of the Archaeology of the Contemporary World*, edited by Paul Graves-Brown, Rodney Harrison and Angela Piccini (Oxford: oxford University Press, 2013), 1–26.

Gillings, Mark, Hacıgüzeller, Piraye and Lock, Garry (eds), *Re-Mapping Archaeology: Critical Perspectives, Alternative Mappings* (London: Routledge, 2018).

Hacigüzeller, Piraye[a], "Collaborative mapping in the age of ubiquitous internet: an archaeological perspective", *Digital Classics Online*, BD. 3(2) (2017): 5–16.

Hacigüzeller, Piraye[b], "Archaeological (digital) maps as performances: towards alternative mappings", *Norwegian Archaeological Review* 50(2) (2017): 149–171.

Hamilton, Sue, Whitehouse, Ruth, Brown, Keri, Combes, Pamela, Herring, Edward, and Saeger-Thomas, Mike, "Phenomenology in practice: towards a methodology for a 'subjective' approach", *European Journal of Archaeology* 9(1) (2006): 31–71.

Harrison, Rodney 2011. "Surface assemblages: towards an archaeology in and of the present", *Archaeological Dialogues* 18(2): 141–161.

Hewison, WS (ed.) *The Diary of Patrick Fea of Stove, Orkney 1766-1796* (East Linton: Tuckwell Press, 1997), 11.

Holtorf, Cornelius, *From Stonehenge to Las Vegas: Archaeology as Popular Culture.* (Lanham, MD: Altamira, 2005), 160.

Holtorf, Cornelius, *Archaeology is a Brand! The meaning of Archaeology in Contemporary Popular Culture* (Oxford: Archaeopress, 2007).

Holtorf, Cornelius. 'Introduction', in Cornelius Holtorf (ed) Forum: "Are we all archaeologists now?", *Journal of Contemporary Archaeology* 2(2) (2016): 217–219.

Ingold, Tim, *Lines: A Brief History* (London: Routledge, 2007).

Kitchin, Rob and Dodge, Michael, 'Rethinking maps', *Progress in Human Geography* 31(3) (2007): 331–344.

Lee, Daniel HJ[a], "Map Orkney Month blog posts" accessed 04/10/20 https://publicarchaeology2015.wordpress.com/category/map-orkney-month.

Lee, Daniel HJ[b], "Map Orkney Month Week 4…" 2015, accessed 04/10/20, https://publicarchaeology2015.wordpress.com/2015/03/29/map-orkney-month-week-4-winding-the- clock-drawing-stromness-and-imaginary-journeys-from-elsewhere/.

Lee, Daniel HJ[c], "Imaginary Tours of Orkney from Elsewhere…" 2015, accessed 04/10/20, https://publicarchaeology2015.wordpress.com/2015/03/31/imaginary-tours-of-orkney-from- elsewhere-mapping-archipelagos-in-east-london/.

Lee, Daniel HJ "Map Orkney month: imagining archaeological mappings", *Livingmaps Review* no. 1 (2016).

Lee, Daniel HJ "Experimental mapping in archaeology: process, practice and archaeologies of the moment", *Re-Mapping Archaeology: Critical Perspectives, Alternative Mappings,* edited by Mark Gillings, Piraye Hacıgüzeller and Gary Lock (London: Routledge, 2018), 43–176.

Lee, Daniel and Coppes, Jasper, "Flux Tower", *Livingmaps Review* no. 7 (2019).

Lee, Daniel and Thomas, Antonia "Archaeologists-in-residence at Papay Gyro nights: Experience, expectations and folklore-in-the-making", Paper presented at *Contemporary and Historical Archaeology in Theory* (CHAT) Conference 2013, UCL London 09/11/13.

Lee, Daniel and Thomas, Antonia, "Archaeologies in between: walking (in)visible lines", *Flettverk* #2 (SCENE) (2015): 4–13.

Lucas, Gavin, *Critical Approaches to Fieldwork: Contemporary and Historical Archaeological Practice* (London: Routledge, 2001).

Lucas, Gavin, *Understanding the Archaeological Record* (Cambridge: Cambridge University Press, 2012).

O'Rourke, Karen, *Walking and Mapping: Artists as Cartographers* (Cambridge, MA: Massachusetts Institute of Technology Press, 2013).

Pearson, Mike and Shanks, Michael, *Theatre/Archaeology* (London: Routledge, 2001).

Pearson, Mike, *In Comes I: Performance, Memory and Landscape* (Exeter: University of Exeter Press, 2006).

Pearson, Mike, *Marking Time: Performance, Archaeology and the City* (Exeter: University of Exeter Press, 2014).

Richardson, Lorna, "A digital public archaeology?", *Papers from the Institute of Archaeology* (23)1: 10 (2013): 1–12.

Shanks, Michael, *The Archaeological Imagination* (Walnut Creek, CA: Left Coast Press, 2012).

Thomas, Antonia, Lee, Daniel, Frederick, Uursula and White, Carolyn, "Beyond art/archaeology: research and practice after the 'creative turn'", *Journal of Contemporary Archaeology* 4(2) (2017): 121–129.

Tilley, Chris, 'Phenomenological approaches to landscape archaeology', in *Handbook of Landscape Archaeology,* edited by Bruno David and Julian Thomas (Walnut Creek, CA: Left Coast Press, 2008), 271–276.

Whitmore, Chris, "The world on a flat surface: maps from the archaeology of Greece and beyond", in *Representing the Past: Archaeology through Text and Image*, edited by Sheila Bonde and Stephen Houston (Oxford: Oxbow Books, 2013), 127–152.

Chapter 18

Field Drawings

Emma McNally

Choral Fields 1-12 (figures 18.1–18.6). Installation of large-scale drawings as part of the *Embassy of the Real*, Cockatoo Island. *20th Biennale of Sydney* (2016). Artistic Director: Stephanie Rosenthal

I think of these drawings as fugitive, heterogeneous grey areas. They are the turbulence between noise and signal. They are a space of difference and deferral, a weather system of graphite. They are also broadband realms where signals at multiple frequencies are being transmitted and received – including those not usually within our 'range': sonar, ultraviolet, the very fast, and the very slow. I am constantly trying to disrupt the figure-ground relationship to make blurred areas where the conditions of focusing are undone.

I mine all sorts of ways of thinking visually about space and time: the spiral paths of particles in bubble chambers, which are infinitely fast and small; images of cellular mitochondria; the Hubble Deep Field images that probe deep time, where all time is held in the surface of the image but cannot be reached. I like looking at images that show fleeting events and images of aerial views of cities at night – all the emergent formations at a macro scale that look like deep-sea organisms in the dark water. I also love aerial images of airports, both in use and obsolete, as well as the Nazca Lines.[1]

I constantly listen to sound when I draw – the white noise of rainfall; field recordings from all environments; the humming and buzzing of Francisco López's album Buildings [New York]; the transmissions from the hydrophones under the Antarctic ice, streamed live on the Internet; as well as all kinds of music. I try to attend as closely as possible to the sound, and to transcribe the rhythms into the drawing, to make a sort of seismograph. Marks that are suggestive of the airborne or the sub-oceanic, for example, can come into relation with marks, lines, traces, and paths suggestive of circuitry, telecommunications, Morse code, molecules, stars, shoals, electronic

Figure 18.1 Choral Fields 1-6′ Installation view 'Mirrorcity' Hayward Gallery, London 10/2014 - 01/2015. Author's Image.

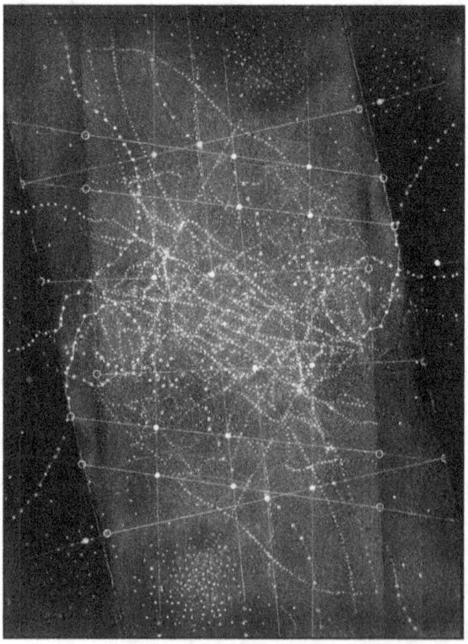

Figure 18.2 CP3 (Carbon on paper 2013). Author's Image.

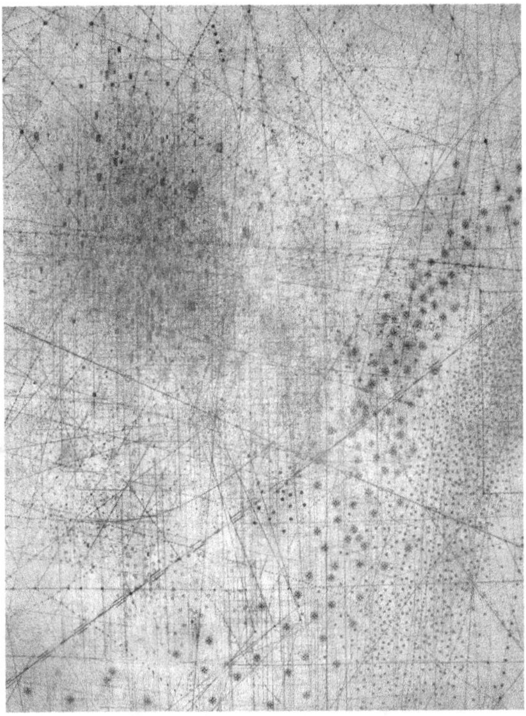

Figure 18.3 Field 4 Graphite on paper. 2011 (150cm 250cm). Author's Image.

Figure 18.4 Space text(ile) (detail) graphite on paper (70c x 100cm) 2015. Author's Image.

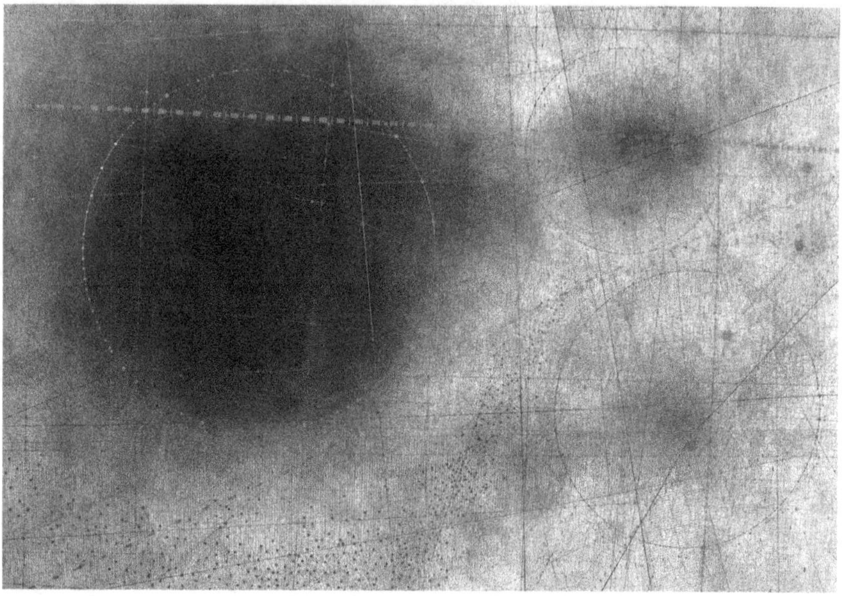

Figure 18.5 Field 12 (detail) Graphite on paper 305cm x 215cm 2013. Author's Image.

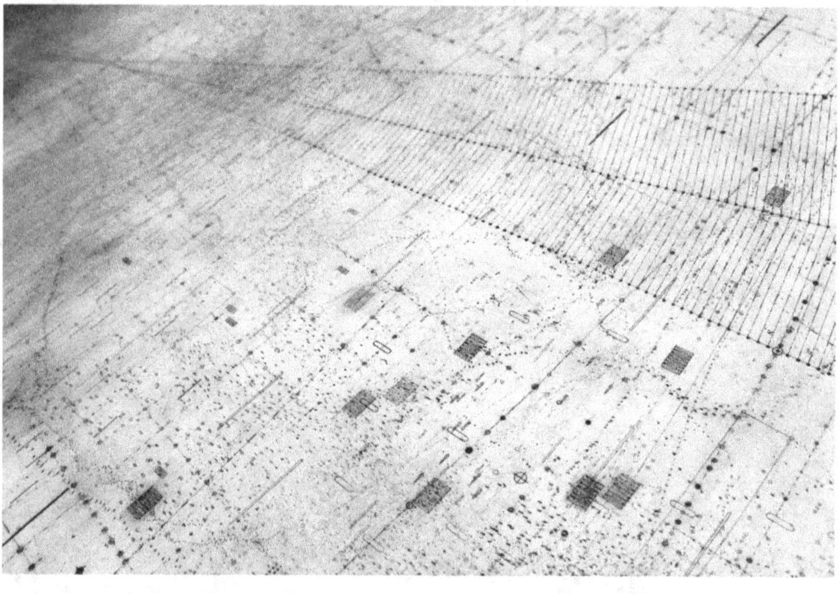

Figure 18.6 CF 2 (detail) Graphite on paper (2014). Author's Image.

pulses, particles, networks. These sorts of 'readings' are at the centre of my drawings.

Graphite is a medium that lends itself perfectly to this practice of rhythmic making and unmaking. The dense graphite areas act as engines in the drawing, emitting dark signals of loss, desire, longing, separation, reaching – they are the material 'heat'. I also like to think of carbon – a material that is both an insulator and a conductor – in different states: coal, diamond, smoke, black oil; as well as water in all its states: ice, snow, mist, rain, vapour. I want the works to be humming graphite soundfields: vibratory, oscillatory, multivoiced assertions and hesitations, yet also full of silences, voids, ghosts, residues, and remainders.

NOTE

1. The Nazca Lines are a series of ancient geoglyphs located in the Nazca Desert in southern Peru.

Chapter 19

Coda

Mapping the Pandemic: Cartesian Cartography and its Other Scene

Phil Cohen and Mike Duggan

Cartography is more popular than ever. Maps are everywhere and adopted by everyone. Location is king and location-based services are the future. Or so we are told. This rhetoric has been perpetuated further by our current moment, somewhere far from the beginning of the global pandemic, but still very far from the end,[1] when maps continue to shape our understanding and imagination of the disease, who has it and where it's coming from. Most, but not all, of the chapters in this collection detail a radical cartography that preceded the pandemic. We have, therefore, written this coda for the book as a response to the events that have unfolded around us as we put the collection together. We offer our take on pandemic mapping and how it might be shaped by some of the radical and counter cartographies outlined in the book.

One of the most important moments in the development of Western cartography occurred when an obscure doctor, John Snow, produced his so-called ghost map of the Cholera outbreak in London's Soho in 1854 (see figure 19.1).

By pinpointing the location of cases, and showing where they clustered, the doctor was able to identify the water pumps that were responsible for the outbreak and consign the air-borne 'miasma' model of transmission to the dustbin of medical history. Ever since, map-making has played a significant role in epidemiology, with ever more sophisticated cartographical techniques being used to capture and represent the big data of disease.

But the story of disease maps does not start here. There is evidence to suggest that maps have been used to shed light on where microscopic diseases have spread for centuries. As far back as 1692, Filippo Arrieta, a royal auditor of Italy, was tasked with visualising the spread of plague in what is now Puglia. His maps showed the areas around the city of Bari that were most

Figure 19.1 John Snow's Cholera Map of Soho, 1854. Wikimedia Commons (CC).

affected, as well as the quarantine zone to south of the city, set up by the military to curb the spread of the disease (see figure 19.2). Though not nearly as intricate, nor as accurate, as Snow's maps of London, and nowhere near the levels of sophistication seen in the digital maps depicting Covid-19, Arrieta's maps were nonetheless used to visualise disease and help the decisions made by those in power over what to do about curbing the spread of infection.

Because the cholera bacterium is transmitted through infected water (and food), and not directly from person to person, it is relatively easy to trace its sources and take appropriate remedial action. Not so, of course, with other infectious diseases from colds, flu, measles and mumps to Smallpox, Ebola, and now, of course, Covid-19. It is the fact that it is impossible to track all the movements of its human carriers, who may remain symptom-free,

Figure 19.2 Filippo Arrieta's Plague Map of Puglia, 1692. New York Academy of Medicine (CC).

that generates such pervasive contagion anxiety around this new virus. The deployment of aerial mapping via the use of drones may be useful in policing lockdown regulations, but the detection of actual chains of human transmission have defeated all but the most totalitarian regimes of track and trace. Meanwhile the cartograms which have been devised to capture the spatial distribution of reported cases, while they may provide some rough rule-of-thumb measurements of how public health policies are, or are not, working, for example, allowing comparisons to be made between countries across the globe, these cartographic practices singularly lack the granularity required to capture the locally situated epidemiologies which shape what is happening on the ground.[2]

If it may not always be possible to map the 'known unknown' (i.e. the full extent and patterns of transmission), how much more difficult is it to represent what the psychoanalyst, Christopher Bollas has called the 'unthought known', the deeper more unconscious responses to the pandemic, sometimes repressed, sometimes emerging in coded forms. We are dealing here with emotional geographies and how they shape perceptions of risk, which belong to what Freud called the 'other scene' of everyday life, where rational calculations of self-interest and limited altruism give way to structures of feeling and belief dominated by often disavowed fears and fantasies of 'the other'.[3]

Where we and 'the other' are in relation to this invisible killer matters to us, and it is maps that visualise these relations in a way we easily understand. Perhaps the most well-known of these maps are those produced by the Johns

Figure 19.3 Johns Hopkins University Covid-19 Global Cases Map and Infographic. Author's Image.

Hopkins University of Medicine (see figure 19.3), who collect and map data about infections, deaths, and recoveries from official sources all over the world. Depending on where we are and who we are, these maps, used by media outlets from Fox News to *Forbes* and the *Financial Times*, have quickly become part of the narrative of how the story of Covid-19 is being told. As such, they are not simply graphics that convey dry spatial information; they have become intertwined with our personal experiences and perspectives of the global crisis. They give us a sense of relief that events are happening far away or feelings of anxiety as events come ever closer to affecting us or our loved ones.

Social distancing and other lockdown measures have made us all other to the other as potential carrier of a lethal (i.e. hidden and life threatening) disease. This social seriality is fertile ground for the popular imagination of disease. These denizens of our inner world inhabit a country of the mind which often remains foreign to us, but which nevertheless we can still own and, after a fashion, assemble into some kind of cognitive map. But how?

What makes a disease like Covid-19 communicable is not just its mode of physical transmission, but its mode of *media and cultural* transmission, in which maps, of course, have played an important role. Covid-19 is, in one sense, news from nowhere; its arrival was not predictable, even if it was anticipated by some epidemiologists; its sudden advent, of course, made it instantly newsworthy. However limited their forensic function, Cartesian cartographies have a major rhetorical role to play in underpinning public health messaging about the pandemic, in the attempt to persuade citizens that despite all appearances to the contrary, the state has the situation under control.

Yet, in another sense, a pandemic is always the same old story; however new the actual bug, the unfolding of the plot line follows a familiar course: It is always and already 'bad news' and gets worse before it hopefully gets better. That combination of repetition and disjuncture is its paradoxical time signature. As such the pandemic takes on a particular resonance within the framework of the media spectacle. This 'infodemic' story is one that always leaves us in suspense, in fearful or excited anticipation of what comes next: is the disease terminable or interminable, will the exit strategy work or produce a new infection spike? Will the vaccinations work? Stay tuned. In other words, the narrative has the structure of a TV soap opera, and this serialisation integrates the pandemic closely with the constant revolution of commodity production and consumption – the quest for permanent novelty that keeps the wheels of the creative and not so creative consciousness industries turning.[4]

In global terms, the news of the virus travelled faster via social media than the virus itself, and in that sense, it was always ahead of itself. But its uncertain incubation period becomes part of a serialised drama which ensures that public responses are invariably behind the curve of events. It is this *double seriality*, of sociality and storying, that makes Covid-19 the first post-modern pandemic, an unreliable narrator of its own demographic trajectory, never coinciding with itself. The role of Cartesian cartography has simply been to contribute to this game of catch up by adding its own visual rhetorics including colour coding: the bigger the red blob on the map, the bigger the existential threat posed by the disease, the bigger the blue blob, the happier we should be that the worst is over. In fact, the dominant iconographic map of the pandemic is not, strictly speaking, a map at all – it is the graphs shown on TV and in mass circulation newspapers, which chart the progress of reported cases and deaths along a time line. And here a curious rhetorical effect occurs, whereby the visual metaphor of peaks and troughs is reversed – the high points of the graph signifying catastrophic rates and making our spirits sink, while the low points provide us with reasons to be more cheerful.[5]

While Cartesian cartography has been quick off the mark to dramatise the spatial distribution of Covid cases and mortality rates, the correlation of this information with a range of socio-economic variables, such as population density, residential patterns, working and housing conditions, household income and resources, public health standards and local situational factors, all of them linked to structural inequalities of class, gender, generation, and race, that is still very much work in progress, although it is to be hoped that the necessary degree of international co-operation will be forthcoming to ensure the standardisation and quality of datasets. In any case, the processes whereby information is locally produced need to be made transparent, since they are rendered as invisible as the algorithms used to give them cartographic form on public media platforms. What remains to be tracked and traced, and then deconstructed is this hidden infrastructure of cartographic reason as it has been mobilised in creating an official narrative around the pandemic. This is certainly an important task for critical theory.[6]

As for counter-mapping the pandemic, so far this has largely been the province of artists who have explored the personal impact of lockdown, and resulting patterns of social confinement and atomisation on their own aesthetic practice and everyday lives. Both Bloomberg and Livingmaps have curated collections of such material;[7] but however interesting it is to see fictional or imaginative mappings of these experiences, filtered through individual creative sensibilities, it cannot really provide much insight into the emotional geographies which shape and inform the perceptions and behaviour of different sections of the population, especially those who are most impacted by the pandemic.[8]

Lockdown regimes, with their one-size-fits-all categorical imperatives about social distancing, take little or no account of the customary social proxemics which characterise particular cultures and communities.[9] For example, affluent suburban neighbourhoods whose social networks are highly individualised and spatially distributed have little difficulty in conforming to the stay-at-home rules; in contrast, certain minority ethnic communities, which have vibrant and economically important street cultures, and patterns of intense social congregation that are central to their shared sense of place, identity and belonging find themselves subject to media-led moral panics about 'super spreaders' and much more intensive police surveillance.[10] It is precisely these hyper-local geographies which need to be recognised and mapped to develop more context-sensitive lockdown strategies. Fortunately, there are well-developed methodologies at hand for this purpose, some of which are outlined throughout this book, that are sensitive to the specific circumstances of people's lives.

It is only to be hoped that participatory mapping will be used in ethno-cartographic studies into the pandemic's impact so that the stories of the young and the old, men and women, the sick and the healthy, majorities and minorities can be put on maps of their own shared making. Such initiatives can provide a more adequate evidential platform for developing public health policies which are genuinely pro-active and precautionary and address the glaring social inequalities which the pandemic has rendered so starkly visible.

We have been offered various 'road maps' out of the present public crisis, the assumption being that societies are like car drivers and can somehow move in a more or less straight line from A (danger) to B (safety) by following the travel instructions and road safety rules issued by government. Unfortunately, it turns out that the situation is somewhat more complicated than that; asking people to give up their own personal lines of desire to follow the straight and narrow path to deliverance can no longer rely on the shared values once provided by the religious journey, whose hand-me-down maps once guaranteed salvation at the end of the road. Science, with its uncertainty principles and empirical scepticism, has proved no substitute, especially given that the Enlightenment belief in the application of scientific rationality to human affairs as guaranteeing social progress has fallen so short of expectation. In this vacuum, conspiracy theories have flourished providing an ever more seductive map of the *terror incognita* that is Covid-19. *There be monsters*, the movie-cum-video game, starring Big Pharma and produced by Hidden Hands Inc.

In general, the pandemic has exposed the gulf which still exists between the two cultures of cartography – the quantitative and qualitative, the macro view from above and the micro view from below. In principle, actor-network theory, with its commitment to flat ontologies and replacing the global with translocal relays, is best placed to track and trace the multiple impacts of Covid-19, but this approach has yet to develop a distinctive cartographic methodology applicable to a pandemic. The challenge which Covid-19 poses for radical cartography is how to scale up its locally mappable knowledges without losing focus: the granularity problem.

Against this background, our aim has been to explore the limits and conditions of current practices in this field and to suggest some possible ways forward in meeting the multiple challenges of the present conjuncture. Livingmaps Network is committed to an open dialogue, not just between different disciplines and approaches to mapping in the arts, sciences, and humanities, working away in the backyards but to support communities and activists who are on the front lines of the many struggles for social and environmental justice, which are now accelerating across the world. We are always keen to hear of new projects and to learn from them and hope this book will encourage an ever more diverse multitude of people to put themselves and their concerns on their own maps.

NOTES

1. At the time of writing, March 2021.
2. The current state-of-the-art atlas of the pandemic is provided by John Hopkins university's Covid-19 platform.
3. For a discussion of this point and the role of emotional geographies in shaping popular perceptions of the pandemic, see Phil Cohen, *There must be some way out of here: mapping the pandemic from Left Field* (Compass, 2020), available to download for free from www.compass.org.uk.

4. See Hans Magnus Enzenberger, *The consciousness industry* (Continuum Books, 1974) and Guy Debord, *The Society of the Spectacle* (Zone Books, 1994).

5. See E. Bowe, E. Simmons, S. Mattern, Learning from lines: Critical COVID data visualizations and the quarantine quotidian. *Big Data & Society*. July 2020. doi:10.1177/2053951720939236.

6. See the very useful critical overview of current cartographic approaches to the pandemic by Andrea Pase, Laura Lo Presti, Tanioa Rossetti, and Giada Peterle, Pandemic cartographies: a conversation on mappings, imaginings and emotions, *Mobilities* (January 2021).

7. See Bloomberg Map Lab Mapping the Pandemic. bloomberg.com/news/newsletters/2020-03-18/maplab-visualizing-a-pandemicand www.livingmaps.org/lockdown-open-call and www.livingmaps.org/ Drawing Maps and Pandemic Storytelling (22/07/2020) especially the presentation by Joel Seath on Children's lockdown maps. This initiative is being developed further in the Autumn issue of Livingmaps Review 2021, and in the Young Citizens Atlas.

8. For an attempt to map these emotional geographies in terms of a theory of environmental risk perception see Phil Cohen, *There must be some way out of here: mapping the pandemic from Left Field* (Compass, 2020).

9. For an interesting historical and cross cultural study of customs of social distancing, see Lily Scherliss, 'Distantiated Communities: a social history of social distancing', *Cabinet* 66, 2020.

10. See, for example, the work of Zuleyka Zevallos on the pandemic, race, class, and moral panics. https://othersociologist.com/2020/07/05/pandemic-race-and-moral-panic/.

BIBLIOGRAPHY

E. Bowe, E. Simmons, S. Mattern. Learning from lines: Critical COVID data visualizations and the quarantine quotidian. *Big Data & Society*. July 2020. doi:10.1177/2053951720939236.

Phil Cohen. 2020. *There must be some way out of here: mapping the pandemic from Left Field*. Compass. London.

Hans Magnus Enzenberger. 1974. *The consciousness industry*. Continuum Books. London.

Guy Debord. 1994. *The Society of the Spectacle*. Zone Books. New York.

Andrea Pase, Laura Lo Presti, Tania Rossetto and Giada Peterle. 2021. Pandemic cartographies: a conversation on mappings, imaginings and emotions. *Mobilities*, 16(1): 134–153.

Lily Scherliss. 2020. Distantiated Communities: a social history of social distancing. *Cabinet*, 66. Available at: https://www.cabinetmagazine.org/kiosk/scherlis_lily_30_april_2020.php.

Chapter Summaries

INTRODUCTION

Part 1: Are We that map?

Chapter 1: The Cultural Life of Maps: Everyday Place-Making Mapping Practices
Mike Duggan

Maps don't simply go out into the world as finished artefacts. Rather, maps become bound up in the practices of everyday life. This is to say that maps have a social and cultural life that extends well beyond the purposes that they were originally designed for. In this chapter, I draw from an ethnographic study of map users to show how both paper and digital maps become intertwined with everyday practices of place-making. In the first instance, I highlight how the popular A–Z pocket atlas has become embedded into the place-making navigational practices of two London residents. In the second instance, I show how digital-mapping software has become integral to the place-making navigational practices of a road cyclist. In documenting these cases, I shed light on the specific ways that paper and digital maps may be annotated and how such annotations matter as they become folded into the social and cultural nuances of everyday place-making process.

Chapter 2: This Noise Matters: Participatory Soundmapping and the Auditory Experience of Homelessness
Paul Tourle

Crowdsourced online soundmaps, developed since the mid-1990s, function ambivalently – both as potent counter maps that subvert and complicate visual cartographies and as massive reservoirs of informational noise: monumental, data-heavy archives that highlight the difficulty of achieving meaningful dialogue or communication in participatory projects. Departing from a brief consideration of the processes through which many contemporary soundmaps are made, and of the often narrow forms of listening they bring to bear on the world, this chapter then describes two collaborative listening projects, which aimed at highlighting auditory experiences of homelessness in London. In documenting sounds and modes of audition born of displacement, makeshift architectures, isolation and bureaucracy, the chapter affirms the value of listening in mapping practices. At the same time, in exploring the limits of the projects in terms of noise, it calls into question an ideal of inclusion frequently embedded in traditions of participatory cartography.

Chapter 3: Mapping the Right to the City: City Perception as a Shaping Force
Giulia Carones

This chapter engages with political activists' perception of the urban environment. It examines how participation in an urban social movement advocating the citizen-led creation and management of the urban commons can shape individuals' perception of the city. The research reported here is based on an innovative cognitive mapping tool used as part of semi-structured interviews with members of the activist collective Laboratorio Universitario Metropolitano (LUMe) in Milan. The maps served to open up and structure participants' reflections on the city, and its spatial politics, in particular focussing on emergent landmarks in the urban fabric linked to sites of political activism considered as a generative, necessary rupture of the status quo.

Part 2: Reclaiming the Territory

Chapter 4: Talking Maps *and Diasporic Community*
Jina Lee

Th is account of an ethno-cartographic research project embeds its methodology in a specific drawing practice. The chapter begins by examining contemporary drawing practices as a form of mapping, and Lee introduces a new method that she named the *talkingmap*. Regarding Joseonjok people living in London as subalterns, I look in detail how the process of the author's engagement with the Joseonjok people in New Malden can be mapped in a different way. The use of drawing as way of notating material emerging from the research interviews is discussed with numerous examples. The chapter concludes by looking at the role that this kind of cartography can play in opening up a space of representation for marginalised communities.

Chapter 5: Stories of the Unmappable
Marija Biljan

This chapter shows how maps and mapping can be used as a tool in work with communities for creating a dialogue between generational and cultural groups. Through different examples, I show how maps and their subsequent nature of representation can reveal much more about one's personal identity. Using maps in different ways besides simply navigating through the city can be an empowering exercise that gives us the feeling of appropriation and ownership of the wider physical, monumental, and impersonal spaces that we often perceive the city to be made up of. I highlight how touching, modifying, making graphical or other interventions on the map can awaken our sense of space and our connection with the place.

Chapter 6: Former Fresnans: Mapping Home through a Memory Palace
Blake Morris

Former Fresnans (2012–on-going) is a memory palace sited in Fresno, California's Tower District. Memory palaces link symbolic images to specific locations. To create a memory palace, one chooses a specific space and imagines vivid symbolic images throughout it. The more absurd the image, the easier it is to recall. To retrieve the memories, one imagines walking through the space and looking at the different images. Unlike an object created in response to a walk, a memory palace has to be imagined, and each participant manifests the images differently depending on their individual interpretation. Since 2012, I have been tracking down former Fresnans to join me for a walk. My instructions were simple: take me on a walk between two places you consider home. I have walked with twenty-one former Fresnans in New York, London, Oakland, and San Francisco. What we all had in common was the place we no longer lived – Fresno. The walks are documented through the creation of an image for a memory palace, sited in Fresno, California. This chapter discusses the construction of that memory palace and the stories it brings to life.

Chapter 7: The Busyness of Button-Mapping: Exploring Children's Everyday Politics in Belfast
Amy Mulvenna

Inspired by feminist materialist thought that takes seriously issues of the material, socio-economic, and crucially, geopolitical narratives within and beyond children's geographies, my recent research engages with creative mapping approaches with the aim to give visibility to ways in which geography matters in *less overt* and *less obvious* ways for children living in Belfast, Northern Ireland. This chapter outlines one of those mapping approaches – *button-mapping* – co-developed with participants aged 7–11 across four field sites in Belfast, Northern Ireland. 'Button mapping' refers to a more-than visual and intra-active mode of mapping that represents one's city using tiny *things* like buttons, shells, paperclips, marbles, bits of straws, cocktail sticks, beads, or anything else to hand. In this chapter, I present an empirical example of this approach in order to focus on the complex and open-ended ways that participants practiced everyday minor politics through mapping. In doing so, I seek to highlight researcher responsibilities when enrolling 'creative' and 'playful' approaches to place-mapping, which can include sensitive questions around identity, belonging, and security.

Chapter 8: Mapping the Overlaid Life of Places of Play
Joel Seath and Kelda Lyons

This chapter is concerned with children's place within, and right to freely enjoy and use, the built environment, which is mainly designed for adults rather than children. We believe that more consideration towards children and play should be given when planning, designing, and rearranging built environments, particularly the public realm. Working from a starting position of play for the sake of playing, the chapter discusses and reflects upon research that Joel Seath undertook between 2015–2016 on and around an adventure playground in west London. Joel worked with children and older residents who live near the playground to discover where and how children play, and have played, over the last few generations in the area. Subtle listening and creation of artefacts and maps were used to understand and explore the playful use of the local area. Thinking is developed on the idea of places as lived experiences, and how their forms are created through the given names and stories that happen when children play; observations and stories told by older generations, of destinations (such as parks), in-betweens (such as streets, steps, kerbs, and alleys), various edges, zones, and routes. The process of engaging adults in this research contributed towards raising the tolerance for play, locally, instigating the idea that mapping play can be the catalyst for a vision of a more equitable built environment. By exploring children's expressions about the importance of place and their experiences in particular places, their inclusion in the built environment, and adults' understanding of children's valid belonging, might all be more fully recognised. After considering the current role and status of children in the UK town planning context, an enhanced vision for child-friendly environments is offered, particularly in the public realm, for design and planning that better acknowledges children, play, and playful histories.

Chapter 9: 'Like the Palm of my Hand': Children and Public Space in Central Athens
Christos Varvantakis

Urban public spaces are noted in literature as being highly important spaces for children and yet children's accessibility to urban public spaces remains a highly contested subject. Children's perception of and accessibility to public space is also, I argue, relevant to their emergent and developing understandings of citizenship. In this chapter, I address uses of public space that children make in central Athens and how this may transgress both limitations that are set forth by urban planners as well as the imaginary borders between the public and private sphere. Focusing on Iason, a 10-year-old boy living in the highly contested and densely politicized urban neighbourhood of Exarcheia, this chapter explores the mismatches between his own and the official's perspectives on the neighbourhood. Iason's views and ways of being in his neighbourhood provide a highly critical outlook on conservative and institutional understandings of children's inclusion in urban public life.

Part 3: Watch this Space

Chapter 10: Just Mapping for Civic Action: Inclusive Neighbourhood Planning in the Elephant and Walworth
Barbara Brayshay and Nicolas Fonty

In this chapter, we discuss the findings from a community-based mapping project in which we explore the role of participatory mapping as a tool for facilitating local community engagement and civic action in an urban planning and redevelopment context. The chapter is divided into four sections, the first introduces JustMap and the project setting in the context of communities experiencing regeneration in London. This is followed by an overview of the theoretical and methodological frame underpinning our approach to participatory collaborative mapping. We then illustrate this with a case study, a project commissioned by the Elephant and Walworth Community Forum, in Southwark, south London, to assist with their development of a Neighbourhood Plan for the area. We conclude with a summary of the challenges and rewards of participatory mapping as a tool for activists and groups campaigning for a say in the (re)development of their neighbourhoods.

Chapter 11: Empathy Walks
Leticia Sabino, Sofia Croso Mazzuco, Julie Plichon, Debanil Pramanik, and Sonja Baralic

Empathy Walks is an open source project which aims to build empathy through walking together. Walks are led by socially and economically vulnerable and underrepresented communities and people such as migrants, maintainers, unemployed, or facing displacement. The walks are used as a tool for people to get closer and understand other realities and layers of coexistence in the same city. The aim is to unlock narratives and open a space for diverse voices to be heard and paths to be walked, making citizens more aware of each other's perspectives and presence. By walking paths together that other people take during the course of their day-to-day lives, we are able to understand better each other, building empathy and more meaningful connections which can lead to respect and collective action.

Chapter 12: WILD CITY | FIADH-BHAILE | ORASUL SALBATIC: Mapping the Wild in the Greeny Howe of Glasgow
Alec Finlay, Deirdre Heddon, and Misha Myers

These poem maps and word-drawings were produced for WILD CITY | FIADH-BHAILE | ORASUL SALBATIC, a collaboration with *The Walking Library,* mapping urban rewilding in the city of Glasgow, in 2018. The accompanying texts are from the Wild City Manifesto, another of Finlay's contributions to the project. A paperback book documenting the entire projects, including communal walks, is available.

Chapter 13: Performing Cartographies: Getting Inside and beyond the Map
Misha Myers and Lucy Frears

This chapter focuses on embodied geographies of landscape, *deep mapping* and embodied forms of mapping, wayfinding, performance and art, including locative media art or remotely transmitted performance, performances involving live or recorded performers mediated by or present through technology. Two case studies of cartographic performance works are presented, *Take me to a place,* a song map for an audio walk created by Misha Myers in 2004 with diverse inhabitants of Plymouth, the United Kingdom, and *Hayle Churks App*, a multimedia geolocated app made for smartphone by Lucy Frears in 2013. Each project represents a different period of time in the development of locative media art technologies, but employs the capacities that forms of digital storytelling offer for immersive and lived cartographies. The latter project was developed as part Frear's interdisciplinary practice-based PhD studentship led by Dr. Misha Myers and co-supervised with Professor Caitlin DeSilvey of Exeter University with funding from the *European Social Fund II.*

Myers is co-author of an additional contribution to the volume with Dee Heddon and Alec Finlay, *WILD CITY | FIADH-BHAILE | ORASUL SALBATIC mapping the wild in the greeny howe of Glasgow,* on their collaboration on the *Wild City* project.

Chapter 14: Unmapping Space (Lines, Smudges, and Stories)
Kimbal Quist Bumstead

This chapter explores the process of collaborative map-making as a tool to visualise and contemplate subjective and *felt* experiences of personally significant places. Using examples from my artistic research project 'Unmapping', I discuss mark making as an embodied practice, the visceral and tactile nature of which can be a tool to explore and capture the complex relationships between body, space, and experience. The text demonstrates how the bodily act of making marks can be used as a way of tuning-in to the sensory and emotional aspects of remembering and imagining places, and consequently how subjectivities and *imaginings* can be visually represented.

Part 4: New Scopes, New Scales

Chapter 15: Cartographic Care, or Caretographies: From London to Hong Kong
Sam Hind

This chapter looks to several digital mapping platforms – some defunct, others active – that provide a blueprint for, and a validation of, a latent cartographic politics. From 2010 to 2012, 'Sukey' was used by protesters in London to avoid police containments at protest events. It became known as the 'anti-kettling' app. For this brief period, its capacities confounded the police and excited the media. More recently, in Hong Kong, two strikingly similar protest mapping projects – 'Hkmap.live' and '103.hk' – were launched to help protesters demonstrate against Chinese interference in Hong Kong affairs. In a navigational sense, maps aid wayfinding, caring for the lost, disoriented, or in danger. They act as assistants, rendering care in material form, offered through navigational practice. As such, I present the Sukey platform as a possible blueprint for, and the Hong Kong platforms as a validation of, cartographic care; drawing on feminist Geographical Information Science (GIS) that has considered how maps routinely leave spaces of care 'off the map'. I build on this work by suggesting a need to consider careful cartographic practice and how protesters and mapping volunteers 'act caretographically' during protests. I use the work of Joan Tronto to suggest protesters in London were engaged in four 'phases' of care, in which an *attentiveness* to care needs, a *responsibility* towards those who required care, *competence* in care-giving, and a *responsiveness* to how care was received, was in evidence. In Hong Kong, I suggest three additional phases were also witnessed, concerning the provision of *self-care*, the *resilience* of caring infrastructures, and the *expiration* of care needs. I refer to cartographic projects that engender caring actions as 'caretographies'.

Chapter 16: Mapshop: Learning to Map, Mapping to Learn
Emily Barrett and Matthew W. Wilson

The emergence of participatory and public participation GIS in North America in the late-1990s have led to a series of productive conversations, disciplinary discussions, and university initiatives broadly termed, 'community geographies'. These efforts are only partially about research methods and have instead drawn in debates around the role of the contemporary public university. In this chapter, we discuss one such modest initiative at the University of Kentucky: Mapshop. Attempting to leverage the technical resources and the expertise of students and faculty in the Department of Geography, Mapshop supports community partners with mapping and visualization needs and offers training opportunities for campus and community members. We outline some of our findings from these efforts including new engagement attempts with Instagram. In doing we, we draw attention to the ways that, for us, learning to make maps and learning from those mapmaking endeavours are sutured, if always a balancing act.

Chapter 17: Map Orkney Month: Imagining Archaeological Mappings
Daniel H. J. Lee

Map Orkney Month proposes new forms of creative mapping for archaeology. When volunteers were asked to map their world for a day, the idea was to create a new collaborative map of the Orkney archipelago based on everyday journeys and places; a kind of countywide archaeological walkover survey with a twist. In the process, the project challenged traditional archaeological power structures, destabilised the way archaeological knowledge is produced by using non-specialists, and experimented with new modes of archaeological mapping. In the end, each contribution became its own map without the need for traditional archaeological cartography. In particular, the role of imagination in both traditional and experimental mappings became an important theme. The project resisted the need for an overall final map and allowed the contributions to speak for themselves. Above all, mappers were challenged to think about archaeology and their relationship with space, place, and landscape in a new way through their everyday routines. This chapter summarises the month-long mapping project interspersed with original contributions from participants.

Chapter 18: Field Drawings
Emma McNally

In this chapter, Emma McNally their mapwork, Choral Fields 1-12. Installation of large-scale drawings as part of the *'Embassy of the Real'*, Cockatoo Island. *20th Biennale of Sydney* (2016). Artistic Director: Stephanie Rosenthal.

Chapter 19: Coda: Mapping the Pandemic: Cartesian Cartography and its Other Scene
Phil Cohen and Mike Duggan

Cartography is more popular than ever. Maps are everywhere and adopted by everyone. Location is king and location-based services are the future. Or so we are told. This rhetoric has been perpetuated further by our current moment, somewhere far from the beginning of the global pandemic, but still very far from the end, where maps continue to shape our understanding and imagination of the disease, who has it and where it's coming from. Most, but not all, of the chapters in this collection detail a radical cartography that preceded the pandemic. We have, therefore, written this coda for the book as a response to the events that have unfolded around us as we put the collection together. We offer our take on pandemic mapping and how it might be shaped by some of the radical and counter cartographies outlined in the book.

Index